Nucleic Acid Probes

Editor

Robert H. Symons, Ph.D.

Professor of Biochemistry
University of Adelaide
Adelaide, S. A., Australia

CRC Press, Inc.
Boca Raton, Florida

Library of Congress Cataloging-in-Publication Data

Nucleic acid probes.

Includes bibliographies and index.
1. DNA probes. 2. DNA probes—Diagnostic use.
3. Genetic disorders—Diagnosis. I. Symons, Robert H.
[DNLM: 1. Nucleic Acids—diagnostic use. QU 58 N9638]
QP624.5.D73N83 1989 616.07′57 88-16745
ISBN 0-8493-4942-7

Direct all inquiries to CRC Press, Inc., 2000 Corporate Blvd., N.W., Boca Raton, Florida, 33431.

International Standard Book Number 0-8493-4942-7

Library of Congress Card Number 88-16745
Printed in the United States

PREFACE

Nucleic acid hybridization is arguably the most important single phenomenon to be exploited in the amazing revolution in biology that has been occurring for the past 30 years. The ability of complementary nucleic acid sequences to form double-stranded hybrids with high efficiency and immense specificity in the presence of a mixture of many different, noncomplementary nucleic acids was first reported in detail in 1961 by B. D. Hall and S. Spiegelman (*Proc. Natl. Acad. Sci. U.S.A.*, 47, 137, 1961). Since then, the technique has been finely characterized and used extensively in all its variations.

Apart from their daily use in all gene technology laboratories, there is widespread interest in, and great potential for, the application of nucleic acid probes in the specific, rapid diagnosis of human, animal and plant pathogens, in the detection of specific genes in animal and plant breeding, and in the diagnosis of human genetic disorders. In addition, there is an increasing requirement to replace radioisotopes with nonradioactive procedures for the detection of nucleic acid probes. Many of these areas are addressed in this book.

The six chapters provide a comprehensive coverage of the enzymatic and chemical techniques for labeling nucleic acid probes with radioisotopes and with nonradioactive ligands and haptens together with associated methods for their detection. In addition, the use of these probes is described in the diagnosis of human and microbial pathogens, plant viruses and viroids, and human genetic disorders, as well as in the detection of nucleic acids in tissues and cells by *in situ* hybridization. Because the overall field is developing and expanding rapidly, the authors have aimed to give a sound background to their topics as well as to provide some insight into future directions.

I would like to express my deep gratitude to all the authors who have made this volume possible. I am very pleased that we have the benefit of the extensive experience of Dr. Timo Hyypiä and his colleagues in Finland and Sweden in Chapter 3 on the use of nucleic acid probes in the diagnosis of human microbial pathogens. Likewise, Dr. Chris Burrell and Dr. Grant Sutherland and their colleagues in Adelaide have provided two chapters based on their widely recognized areas of expertise. Finally, I would like to acknowledge my indebtedness to my colleague, Dr. Jim McInnes, who most willingly bore the brunt of the work in the preparation of the three chapters which we co-authored.

THE EDITOR

Robert H. Symons, Ph.D., is a Professor of Biochemistry at the University of Adelaide, South Australia.

Professor Symons received his B.Ag.Sc. in 1956 at the University of Melbourne, Victoria, Australia, and his Ph.D. in 1963 from the same University for a project on ketone body formation from butyric acid by rumen and omasum tissue of the sheep. His introduction to nucleic acids was in February, 1961, when he joined Dr. Roy Markham at the Agricultural Research Council, Virus Research Unit, Cambridge, England. During the next two years he developed his lifetime interest in the structure and properties of plant virus nucleic acids. In 1962, he was appointed to a lectureship in the Biochemistry Department of the University of Adelaide where he has remained except for study leave periods at Stanford University in California and at the Medical Research Council Laboratory of Molecular Biology in Cambridge. He was promoted to Senior Lecturer in 1967, Reader in 1973, and to a Personal Chair in 1987. In 1983, he was elected Fellow of the Australian Academy of Science, and in 1988, a Fellow of the Royal Society.

Professor Symons has had diverse research interests in addition to his main one on the structure, function, and replication of plant viruses, viroids, and satellite RNAs. During the 1960s he developed a widely used procedure for the preparation of ^{32}P-labeled nucleotides, while, for about 10 years from 1966, he investigated the use of low molecular weight analogues of puromycin to characterize the peptidyl transferase reaction on the ribosome. For the past few years he has been involved in the development of nonradioactive nucleic acid probes and related techniques for the rapid diagnosis of pathogens of plants and animals.

In addition to these research interests, Professor Symons was Chairman of Directors from 1983 to 1987 of Biotechnology Research Enterprises S. A. Pty. Ltd. (BRESA), a company set up in 1982 by the Department of Biochemistry and wholly owned by the University of Adelaide to market products arising from research developments in gene technology and biotechnology. BRESA was converted in 1987 to a public, unlisted company, Bresatec Ltd., of which he is a director.

CONTRIBUTORS

C. J. Burrell, Ph.D.
Senior Director
Division of Medical Virology
Institute of Medical and Veterinary Science
Adelaide, South Australia, Australia

Eric James Gowans, Ph.D.
Department of Medical Virology
Institute of Medical and Veterinary Science
Adelaide, South Australia, Australia

Martin Holmberg, M.D.
Department of Medical Genetics
Biomedical Center
Uppsala, Sweden

Pentti Huoyinen, M.D.
Assistant Professor
Department of Medical Microbiology
University of Turku
Turku, Finland

Timo Hyypiä, Ph.D.
Department of Virology
University of Turku
Turku, Finland

Allison R. Jilbert, B.Sc.
Division of Medical Virology
Institute of Medical and Veterinary Science
Adelaide, South Australia, Australia

James L. McInnes, Ph.D.
Department of Biochemistry
University of Adelaide
Adelaide, South Australia, Australia

John C. Mulley, Ph.D.
Department of Histopathology
Adelaide Children's Hospital
North Adelaide, South Australia, Australia

Ulf Pettersson, M. D.
Professor and Chairman
Department of Medical Genetics
Biomedical Center
Uppsala, Sweden

Grant R. Sutherland, Ph.D.
Chief Cytogeneticist
Department of Histopathology
Adelaide Children's Hospital
North Adelaide, South Australia, Australia

R. H. Symons, Ph.D.
Professor
Department of Biochemistry
University of Adelaide
Adelaide, South Australia, Australia

TABLE OF CONTENTS

Chapter 1

ENZYMATIC AND CHEMICAL TECHNIQUES FOR LABELING NUCLEIC ACIDS WITH RADIOISOTOPES

J. L. McInnes and R. H. Symons

TABLE OF CONTENTS

I. INTRODUCTION

The more recent techniques of gene cloning[1-3] and oligodeoxyribonucleotide synthesis[4,5] have led to a new era in the preparation of nucleic acid probes. These probes have found widespread application in gene structure and gene function studies, the diagnosis of human, animal, and plant pathogens, and the detection of human genetic abnormalities. Chapters 3 to 6 in this volume describe such usage of suitably labeled DNA and RNA probes and oligonucleotides.

The basis for studies involving nucleic acid probes is the technique known as nucleic acid hybridization analysis. Briefly, the nucleic acid probe, either labeled with a radioactive isotope (this chapter) or nonisotopically (Chapter 2), hybridizes to form a duplex only with target nucleic acid (e.g. DNA or RNA) which is exactly complementary to itself; no such double-strand hybrid is formed with other nucleic acids. The technique is extremely popular due to its speed, specificity, and sensitivity. Readers are referred to two excellent publications[6,7] for details of the physicochemical basis of hybridization and specific technical protocols.

The choice of nucleic acid material to be used as probe is dependent upon several factors. These include the availability and source of the material for labeling, the degree to which the nucleic acid can be labeled (by the various methods outlined in this chapter), and the particular hybridization strategy. The advent of recombinant DNA technology has allowed great flexibility in the preparation of nucleic acid probes. Essentially any nucleic acid can now be cloned into a phage or bacterial vector and be produced in large quantities. It is also possible to derive segments from the original clone by subcloning fragments. In this way probes can be selected which represent either unique sections of a genome or sections which are shared by other members of, for example, a virus family. Details of methods available for the isolation, preparation, and cloning of nucleic acid molecules for probe usage are beyond the scope of this chapter and readers are referred to Maniatis et al.[1] and Glover[2,3] for specific protocols.

One important limitation with the use of cloned probes lies in the nature of the vector since this may contain sequences which cross-hybridize with other nucleic acids in the sample under test. This limitation, known as the vector homology problem,[8] is difficult to overcome, and the concomitant use of a control probe of vector alone in order to demonstrate the absence of material with vector homology in the test sample is recommended. To circumvent the problem, the isolated insert (excised from the vector), RNA transcripts derived from cloned nucleic acids *in vitro* (Section III.A), or a chemically synthesized deoxyoligonucleotide (Section IV) could be used as probe.

Gene fragments for probe usage can also be produced by the construction of synthetic oligonucleotides.[4,5] Automated "gene machines" are now available commercially which allow the rapid synthesis of single-stranded oligonucleotides of user-defined sequence. These synthetic oligomers with lengths over 50 nucleotides (normal range, 15 to 25 nucleotides) can be produced in microgram to milligram quantities for use as diagnostic probes in much the same fashion as cloned gene fragments.

Table 1 lists the main properties of the radioactive isotopes that have been employed for labeling nucleic acids. When used as labels in hybridization studies in conjunction with autoradiographic means of detection, the sensitivity and resolution achieved is directly related to the physical properties of the radioisotopes. ^{32}P, for example (Table 1), remains the isotope of choice for both filter and solution hybridization analyses. ^{32}P has two major advantages; it is available at high specific activity and it emits highly energetic β-particles — both lead to short autoradiographic exposure times. Its major disadvantage, however, is its short half-life. ^{35}S, on the other hand, is attainable at relatively high specific activity but is a low-energy β-particle emitter. The sensitivity of ^{35}S is thus lower than that of ^{32}P, but its resolution on X-ray film is correspondingly higher.

Table 1
PROPERTIES OF ISOTOPES USED FOR LABELING NUCLEIC ACIDS

Isotope	Half-life	Approx. specific activity at 100% isotope enrichment (Ci/mmol)	E_{max} of radiation (keV)	
			β	γ
^{3}H	12.1 years	29	18.5	—
^{14}C	5100 years	62	156	—
^{32}P	14.3 d	9,120	1,710	—
^{35}S	87.1 d	1,490	169	—
^{125}I	60 d	2,400	34.6	35.4
^{131}I	8.6 d	16,100	608	365

^{125}I emits both γ-radiation and β-particles. The former readily penetrates film without efficient silver grain production and is best detected using an intensifying screen while the low-energy β-particles are stopped entirely within the emulsion. Thus, high sensitivity may be obtained with a screen and high resolution without. The very weak β-emission of ^{3}H yields the highest resolution, but its low maximum specific activity and low detection efficiency necessitate long exposure times. The traditional isotope of choice for *in situ* hybridization (Chapter 5) has been ^{3}H, since its low energy results in low backgrounds. However, ^{35}S and even ^{32}P are being increasingly used since they allow shorter time periods for autoradiographic exposure of slides. The unique emission characteristics of ^{125}I (Table 1) suggest an as yet largely unrealized role in high sensitivity, high resolution *in situ* hybridization.

This chapter discusses a range of methods available for probe labeling by both enzymatic and chemical means. For sensitive detection of hybrids, probes (either DNA or RNA) need to be labeled isotopically to high specific activities. As will be seen, some methods are better than others with respect to producing high specific activity probes. A chapter of this type is never complete; new methods for labeling nucleic acid fragments radioisotopically are constantly being developed.

II. ENZYMATIC LABELING OF DNA

A. Nick Translation

Without question, this is the most popular method for the preparation of ^{32}P- labeled double-stranded DNA. The basis of this method and its name were first described in 1970 by Kelly et al.[9] as part of a comprehensive characterization of *E. coli* DNA polymerase I in Arthur Kornberg's laboratory in the Department of Biochemistry at Stanford University. Certain unique properties of *E. coli* DNA polymerase I allow the preparation of labeled double-stranded DNA by nick translation and this method was soon developed in the same Department and first reported by Schachat and Hogness.[10] Although the method was soon in widespread use, it was not until 1977[11] that the detailed description of the method was published.

The nick translation technique utilizes the ability of *E. coli* polymerase I to combine the sequential addition of nucleotide residues to the 3′-hydroxyl terminus of nicks in double-stranded DNA (generated by low concentrations of DNase I) with the concomitant sequential removal of nucleotides from the adjacent 5′-phosphoryl terminus in what is really a path-clearing function.[1,9,11] When one or more of the incoming dNTPs is radioactive (usually α-^{32}P-labeled), the overall result is movement of the nicks 5′→3′ on each strand of double-stranded DNA with unlabeled nucleotides being replaced by labeled ones. A diagrammatic representation of nick translation is presented in Figure 1.

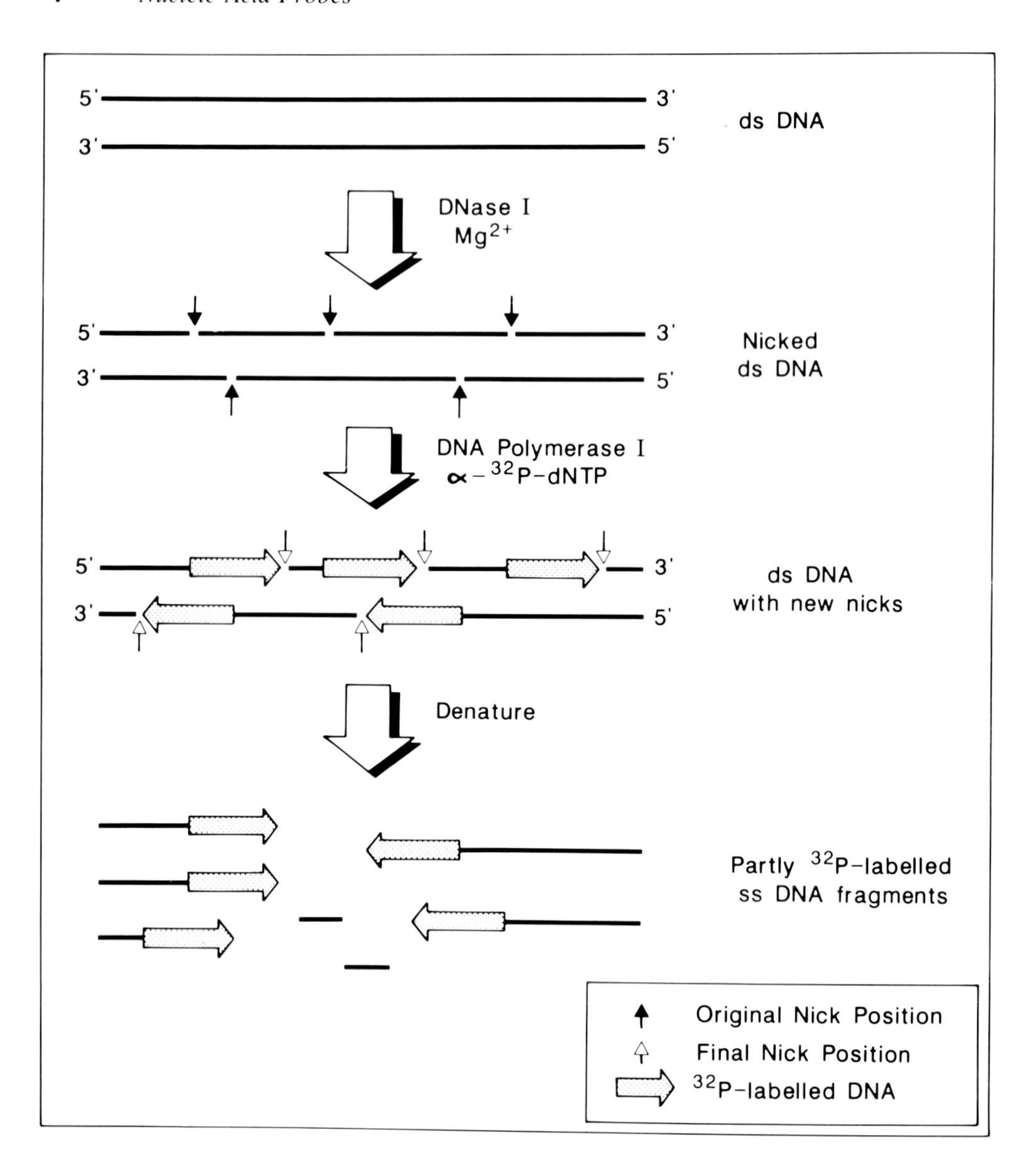

FIGURE 1. Preparation of ^{32}P-labeled DNA probes by nick translation.[9-11]

Double-stranded DNA in any form (linear, nicked, supercoiled, etc.) can be efficiently labeled by this approach. For most purposes, it is adequate to use one α-^{32}P-dNTP (specific activity 1000 to 3000 Ci/m mole) and three unlabeled dNTPS. Conditions are normally chosen that result in incorporation of approximately 30 to 60% of the α-^{32}P- dNTP into the DNA and provide DNA with a specific activity of about 10^6 to 10^9 cpm/μg. The specific activity is calculated from the amount of starting DNA and the total cpm incorporated.

It is important to appreciate that a high incorporation of a high specific activity α-^{32}P-dNTP does not guarantee a high specific activity DNA. For example, where excess DNA is present in the reaction mixture, the α-^{32}P-dNTP may be limiting such that a high incorporation of the total ^{32}P only results in a low percent replacement of unlabeled by labeled nucleotide residues. Hence, in order to obtain maximum specific activity of DNA, it is essential to use limiting amounts of starting DNA and conditions that ensure a high percent replacement of total nucleotides.

In the case of ^{32}P-labeled recombinant plasmid probes, the label is distributed between the cloned insert as well as the plasmid vector. Where the cloned insert is small relative to the vector, then most of the label will be in vector sequences and may be considered wasteful of starting α-^{32}P-dNTPS. In this situation, and also where vector sequences are not required, it is usual to first cleave the insert from the vector with a suitable restriction enzyme and then to label it by nick translation.

Double-stranded DNA probes must first be denatured to single-stranded segments prior to addition to hybridization buffer; this is normally achieved by heating the probe (95°C/10 min) followed by rapid ice chilling. The size of these single-stranded segments is dependent upon the number of nicks introduced originally by DNase I. Ideally, the DNA fragments should be approximately 400 to 800 nucleotides in length. Fragments in this size range produce optimal signal to noise ratios, presumably from their ability to hybridize with each other in overlapping complementary regions to form networks, a process accelerated by dextran sulfate.[6]

Commercial kits are widely available for nick translation. The common protocol for these is to mix together the sample DNA, DNase I, DNA polymerase I, and the labeled and unlabeled dNTPs with both enzymes acting *simultaneously* on the DNA. This procedure is critically dependent on the delicate balance between the three macromolecular components so that variability can occur in probe lengths and the degree of labeling. Koch et al.[12] have presented a modified nick translation protocol whereby the action of the two enzymes are separated. Here, DNase I treatment precedes the incubation with DNA polymerase I; the resulting probes have been found to possess higher specific activities and more constant fragment length than probes prepared by traditional nick translation.

B. Random Priming or Oligolabeling

A technique for conveniently radiolabeling DNA restriction fragments to high specific activity has been described by Feinberg and Vogelstein.[13,14] The DNA fragment is first denatured and hybridized to a mixture of short primer of random sequence which are then used to prime the synthesis of a complementary DNA strand using the large (Klenow) fragment of *E. coli* DNA polymerase I[15] and four dNTPs, one or more of which are α-^{32}P-labeled. This large fragment lacks the 5′→3′ exonuclease activity associated with the complete enzyme and, therefore, cannot carry out the nick translation reaction.

The mixture of short oligodeoxynucleotide primers required for this technique was originally obtained by the method of Taylor et al.[16] where calf thymus DNA is digested to completion with DNase I in the presence of Mg^{2+} and then autoclaved to inactivate the DNase I and completely denature the fragments. Such a limit digest contains a mixture of short fragments only a few residues long and presumably all possible sequences. More recently, a mixture of hexanucleotides prepared by chemical synthesis and containing all possible sequences has proven to prime more efficiently. Such synthetic mixtures are readily available and normally are included in commercial kits for oligolabeling. A practical point that may be important in the use of these hexamers is that the average size of nascent, single-strand DNA products is an inverse function of the primer concentration.[17] A diagrammatic representation of the oligolabeling procedure is shown in Figure 2.

It should be noted that DNA synthesis is initiated at several sites along the DNA template; hence, after a final denaturation step, the radiolabeled DNA prepared in this manner will consist of partial length copies of the original restriction fragment. Using standard protocols, the mean probe length obtained with oligolabeling is between 200 to 400 nucleotides and is more than suitable for use in Northern and Southern blots, colony screening, and *in situ* hybridization. At least 40 to 60% of the added dNTP label will become incorporated within 3 h with as little as 25 ng sample DNA being used. The specific activity of the resulting probes is high ($>1 \times 10^9$ cpm/μg) (Table 2). The method is rapid; DNA isolated from low

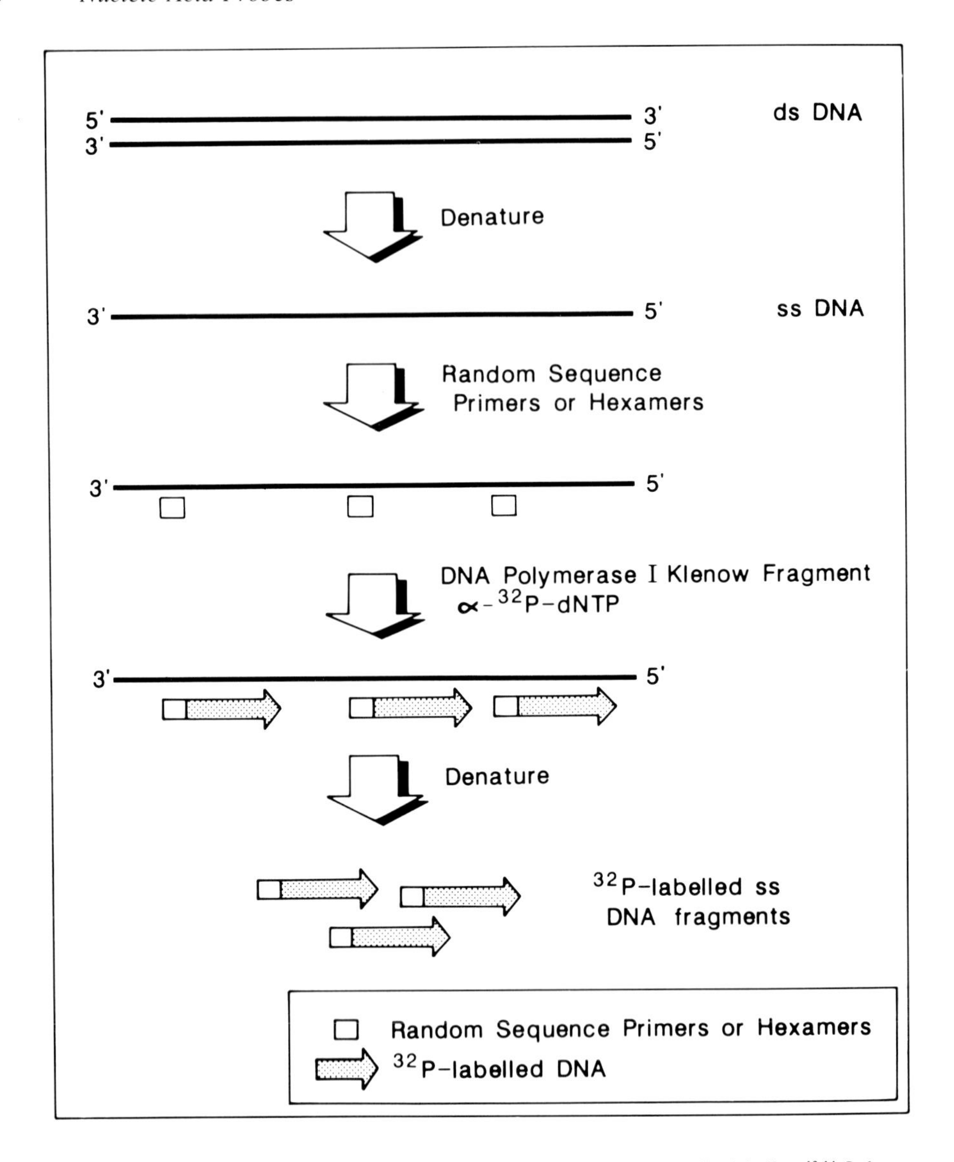

FIGURE 2. Preparation of ^{32}P-labeled DNA probes by random priming or oligolabeling.[13,14] Only one DNA strand is illustrated here as acting as a template; however, both DNA strands can serve as templates.

melting point agarose can be labeled in the presence of the melted gel without any need for further purification.[14]

The oligolabeling method provides several distinct advantages over nick translation for labeling DNA and is becoming an economical and efficient alternative.

1. Labeled complementary DNA to single-stranded DNA can be rapidly prepared; this is not possible with nick translation which requires double-stranded DNA. The obvious limitation is that the single-stranded template DNA itself cannot be labeled; if this is required, then the starting material must be its double-stranded form or a single-stranded cloned complementary copy of it (see Section II.C).
2. The oligolabeling method is easier to perform because it is simpler and does not require treatment of DNA with DNase I, the reaction which causes most problems with the nick translation procedure.

Table 2
MAXIMUM SPECIFIC ACTIVITY OF DNA OBTAINABLE IN RELATION TO THE NUMBER AND SPECIFIC ACTIVITY OF THE α-^{32}P-dNTPs IN ENZYMATIC REACTION MIXTURES

S.A. of each α-^{32}P-dNTP (Ci/mmol)	S.A. (cpm/μg × 10^{-9}) of product DNA[a]			
	No. of α-^{32}P-dNTPs in reaction mix			
	1	2	3	4
500	0.75	1.5	2.25	3.0
1000	1.5	3.0	4.5	6.1
2000	3.0	6.1	9.1	12.1
3000	4.5	9.1	13.6	18.2
4000	6.0	12.1	18.2	24.2
5000	7.6	15.2	22.3	30.3

[a] Calculations are based on 1.0 mCi = 2.2 × 10^{12} cpm and an average nucleotide residue weight of 330.

3. Perhaps the most important advantage is that it can provide ^{32}P-labeled DNA of higher specific activity than nick translation.

As pointed out in Section II.A, the specific activity of nick translated DNA is determined by the percent of total residues which are replaced by the one or more α-^{32}P-dNTPs that are used, as well as the specific activity of the starting α-^{32}P-dNTPs. A low percent replacement therefore decreases the potential advantage of using α-^{32}P-dNTPs of the highest specific activity that are available in the laboratory. In the case of the oligolabeling procedure, the specific activity of the product labeled DNA is determined solely by the specific activity of the α-^{32}P-dNTPs when *single-stranded* DNA is used as the starting material. With *double-stranded* DNA, the final specific activity of labeled product is, as with nick translated DNA, dependent on the extent to which the *total* template DNA is copied. In practice, the ease with which the system can be saturated with oligonucleotide primers ensures a high percent copying of the template and hence higher specific activity DNA.

The relationship of the specific activity of α-^{32}P-dNTPs to the theoretical maximum specific activity of DNA obtainable is given in Table 2. The oligolabeling technique using *single-stranded* DNA should provide labeled complementary DNA with specific activities indicated, whereas oligolabeling and nick translation of *double-stranded* DNA will provide products of lower specific activity. Using one or two labeled dNTPs of greater than 1,000 Ci/mmol in each reaction mixture, it should be possible to readily obtain labeled probes with a specific activity greater than 10^9 cpm/μg (Table 2).

In using α-^{32}P-dNTPs of high specific activity, it is important to appreciate that the actual concentration of the labeled dNTP in the reaction mixture will be very low. Table 3 gives the calculated concentrations of dNTP in a reaction mixture of the range 0.1 μCi/μl to 5 μCi/μl and specific activity from 1000 to 5000 Ci/mmol. For example, if an α-^{32}P-dNTP at 2000 Ci/mmol is used at 1 μCi/μl, then its concentration will be 0.5 μM. This is to be compared with a Km of 8 μM for dTTP in the *E. coli* DNA polymerase I reaction with poly(dAT) as a template.[18] Hence, the concentration of labeled nucleotide will be the limiting component in essentially all the reactions described in this chapter.

C. M13 Bacteriophage Systems

High specific activity hybridization probes can be generated using recombinant vector clones based upon the single-stranded M13 bacteriophage DNA.[19-21] The DNA which ulti-

Table 3
ACTUAL CONCENTRATION OF α-^{32}P-dNTP IN A REACTION MIXTURE IN RELATION TO ITS SPECIFIC ACTIVITY AND ITS RADIOACTIVE CONCENTRATION

Radioactive concentration (μCi/μl)	Concentration of dNTP in reaction mixture (μM)				
	S.A. (Ci/mmol) of α-^{32}P-dNTP				
	1,000	2,000	3,000	4,000	5,000
0.1	0.1	0.05	0.03	0.025	0.02
0.5	0.5	0.25	0.17	0.12	0.10
1.0	1.0	0.5	0.33	0.25	0.20
5.0	5.0	2.5	1.67	1.25	1.00

mately serves as a source of probe must first be cloned into the M13 DNA vector by standard cloning techniques.[1] M13 hybridization probes can then be prepared by upstream priming, downstream priming, or the use of a M13/M13 replicative form (RF) system.

1. M13 Upstream (Universal) Probe Primer

Single-stranded M13 phage DNA containing the specific insert of interest is first hybridized with a small oligonucleotide primer, 16 bases in length (5′-CACAATTCCACACAAC-3′), complementary to the 5′-side of the phage (plus strand) sequences upstream from the insert. Extension of the primer away from the insert using the Klenow fragment of *E. coli* DNA polymerase I in the presence of one or more α-^{32}P-labeled dNTPs will generate hybridization probes of high specific activity. M13 probe preparation using this upstream primer technique is depicted diagrammatically in Figure 3.

It is important to note that, under the experimental conditions specified, synthesis of the ^{32}P-labeled complementary strand does not proceed to completion; thus the inserted probe sequence remains single-stranded and hence available for hybridization (Figure 3). Probe preparation is completed by removal of the low molecular weight components of the reaction mixture by passage through a Sephadex G-50 (medium) chromatography column. Based on our experience,[22] we consider it advisable to phenol extract the reaction mixture prior to purification of the ^{32}P-labeled probe on the Sephadex G-50 column to eliminate the occasional low, nonspecific background on autoradiograms after hybridization of dot-blots on nitrocellulose membranes.

DNA probes prepared by this approach have a specific activity of approximately 1.5×10^8 cpm/μg of total clone.[22] Two cautionary points in using this upstream priming technique are: (1) Ensure that the correct insert orientation has been achieved in cloning into the M13 vector. If, for example, a (+) target viral RNA is under study, then a (−) orientation in the clone must be used for probe manufacture. (2) Probes prepared by this method do not require heat denaturation prior to hybridization.

2. M13 Downstream Probe Primer

By utilizing a different primer from that described above, one that now hybridizes downstream and on the opposite (3′) side of the cloned insert of a M13 recombinant DNA phage, high specific activity hybridization probes can also be generated.[22,23] The technique first involves hybridization of a synthetic oligonucleotide primer (e.g., the 15-mer 5′-TCCCAGTCACGACGT-3′) with the M13 phage clone under study. On incubation with the Klenow fragment of *E. coli* DNA polymerase I and the four dNTPs, one or more of which is α-^{32}P labeled, DNA synthesis extends from the primer across the cloned insert and into the phage

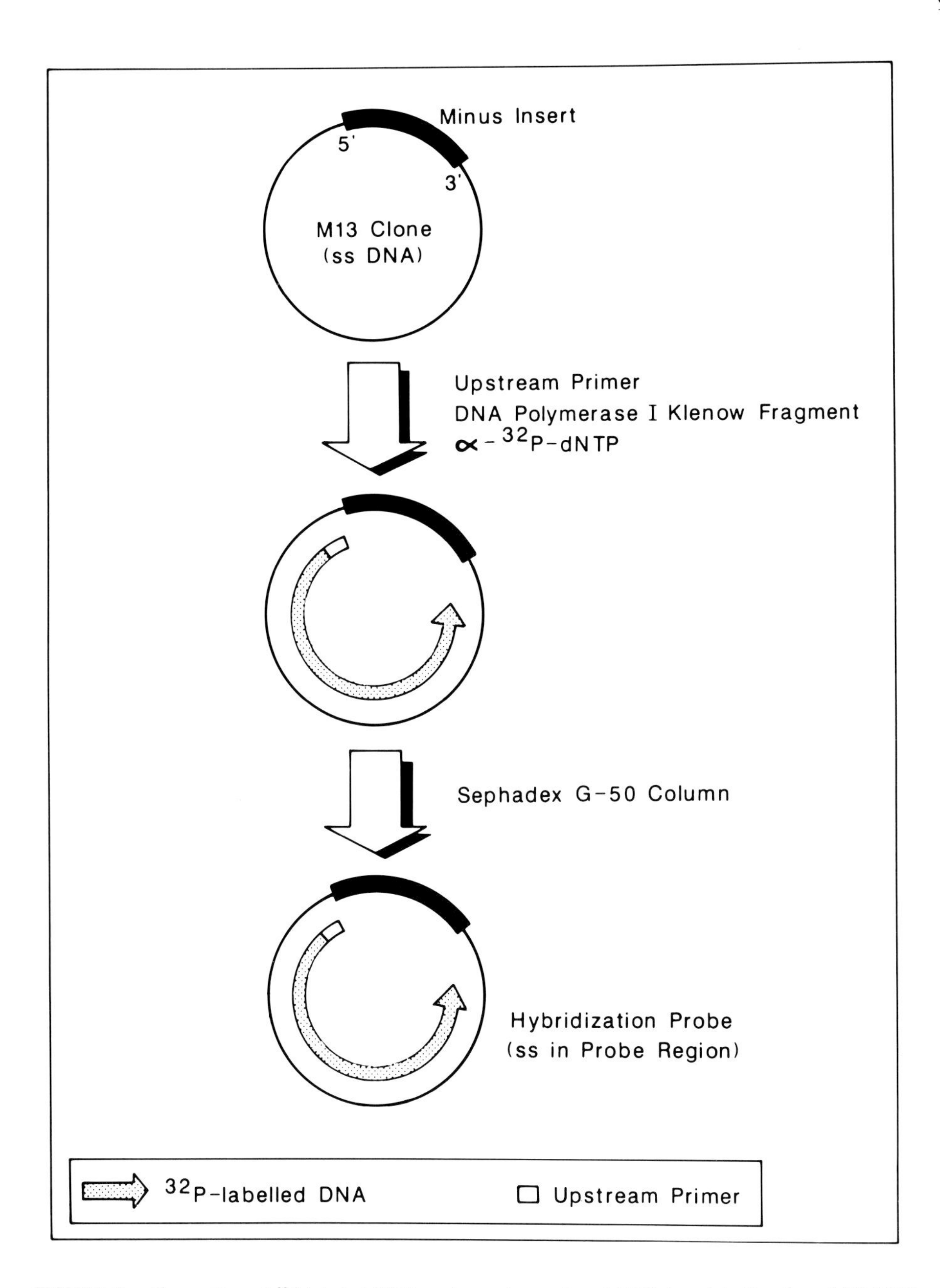

FIGURE 3. Preparation of ^{32}P-labeled DNA probes using a cloned DNA insert in the phage M13 DNA vector[19-21] and a synthetic, upstream (Universal) primer.[22]

vector DNA (Figure 4). The reaction mixture is then incubated with a carefully chosen restriction enzyme in order to cut the double-stranded DNA upstream, i.e., on the 5′-side of the cloned insert. After heat denaturation of the double-stranded DNA, the single-stranded DNA fragments can be purified on the basis of size by polyacrylamide gel electrophoresis under denaturating conditions (Figure 4). The ^{32}P-cDNA single-stranded probe is then eluted from the gel for hybridization usage.

Extensive experience in our laboratory[22,23] has shown that this downstream priming technique for ^{32}P-probe preparation is convenient, rapid and economical in the use of α-^{32}P-dNTPs and gives probes of high specific activity (e.g., approximately 3 × 10^9 cpm/μg

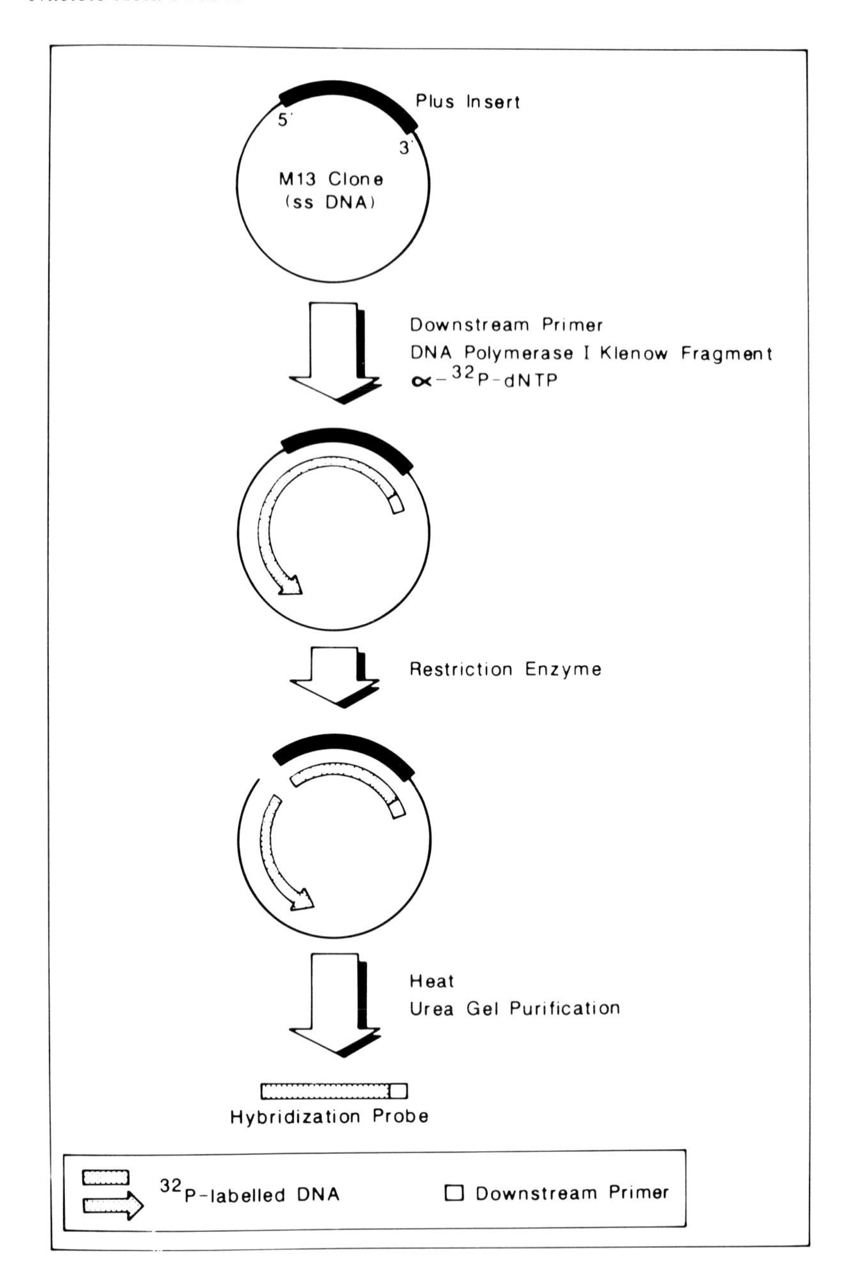

FIGURE 4. Preparation of single-stranded ^{32}P-labeled DNA probes using a cloned DNA insert in the phage M13 DNA vector[19,21] and a synthetic, downstream primer.[22,23]

DNA). A typical autoradiogram of a 5% polyacrylamide/7*M* urea gel used for the preparation of a ^{32}P-DNA probe which detects plus avocado sunblotch viroid (ASBV) sequences[23] is shown in Figure 5. Again, care should be taken in the choice of the clone for labeling: if, as in the previous example quoted (Section II.C.1) the target is a (+) viral RNA, then a (+) orientation of the insert in the clone must be chosen since the technique will produce a (−) orientation labeled ^{32}P-DNA probe.

3. M13/M13 Replicative Form (RF) System

Boyer[24] has outlined a two stage *sandwich hybridization* protocol which utilizes both single-stranded M13 DNA containing the cloned DNA fragment of interest (unlabeled) and

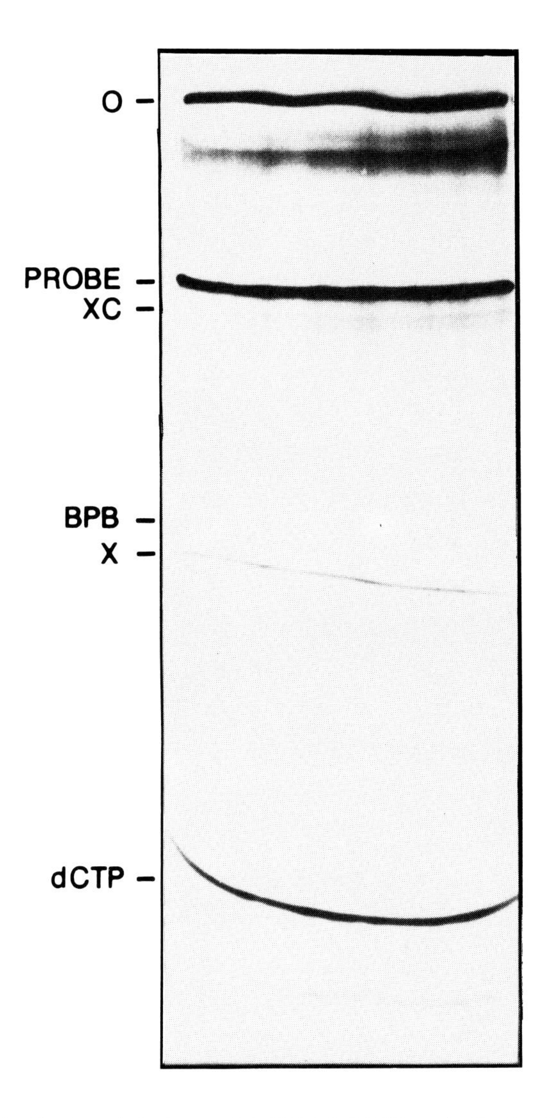

FIGURE 5. Purification of single-stranded ^{32}P-labeled DNA probe for the detection of the plus RNA of avocado sunblotch viroid (ASBV).[23] The reaction mixture (Figure 4)[23] was first denatured by heating (100°C/3 min) in 35% deionized formamide and then electrophoresed on a 5% polyacrylamide gel (40 × 20 × 0.05 cm) for 1.2 h at 30 mA in 90 m*M* Tris-borate (pH 8.3), 2 m*M* EDTA, 7 *M* urea. Autoradiography was for 2 min *Abbreviations:* 0, origin of gel; probe, position of the single-strand DNA probe; XC and BPB, position of the marker dyes, xylene cyanol FF and bromophenol blue; X, unidentified material; dCTP, unincorporated α-^{32}P-dCTP. (From Bruening, G., Gould, A. R., Murphy, P. J., and Symons, R. H., *FEBS Lett.*, 148, 71, 1982. With permission.)

M13 RF DNA (radioactively labeled). The technique is depicted diagrammatically in Figure 6.

In the first stage, unlabeled single-stranded M13 DNA hybridizes with the target nucleic acid by complementary base pairing of the cloned sequence. In the second stage, hybridization is repeated with M13 RF DNA which has been nick-translated with ^{32}P and denatured; labeled sequences complementary to the M13 vector now hybridize with M13 vector tails which protrude from the hybridized target sequences (Figure 6). Using this approach, several probes representing numerous regions along a DNA fragment can be tested simultaneously, but only one, the M13 RF, needs to be radioactive.

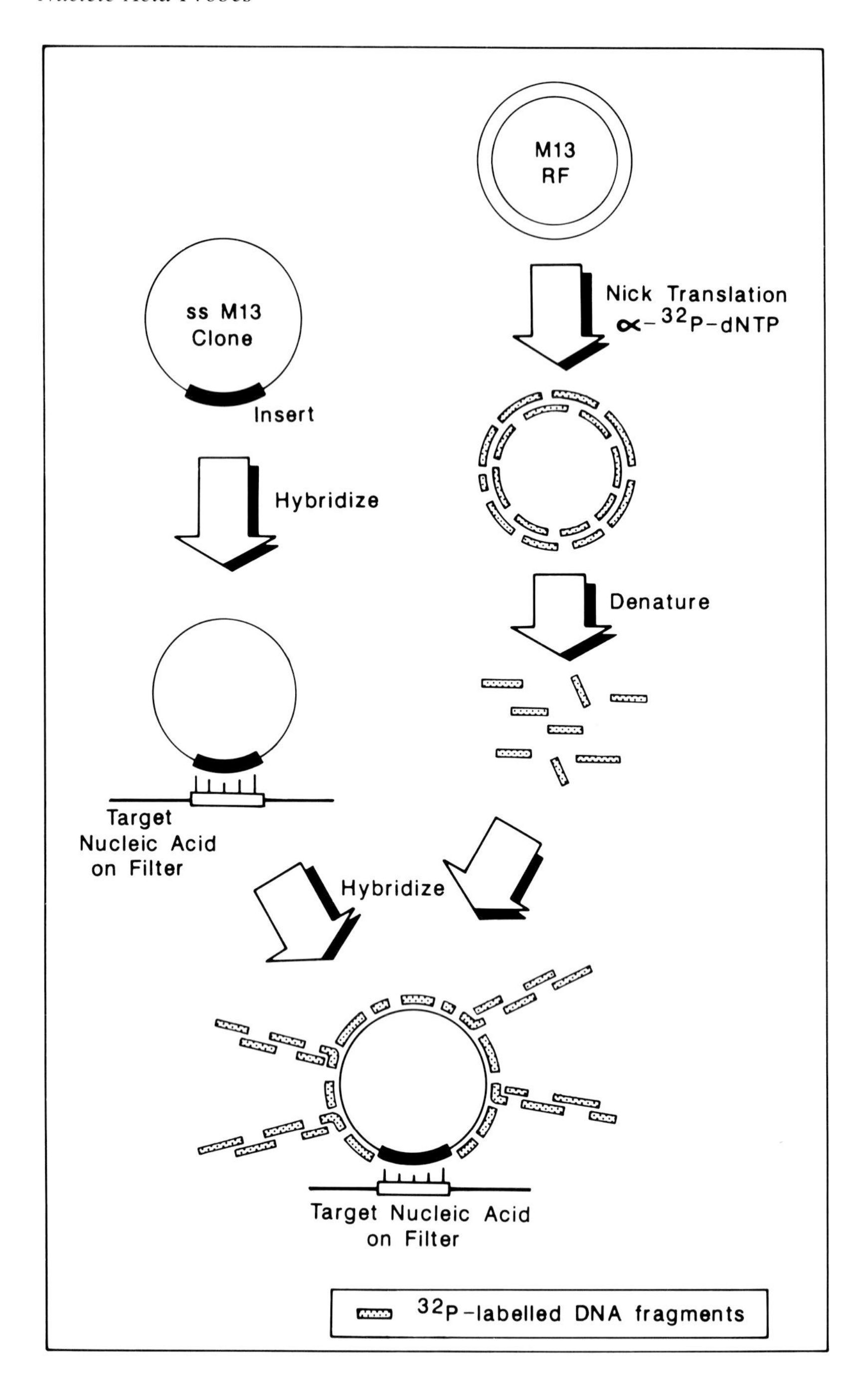

FIGURE 6. A two-stage sandwich hybridization protocol.[24] Stage 1 involves the hybridization of unlabeled single-stranded phage M13 DNA containing cloned DNA insert with the target nucleic acid. Hybridization is repeated in Stage 2 with nick translated (^{32}P-labeled) and denatured phage M13 DNA replicative form (RF).

D. T4 DNA Polymerase

T4 DNA polymerase possesses both a 3′→5′ exonuclease activity as well as 5′→3′ DNA polymerse activity.[25,26] The enzyme can be utilized in a convenient replacement reaction to end label DNA strands for use as hybridization probes using one or more α-^{32}P-dNTPs (Figure 7). In the absence of exogenous dNTPs only the exonuclease is active; it is ap-

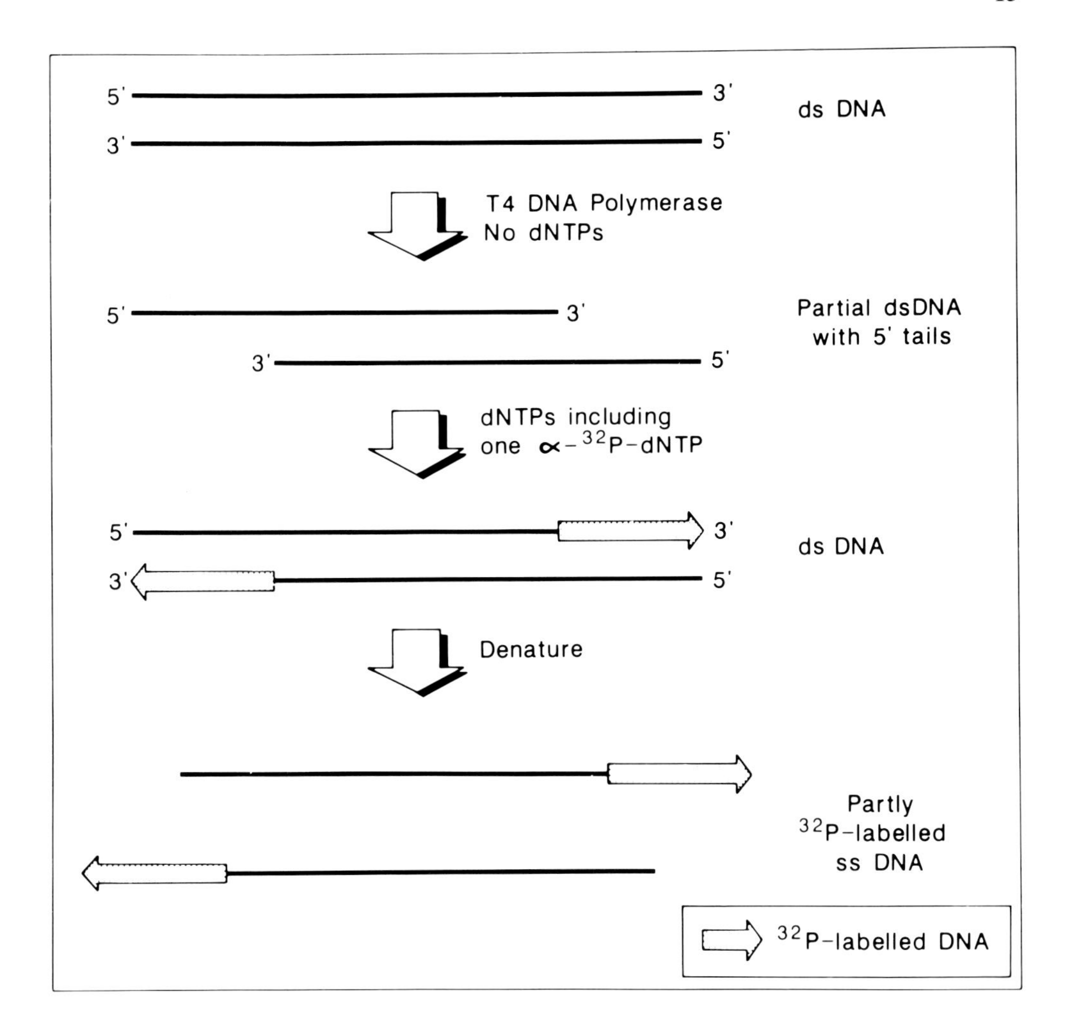

FIGURE 7. Preparation of 3′-terminally labeled DNA probes using T4 DNA polymerase.

proximately 200 times as active as the corresponding exonuclease of DNA polymerase I and, under normal conditions, will remove approximately 25 nucleotides per minute giving rise to DNA molecules with recessed 3′-termini. On subsequent addition of labeled dNTPs, the partially digested DNA molecules serve as primer templates that are regenerated by the polymerase into intact, double-stranded DNA.

Unlike the nick translation reaction, this procedure yields intact double-stranded DNA molecules with no nicks. Probe denaturation will therefore result in intact single-stranded DNA molecules with relatively long ^{32}P-labeled stretches. The method is suited to labeling short DNA fragments of approximately a few hundred bases. Care should be taken if the DNA to be labeled is a circular recombinant plasmid; it must first be linearized by cutting with a suitable restriction enzyme. Commercial kits are available for T4 DNA polymerase labeling.

Because of the asymmetry of the end-labeling, this replacement synthesis method can be used to specifically label one strand or another of a cloned DNA fragment.[1,26] For example, after the end-filling reaction and prior to heat denaturation, the double-stranded fragment can be cut with a restriction enzyme to give two fragments of different size which can then be separated by gel electrophoresis. The labeled strand of each of these heat denatured fragments will then hybridize to opposite strands of a target sequence. Such probes can be extremely useful in defining the direction of transcription by their ability to hybridize to RNA.

E. T4 Polynucleotide Kinase

Labeling of DNA and RNA fragments at the 5′- end with T4 polynucleotide kinase is a method currently employed to obtain ^{32}P-labeled nucleic acids for sequencing purposes. The enzyme transfers the γ- phosphate of ATP to a free 5′-hydroxyl group in either DNA or RNA.[1,27-29]

In the *main (forward) reaction* the phosphate groups at the 5′- ends of the DNA or RNA fragments are first removed with bacterial or calf-intestinal alkaline phosphatase. Polynucleotide kinase then attaches labeled phosphate molecules to the 5′-hydroxyl groups of the fragments when the incubation is carried out in the presence of γ-^{32}P-labeled ATP. Unincorporated nucleotides can be removed by Sephadex G-50 chromatography. The enzyme can also catalyze the exchange of the γ-phosphate group of ATP for the existing 5′-terminal phosphate. This second procedure, known as the *exchange reaction,* needs to be carried out in the presence of excess ADP in order to be efficient. Both reactions give rise to efficient labeling if fragments possess protruding 5′-ends; fragments with blunt or recessed 5′-ends are poorly phosphorylated. The specific activity of the final product is solely dependent on the specific activity of the γ-^{32}P-ATP if the phosphorylation of 5′-hydroxyls goes to completion and cannot be adjusted by addition of more enzyme. Commercial kits are available for T4 polynucleotide kinase labeling.

The labeling of nucleic acid fragments with polynucleotide kinase is extremely sensitive to the presence of certain impurities.[1,30] Agarose, for example, is known to contain sulfate impurities which may inhibit the kinase reaction. Hence, fragments for labeling should preferentially be eluted from acrylamide rather than agarose gels. NH_4^+ ions are also strong inhibitors of polynucleotide kinase; ethanol precipitations of nucleic acid fragments prior to the kinase reaction should therefore be carried out with a sodium salt. Recently, Harrison and Zimmerman[31] have shown that addition of high concentrations of nonspecific polymers (e.g., polyethylene glycol 8000, Ficoll 70) cause a striking increase in kinase activity in both the forward and exchange reactions. This effect was shown to result from a stabilization of the enzyme, presumably due to *macromolecular crowding.*

F. Terminal Deoxynucleotidyl Transferase

Terminal deoxynucleotidyl transferase (terminal transferase) catalyses the polymerization of dNTPs at the 3′-ends of single-stranded DNA.[32] The enzyme can accept double-stranded DNA as primer when Co^{2+} is used in place of the normal cofactor, Mg^{2+}. In the presence of Co^{2+}, all forms of duplex DNA molecules can be labeled at their 3′-ends regardless of whether such ends are single-stranded or base-paired.[33]

The most common protocol for 3′-end labeling with terminal transferase is based on the method of Tu and Cohen[34]; this method utilizes the nucleotide analogue α-^{32}P-cordycepin triphosphate (3′-deoxyadenosine 5′-triphosphate). This compound lacks a 3′-hydroxyl group so that further nucleotide additions are prevented thus resulting in chain termination and the generation of DNA molecules labeled by the addition of a single nucleotide only. Commercial kits are available for terminal transferase labeling using this approach.

It should be noted that, like the labeling of DNA with T4 polynucleotide kinase (Section II.E), DNA labeling with terminal transferase is also sensitive to contaminating salts (e.g., Na^+ ion) and protein.[35] One disadvantage (again, as seen with T4 polynucleotide kinase), is that the specific activity of the final DNA is totally dependent on the specific activity of the α-^{32}P-cordycepin triphosphate. Overall, the efficiency of labeling by this method is in the order: 3′-protruding ends > blunt ends > 3′-recessed ends. This order is probably a result of increasing steric hindrance of the enzyme.[35]

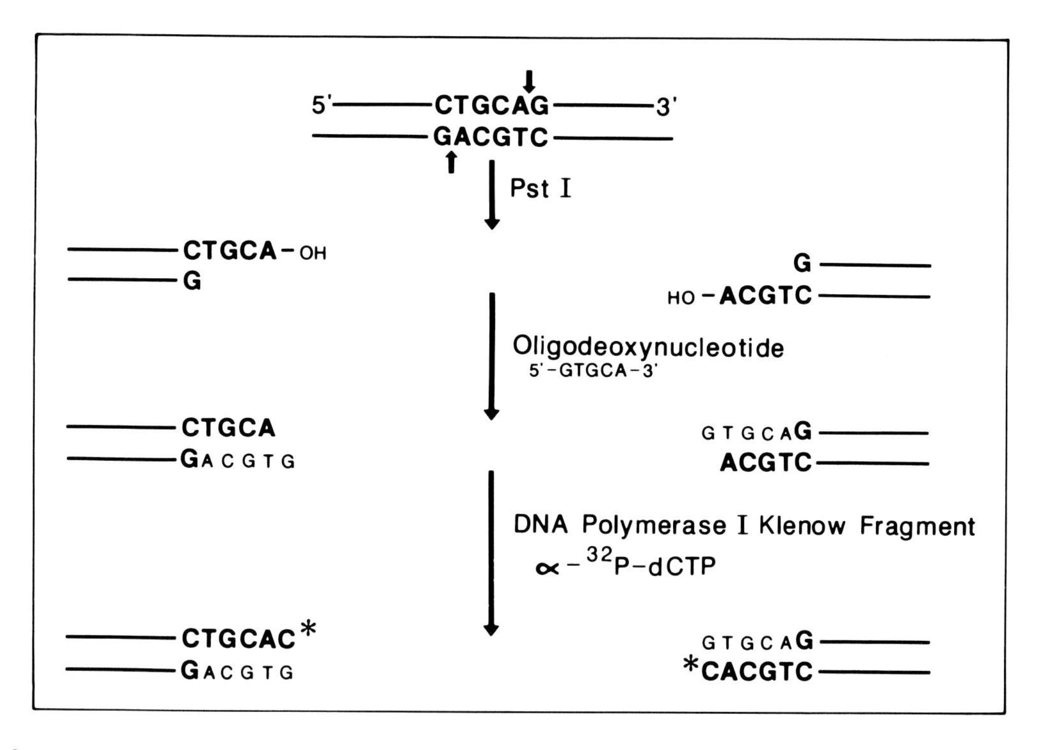

FIGURE 8. A scheme outlining the oligodeoxynucleotide-mediateed ^{32}P-labeling of the 3′ ends of *Pst*I-generated DNA restriction fragments.[36] The * indicates the newly incorporated radioactive nucleotide. (From Chirala, S. S. and Wakil, S. J., *Gene* 47, 297, 1986. With permission.)

A new and simple method has been developed for the efficient ^{32}P-labeling of the protruding 3′-ends of DNA fragments generated by class II restriction enzymes (e.g., Pst I).[36] This method, as an alternative to using terminal transferase, utilizes a synthetic oligodeoxynucleotide, 5′-GTGCA-3′, which is complementary to the protruding 3′-end of the Pst I fragment and hybridizes with it in such a manner that the end of the restriction fragment will now have a protruding, one-nucleotide-long 5′-end. This then serves as a template for the Klenow fragment of DNA polymerase I (see Section II.G) which, in the presence of α-^{32}P-dCTP, will incorporate just one nucleotide at the 3′-end of the restriction fragment. A diagrammatic representation of this approach is given in Figure 8. The method can be utilized in a general fashion to label any DNA restriction fragment with a 3′-sticky end by correctly choosing the appropriate oligonucleotide and α-^{32}P-dNTP.

G. Large (Klenow) Fragment of *E. coli* DNA Polymerase I

The large (Klenow) fragment of *E. coli* DNA polymerase I,[15] as previously stated (Section II.B), possesses the 5′→3′ DNA polymerase activity (and 3′→5′ exonuclease activity) but lacks the 5′→3′ exonuclease activity of the holoenzyme. Radiolabeling with Klenow is therefore most efficient with DNA fragments containing recessed 3′-ends. DNA fragments 3′-end labeled by this technique are suitable for sequencing by the method of Maxam and Gilbert[29] and also for hybridization probe usage.

The principle of 3′-end labeling with the Klenow fragment is shown diagrammatically in Figure 9. Essentially, the fragment is used to fill in the 3′-ends of DNA fragments opposite naturally occuring 5′-extensions or those produced by certain restriction enzymes (e.g. BamHI, EcoRI, HindIII, etc.). The exact choice of dNTPs (including one of which is α-^{32}P-labeled) to be added to the reaction mix is dependent upon the sequence of the protruding 5′-termini at the ends of the DNA fragments.[1,37] For example, to fill in the recessed 3′-ends

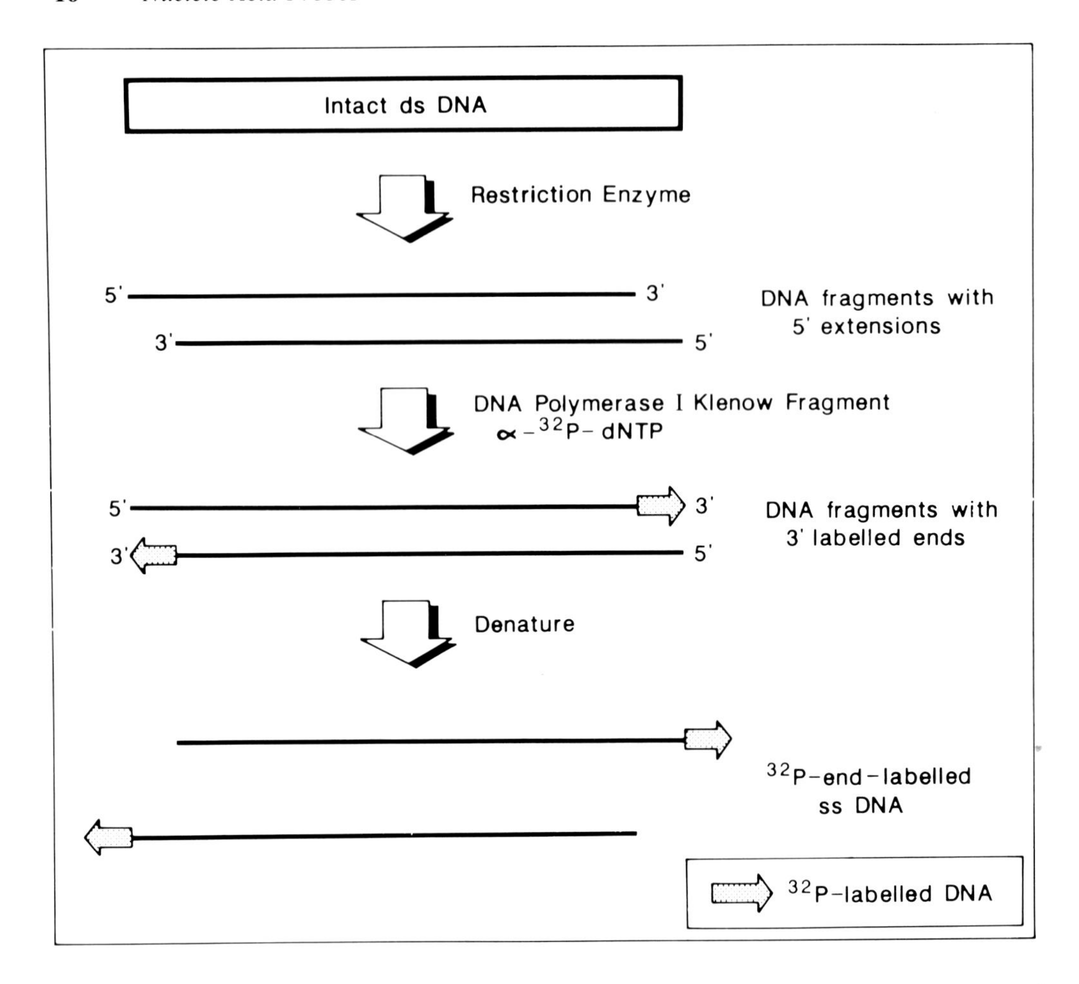

FIGURE 9. Preparation of 3′-end labeled DNA probes by end filling of DNA restriction fragments containing recessed 3′-ends using the Klenow fragment of *E. coli* DNA polymerase I.

created by cleavage of DNA by EcoRI, only dATP and dTTP need be added (since TCGA 5′-overhang is generated). On the other hand, all four dNTPs would be required to fill the recessed 3′-ends created by the restriction enzyme Hind III (since TCGA 5′- overhang is generated).

The labeling of DNA with the Klenow fragment of DNA polymerase I is not as sensitive to inhibition by certain impurities as seen with T4 polynucleotide kinase labeling (Section II.E). Hence, ^{32}P-labeling can be carried out directly on fragments obtained by restriction enzyme digestion and on fragments eluted from agarose gels.[37]

One disadvantage of this labeling approach is that DNA with 3′-overhanging termini is not labeled. If it is necessary to end label DNA which has 3′-overhangs (or blunt ends) and *replacement synthesis* is desired, then an alternative to T4 DNA polymerase labeling (Section II.D) is to use the enzyme exonuclease III. This enzyme, which is a 3′→5′ nuclease specific for double-stranded DNA,[38] can be used in conjunction with the Klenow fragment in a two-step reaction to label the 3′-ends of DNA.[39] First, the 3′-end is chewed back using exonuclease III, leaving the 5′-end protruding as a single-stranded region. The Klenow fragment is then used to refill the gap created by resynthesizing the 3′-end using one or more labeled dNTPs.

H. Reverse Transcriptase

The RNA-dependent DNA polymerase from avian myeloblastosis virus (AMV reverse

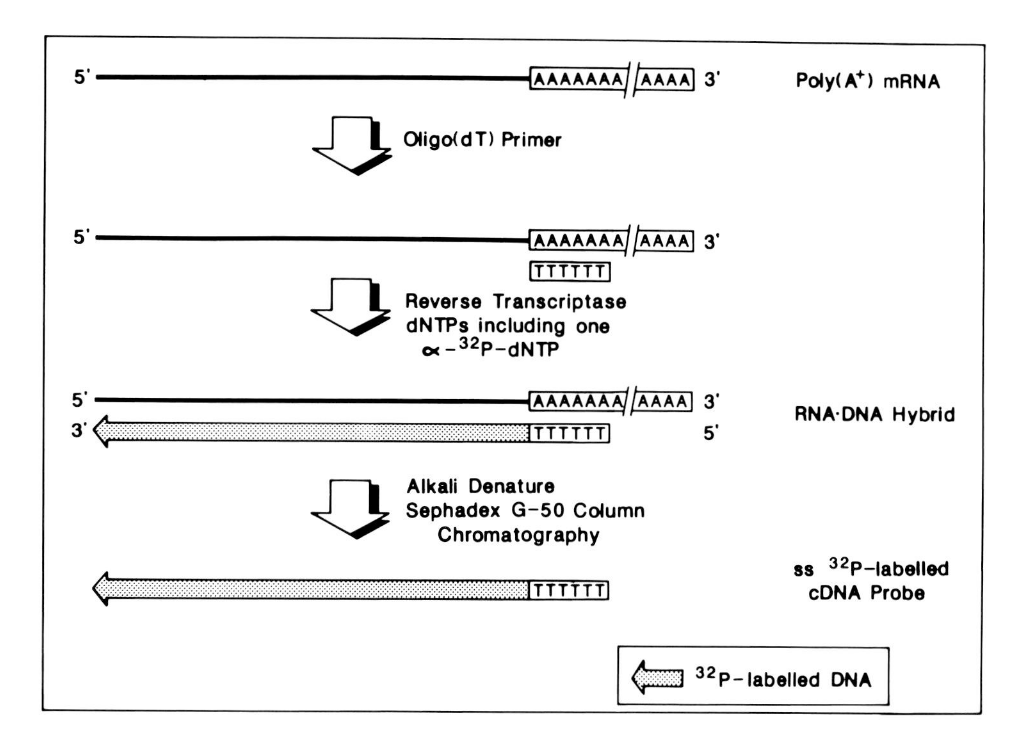

FIGURE 10. Preparation of single-stranded ^{32}P-labeled cDNA probes using RNA and reverse transcriptase. The primer chosen is oligo(dT) since poly (A^+) mRNA is used as template. The alkali treatment hydrolyses the RNA to small fragments as well as rapidly denaturing DNA-RNA hybrids. DNA is stable under these conditions.

transcriptase) displays several enzymatic activities including a 5′→3′ DNA polymerase activity and a degradative (RNase H) activity specific for RNA/DNA hybrids.[40] The enzyme is used mainly to transcribe mRNA into double-stranded DNA (cDNA) which can then be inserted into prokaryotic vectors for cloning purposes. Reverse transcriptase also finds use with either single-stranded DNA or RNA templates to make ^{32}P-radiolabeled probes of high specific activity.

Synthesis of ^{32}P single-stranded cDNA probes can be carried out on any single-stranded DNA or RNA template in an incubation mix containing reverse transcriptase, dNTPs, including one α-^{32}P-labeled, and a suitable primer(s).[1] The primers chosen may be either oligo(dT) if poly(A^+) mRNA is used as template or a collection of randomly generated oligodeoxynucleotides or hexamers where mRNA or DNA is the template.[13,14,16] Whereas oligo(dT) binds only to the 3′-poly (A^+) region of mRNA and so primes the synthesis of DNA that is heavily biased toward 3′-end sequences of the mRNA template, randomly generated oligonucleotides bind to all parts of the template and so enhance cDNA synthesis on all parts of the template. The RNA can be removed as the end of the reaction by alkaline hydrolysis and the labeled DNA separated from unincorporated dNTPs by Sephadex G50 column chromatography. Figures 10 and 11 show the use of reverse trasncriptase for cDNA probe preparation from polyadenylated and nonpolyadenylated mRNA, respectively.

Ashley and MacDonald[41] have presented an alternative strategy involving reverse transcriptase for generating single-stranded DNA hybridization probes. Here, single-stranded M13 DNA containing a cloned cDNA insert is bound to diazobenzyloxymethyl (DBM) - cellulose. After priming with either the M13 universal primer or oligo(dT), a complementary DNA strand is synthesized with reverse transcriptase. Following synthesis, the unincorporated radiolabeled nucleotides can be washed away and the probe is eluted from the DBM-cellulose solid support with formamide.

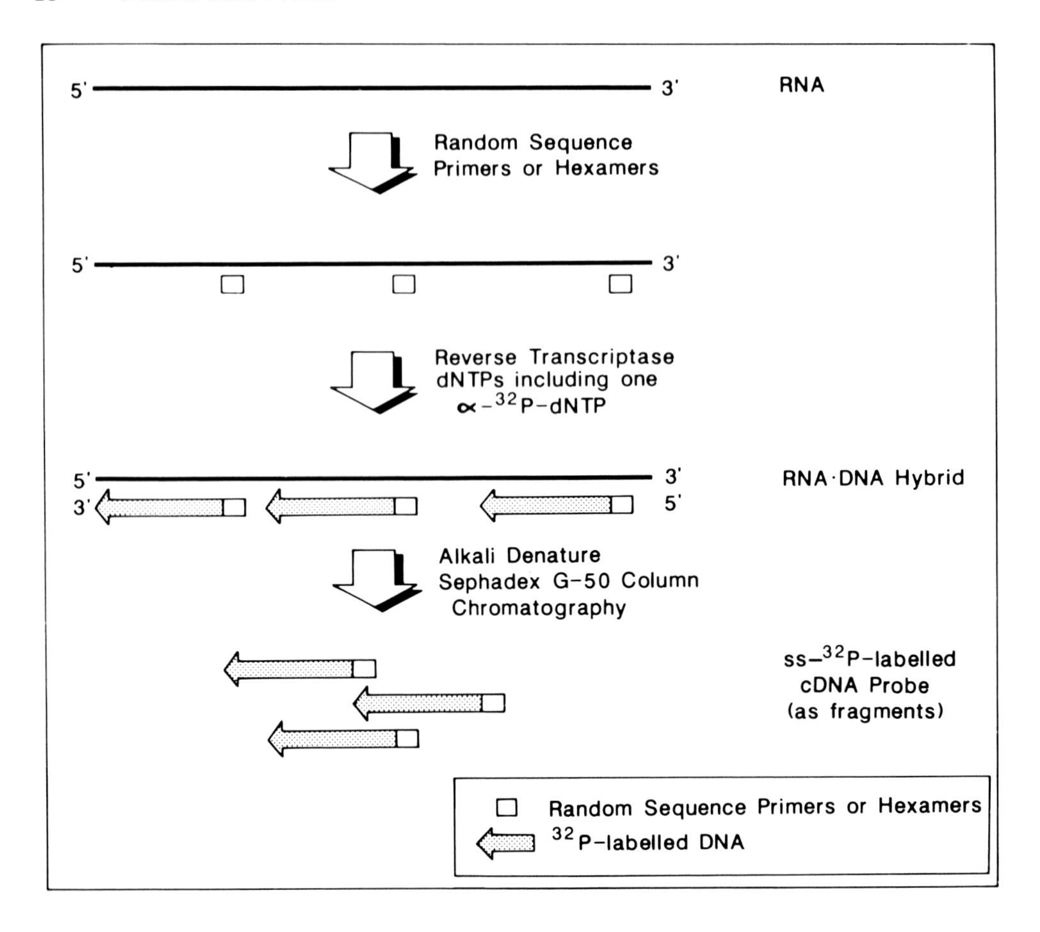

FIGURE 11. Preparation of single-stranded ^{32}P-labeled cDNA probes using RNA and reverse transcriptase. The primer chosen is either a collection of randomly generated oligodeoxynucleotides or hexamers[13,14,16] since non-polyadenylated mRNA is used as template.

To limit the DNA copy to the insert, a preliminary synthesis reaction is performed with unlabeled nucleotides, primer, and enzyme, followed by digestion of the reaction mix with a restriction endonuclease that recognizes a unique site in the recombinant DNA immediately upstream of the 5′-side of the cDNA insert. After elution of the unlabeled synthesized complementary DNA, a second synthesis reaction now yields highly radiolabeled single-stranded DNA that extends only the length of the mRNA insert. One major advantage is that the restriction enzyme-cleaved, cellulose-bound template can be stored and reused repeatedly.

III. ENZYMATIC LABELING OF RNA

A. The SP6 RNA Polymerase System

A simple and efficient method for synthesizing pure single-stranded RNAs of high specific radioactivity has been described.[42,43] This *in vitro* transcription system is based on specific RNA synthesis by bacteriophage SP6 RNA polymerase[44] which exhibits an extremely high specificity for SP6 promoter sequences.[45]. This property can be exploited by cloning the DNA sequence of interest directly downstream from an SP6 promoter. Specialized cloning vectors pSP64 and pSP65[42] have been constructed to contain an SP6 promoter directly upstream of a multiple cloning region to facilitate insertion of DNA sequences downstream from the SP6 promoter. (The pSP64 and pSP65 vectors differ in the orientation of the polylinkers.) The gene for the enzyme has recently been isolated from the DNA of phage SP6 purified from infection *Salmonella typhimurium* and cloned into an *E. coli* expression vector.[46] The enzyme was efficiently expressed and purified. It is expected that commercial preparations of the enzyme will be available in the near future.

In order to obtain RNA transcripts of a defined length, it is first necessary to linearize the recombinant plasmid at a unique site immediately downstream of the inserted DNA sequence. Transcription in the presence of the four NTPs, one or more of which is α-^{32}P-labeled (e.g., α-^{32}P- UTP; specific activity 3000 Ci/mmol), and SP6 RNA polymerase proceeds from the SP6 promoter through the probe sequence, giving rise to a near homogeneous population of *run-off* RNA transcripts of high specific radioactivity. The length of the transcript is defined by the precise relationship of the promoter and DNA inserted and by the *run-off* cleavage site. The transcript will be either mRNA-like or *anti*-mRNA depending upon the orientation of the gene or DNA inserted within the plasmid. An outline of the SP6 RNA polymerase procedure is given in Figure 12.

One advantage of the method is that the labeled single-stranded RNA probes can be prepared without a gel isolation step. This is due to the fact that the transcription reaction yields only a single RNA species (Figure 12). Once the transcription reaction is completed, the DNA template can be removed by treatment with RNase-free DNase. The reaction mixture is then phenol extracted and ethanol precipitated to remove all unincorporated radionucleotides and the digested DNA template.

SP6 derived RNAs have several advantages as hybridization probes. They can be synthesized at very high specific activities (Table 2) and, being single-stranded, an increase in probing efficiency results when compared to nick translated DNA probes since there is no competition for the complementary labeled DNA strands. Single-stranded RNA probes are also potentially more effective than single-stranded DNA probes since RNA/RNA and RNA/DNA duplexes are more stable than DNA/DNA duplexes.[47-49] This extra stability should allow for the formation and washing of hybrids under more stringent conditions, which should increase the signal to noise ratio. Depending upon the reaction conditions chosen, as much as 80 to 90% of the α ^{32}P-NTP can be incorporated into transcripts; typically, 250 ng of a single RNA probe can be synthesized with a specific activity of $> 6 \times 10^8$ dpm/μg.[42]

Other advantages include the stability of the SP6 RNA polymerase enzyme, the means of generating large amounts of single-strand RNA of predefined length, e.g., 10 μg RNA per 1 μg template plasmid DNA can be achieved,[42] and the versatility of available vectors (see Section III.C). RNA probes prepared by this SP6 transcription system have been shown to be especially effective for Northern blots and *in situ* hybridizations where they offer up to a tenfold increase in detection sensitivities over conventional nick translated DNA probes.[43] One disadvantage, however, is their susceptibility to degradation by ubiquitous contaminating RNases.

B. The T7 RNA Polymerase System

T7 RNA polymerase is a DNA-dependent RNA polymerase which has high specificity for T7 promoter sequences.[50] RNA synthesis with the enzyme is dependent upon the presence of the four ribonucleoside triphosphates, Mg^{2+} as well as template DNA. The gene for the enzyme (T7 gene 1) has been cloned[51,52] and can be expressed at a high level in *E. coli*. Several vectors are currently available commercially which contain the T7 promoter sequence for T7 RNA polymerase. These vectors, like those described in the previous section (III.A), can be used to construct recombinant DNA clones which, when used with T7 RNA polymerase, generate high specific activity *run-off* RNA transcripts suitable for probe usage.

Two interesting and novel approaches for generating labeled RNA transcripts in M13

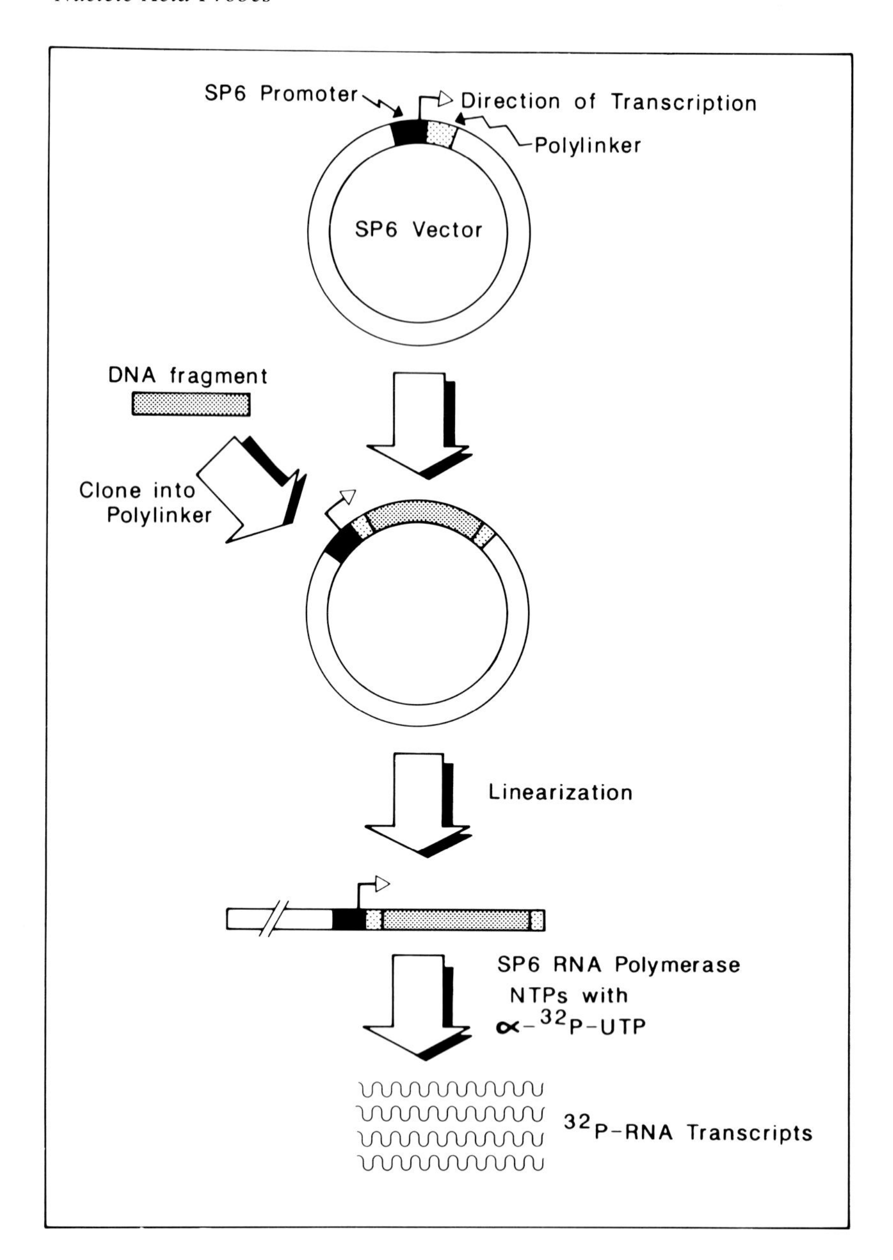

FIGURE 12.. General scheme for the *in vitro* synthesis of ^{32}P-labeled RNA using the bacteriophage SP6 DNA vector system.[42,43] The DNA sequence to be transcribed is cloned into the polylinker site of the SP6 DNA vector which contains the promoter for phage SP6 RNA polymerase. After cutting with a suitable restriction enzyme to generate a linear DNA template, *in vitro* transcription is initiated by addition of buffer, NTPs and SP6 RNA polymerase.

vectors using T7 RNA polymerase have been described.[53,54] In the first of these, Eperon[53] has created a series of new vectors called mICE, by cloning synthetic sequences corresponding to the T7 promoter into specific restriction sites of the polylinker regions of M13mp phage DNA vectors. These new vectors can be utilized in a similar manner as the M13mp vectors for cloning, sequencing and *in vitro* mutagenesis.

In each of these constructs, the T7 promoter has been inserted downstream of the polylinker away from the universal DNA priming site (Figure 13A). To generate labeled RNA tran-

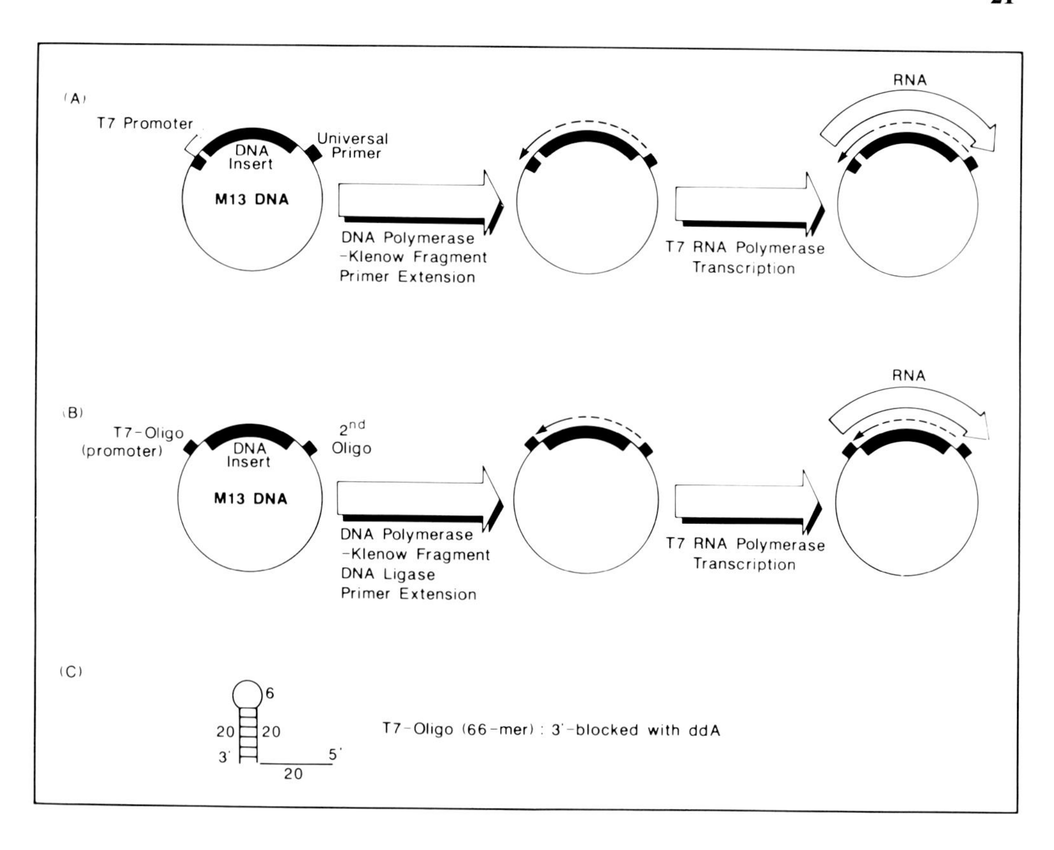

FIGURE 13. (A) Scheme for the *in vitro* synthesis of ^{32}P-labeled RNA transcripts from a specially constructed phage M13 DNA vector which contains a cloned synthetic promoter sequence for phage T7 RNA polymerase.[53] (B) Scheme for the *in vitro* T7 RNA polymerase transcription of a recombinant phage M13 DNA vector, utilizing two oligonucleotides, one of which (T7-oligo) contains the T7 RNA polymerase promoter sequence.[54] (C) Structure of the synthetic oligodeoxynucleotide (T7-oligo) used in (B).[54]

scripts, the single-stranded M13 phage DNA is first used as a template for DNA synthesis by primer extension with Klenow DNA polymerase. (Note: either the universal primer or a different oligonucleotide priming within the cloned DNA fragment can be used to initiate DNA synthesis). After new strand synthesis has crossed the T7 promoter sequence (ensuring promoter is present as double-stranded DNA), transcription is initiated by the addition of NTPs (including one ^{32}P-labeled nucleotide) and T7 RNA polymerase. Transcription will end at the site defined by the primer oligonucleotide (Figure 13A).

A different approach utilizing the M13 vector/T7 RNA polymerase system to generate labeled RNA transcripts has been presented by Krupp and Söll.[54] Here, use is made of a specially constructed oligonucleotide (T7-oligo) which contains the T7 RNA polymerase promoter sequence. In combination with a second oligonucleotide (terminator-oligo), a series of transcripts, initiating and terminating at any chosen positions on a cloned single-stranded DNA in the M13 vector, can be generated.

The following is a brief outline of Krupp and Söll's[54] approach (Figure 13B). The single-stranded M13 phage DNA contains a cloned insert of the same sequence and polarity as the desired transcript; the T7-oligo determines the transcription start site whereas the second oligonucleotide permits the choice of the transcription termination site. The required template DNA strand is first obtained by primer extension of the terminator oligonucleotide (2nd-oligo) with Klenow enzyme, *closing the gap* between the two hybridized oligonucleotides.

DNA ligase is present in the incubation mix to ensure proper joining of the synthesized DNA to the 5′-phosphorylated T7-oligo. This reaction product can be then be used directly as a template for T7 RNA polymerase transcription in the normal fashion (Figure 13B).

The structure of the 66-mer T7-oligo is represented in Figure 13C and contains the double-stranded T7 promoter, a 6-residue loop and a 5′-variable, 20-nucleotide single-stranded sequence which hybridizes to the desired region. A 2′, 3′-dide-oxynucleotide is incorporated at the 3′-end of the T7-oligo to prevent extension by the Klenow DNA polymerase.

C. Paired-Promoter Systems

We have seen previously (Sections III.A and III.B) that both the SP6 and T7 phage RNA polymerases display a high degree of promoter specificity *in vitro*, transcribing selectively from promoter sites on DNA templates. In a similar mode, the RNA polymerase from the bacteriophage T3 also displays such an ability.[55]

In recent years, combination vectors containing promoters for two different RNA transcription systems have become available. Such dual promoter vector systems include pSP6/T7, pSP6/T3 and T7/T3 combinations. For example, the plasmid vector pSP6/T7-19 (2884 bp; BRL) has been constructed from a pUC vector derivative containing the bacteriophage SP6 promoter, the M13mp19 multiple cloning site, and a DNA fragment containing the bacteriophage T7 promoter. The construct is thus designed for the transcription of a DNA insert cloned into the multiple cloning site by use of either of the two RNA polymerases; the SP6 promoter is transcribed in the clockwise direction and the T7 promoter is transcribed in the counter-clockwise direction from the opposite DNA strand. These *in vitro* transcription systems, by utilizing either SP6 or T7 polymerase, thus allow complementary RNA copies of either strand of the inserted DNA to be obtained from a single plasmid construct (Figure 14).

One point of interest with these combination vector systems is that the high promoter specificity of these RNA polymerases should allow synthesis of RNA from either DNA strand with virtually no opposite strand transcription. For example, RNA synthesis by SP6 RNA polymerase is effectively limited to the strand specified by the SP6 promoter and no more than 0.2% of the total transcription product is derived from the opposite strand.[42] However, other reports indicate that care needs to be taken in the use of these paired promoter systems.[56] For example, aberrant transcription from sequences other than the specific promoters can occur. The ends of restriction fragments can provide initiation sites if they contain 3′-protruding ends to the extent of about 3 to 5% of that observed from the promoter sequence. In addition, the T7 and T3 RNA polymerases recognize the heterologous promoter at about 0.5 to 3% frequency. All these effects can be ameliorated by increasing the concentration of NaCl in the reaction mixtures.[56]

D. SP6 (or T7) RNA Polymerase/Oligonucleotide Combination

Wölfl et al.[57] have developed an interesting procedure for the synthesis of highly radioactively labeled RNA hybridization probes from templates involving synthetic single-stranded oligonucleotides and SP6 or T7 promoter-bearing plasmids. Briefly, the technique is as follows (Figure 15).

Oligonucleotides are first synthesized to contain, in addition to the hybridization sequence, four nucleotides at the 3′-ends which are complementary to the 3′-four nucleotide overhang (5′-TGCA-3′) of Pst I restricted DNA. The 3′-end of such a single-stranded oligonucleotide is then annealed to the Pst I generated ends of plasmid vectors pSP64 or pGEM-2 which carry the promoter for SP6 and T7 RNA polymerase, respectively. *E. coli* DNA polymerase I is then utilized in a nick translation reaction to concomitantly translate the nick in the lower strand beyond the promoter sequence and synthesize the complementary DNA strand of the synthetic oligonucleotide. The reaction product is an efficient template for run-off

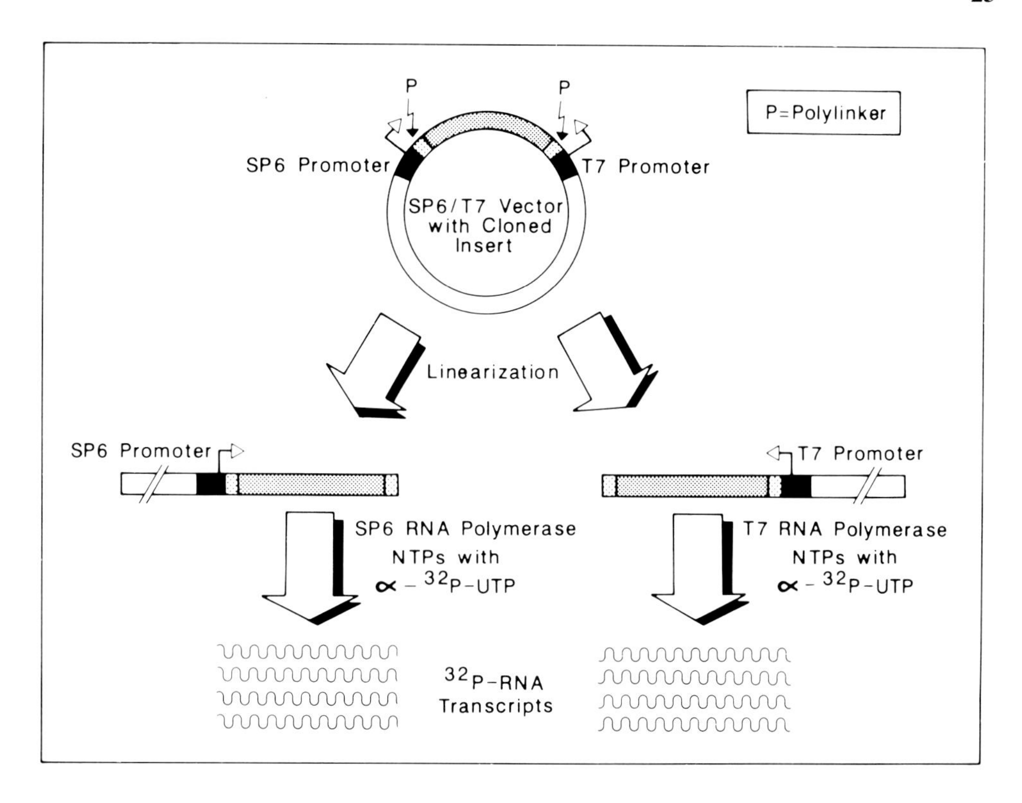

FIGURE 14. General scheme for the *in vitro* synthesis of plus and minus ^{32}P-labeled RNAs using a combination Phage SP6/T7 vector system. The DNA sequence to be transcribed is cloned into the polylinker site of the dual vector which contains phage SP6 and T7 RNA polymerase promoters in opposite orientations. After cutting with a suitable restriction enzyme to generate a linear DNA template, selective transcription of ^{32}P-labeled RNA from either DNA strand can be achieved depending upon the choice of the RNA polymerase.

transcription, by SP6 or T7 RNA polymerse, of RNA hybridization probes at $> 10^9$ cpm/μg.

E. T4 RNA Ligase

The enzyme T4 RNA ligase catalyzes the ATP-dependent formation of an internucleotide phosphodiester bond between an oligoribonucleotide molecule with a 5′-terminal phosphate and an oligoribonucleotide acceptor molecule with a 3′-terminal hydroxyl.[58] For a comprehensive review of the enzyme, readers are referred to Uhlenbeck and Gumport.[59] Although the oligoribonucleotide with the 3′-hydroxyl (the acceptor) must be three or more residues long, it can be joined to an oligoribonucleotide bearing a 5′-phosphate (the donor) as short as a nucleoside 3′,5′-biphosphate.[60]

The above reaction provides a convenient *in vitro* method for labeling the 3′-end of RNA molecules.[60,61] By using a 5′-^{32}P-nucleoside 3′,5′-biphosphate as a donor and an RNA molecule as an acceptor, the product of the reaction is an RNA molecule one nucleotide longer, with a 3′-terminal phosphate and a ^{32}P-phosphate in the last internucleotide linkage. This reaction is:

$$\text{RNA-OH} + {}^{*}\text{pNp} + \text{ATP} \rightarrow \text{RNA}-{}^{*}\text{pNp} + \text{AMP} + \text{PPi}$$

This 3′-terminal labeling method using T4 RNA ligase complements the 5′-terminal labeling protocol outlined previously (Section II.E), which makes use of polynucleotide kinase and γ-^{32}P- labeled ATP. For detailed protocols on the 3′-terminal labeling reaction

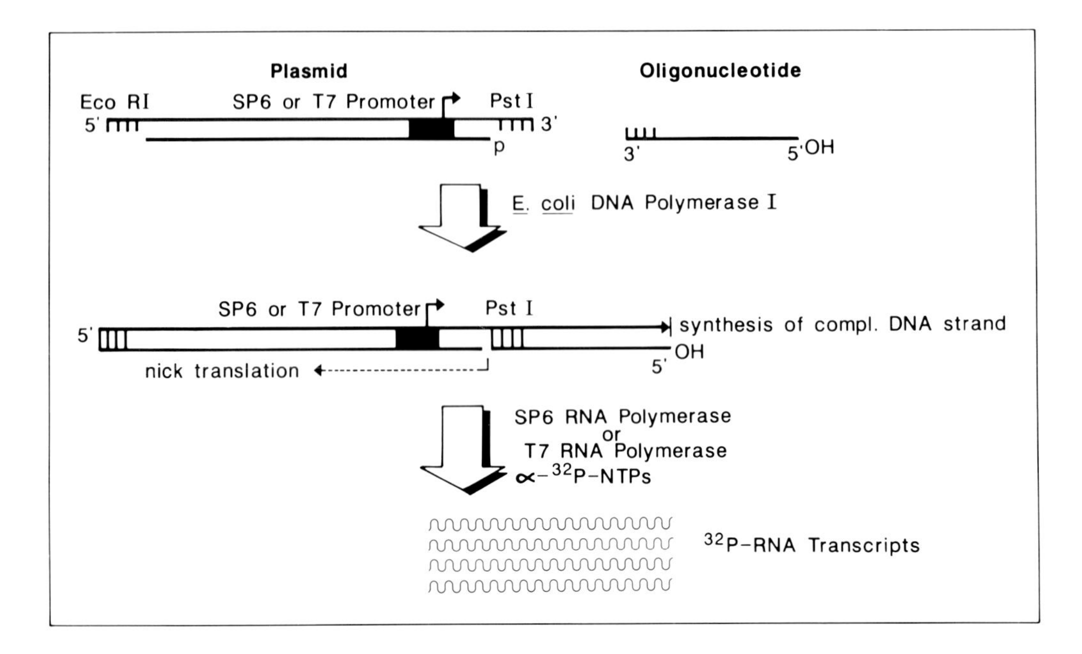

FIGURE 15. Schematic outline of a method[57] for the synthesis of ^{32}P-labeled RNA hybridization probes utilizing synthetic single-stranded DNA oligodeoxynucleotides and plasmid vectors containing a promoter for either phage T7 or SP6 RNA polymerase. (Modified from Wölfl, S., Quaas, R., Hahn, U., and Wittig, B., *Nucleic Acids Res.*, 15, 858, 1987. With permission.)

with T4 RNA ligase and the preparation of 5′-^{32}P- labeled nucleoside - 3′,5′-biphosphates, readers are referred to England et al.[62]

IV. ENZYMATIC LABELING OF OLIGONUCLEOTIDES

Improvements in DNA probe technology have been made possible with the introduction in recent years of automated DNA synthesis machines which can rapidly generate large amounts of oligonucleotides, each made to a specific sequence.[4,5] As hybridization probes, these oligonucleotides, normally 15 to 25 bases in length and labeled with ^{32}P, can distinguish a perfect match in target sequences from various single base-pair mismatches.[63-65] Hence, they have been used to distinguish between normal genes and abnormal ones arising from known point mutations.[65] In addition, synthetic oligonucleotides have been used to introduce mutations in DNA at specific sites[66] and to identify recombinants of interest.[67] Currently, three general enzymatic approaches are used to label synthetic oligonucleotides; these involve the enzymes T4 polynucleotide kinase, the Klenow fragment of DNA polymerase I and terminal deoxynucleotidyl transferase.

A. T4 Polynucleotide Kinase

A convenient method for labeling synthetic oligonucleoides at their 5′-hydroxyl termini is to utilize the enzyme T4 polynucleotide kinase (see Section II.E) and γ-^{32}P-ATP. In a typical reaction,[68] a solution of the oligonucleotide (100 ng in30 μl water) is first heated at 65°C for 5 min and then quick-chilled. Five μl of 10x kinase buffer (0.5 *M* Tris-HCl, pH 7.5, 0.1 *M* $MgCl_2$, 50 m*M* dithiothreitol, 1 m*M* spermidine, 1 m*M* EDTA), 100 μCi γ-^{32}P-ATP (7,000 Ci/mmol, 2 μl kinase (10 units/μl) and water to a final volume of 50 μl are added. The solution is incubated at 37°C for 20 min; another 20 units (2 μl) of enzyme are added and incubation proceeds for a further 20 min before placing the reaction mix on ice.

The labeled oligonucleotide is separated from unincorporated labeled nucleotide by Sephadex G-50 column chromatography. Oligonucleotide probes of specific activity 1 to 4 $\times$ 10^8 cpm/μg can be routinely achieved by this protocol.

One limitation of this 5′-end labeling method is that it permits a maximum addition of one radioactive phosphate per probe molecule. In a theoretical calculation involving hybridization to DNA targets, Berent et al.[68] showed that one could expect end-labeled oligonucleotides to be 40 to 80 fold less sensitive than nick translated probes. (This result was more or less confirmed in the studies reported in their paper.) To prepare an oligonucleotide probe radiolabeled to high specific activity, the primer extension method involving the use of the Klenow fragment of DNA polymerase I (Sections II.G ad IV.B) is recommended.

B. Large (Klenow) Fragment of *E. coli* DNA Polymerase I

Studencki and Wallace[69] have reported a primer extension method for introducing multiple-labeled nucleotides into an oligonucleotide probe. This approach makes use of a target-sense 19-mer oligonucleotide as a template and DNA polymerase I (Klenow fragment) plus a 9-mer oligonucleotide primer for extension across the template in the presence of all four α-^{32}P-dNTPs. Thus, ten residues of the final 19-mer probe are labeled with ^{32}P.

The following is a typical primer extension reaction according to the method of Studencki and Wallace.[69] A 5 μl reaction mix contains 0.8 m*M* template oligonucleotide, 2.7 μ*M* primer, 2 to 3 μ*M* of each of the four α-^{32}P- dNTPs (specific activities about 3000-5000 Ci/mmol), 50 m*M* NaCl, 25 m*M* Tris-HCl, pH 7.5, and 1.5 to 3.0 units *E. coli* DNA polymerase, Klenow fragment. Before initiation of synthesis with the enzyme, the reaction mix is cooled to 0°C; incubation then proceeds in the presence of the enzyme at 0°C for a 1.5 to 3.0 h period. The extended labeled primer DNA (probe) is then separated away from the template by polyacrylamide gel electrophoresis under denaturing conditions. Oligonucleotide probes of extremely high specific activities (up to 1 to 2 $\times$ 10^{10} dpm/μg) can be obtained by this procedure (Table 2).

Dattagupta et al.[70] have recently reported a variation of the primer extension method of Studencki and Wallace.[69] Briefly, an overlapping primer-template pair can be designed for most probe sequences such that only the primer and not the template is extended by the Klenow fragment of DNA polymerase I or reverse transcriptase. For example, a 19-mer can be synthesized from a pair of overlapping 14-mers by aligning the overlap such that extension of one oligonucleotide requires a different set of dNTP precursors than the other:

```
5'   GCAGACTTCTCCTC
              GAAGAGGAGTCCTC   5'
```

Only the upper strand will be extended and labeled if α-^{32}P-dATP and α-^{32}P-dGTP are used as dNTP substrates for either DNA polymerase (Klenow fragment) or reverse transcriptase.

A major feature of this labeling reaction[70] is that the probe and template are of different sizes, 19- and 14-nucleotides, respectively. This greatly facilitates purification of the probe from the template, if desired, and from unincorporated precursors. An interesting observation was that reverse transcriptase was more reliable than the Klenow fragment of DNA polymerase in the extension reaction; the latter enzyme frequently gave a lower molecular weight species, presumably due to incomplete extension.

C. Terminal Deoxynucleotidyl Transferase

Terminal deoxynucleotidyl transferase (see Section II.F), can be utilized to add labeled dAMP residues to the 3′-end of oligonucleotide probes. The following protocol is from Collins and Hunsaker.[71] In a typical reaction volume of 10 μl, 10 ng of oligonucleotide is incubated at 37°C for 60 min in 100 m*M* potassium cacodylate, pH 7.0, 1 m*M* $CoCl_2$, 1 m*M* β-mercaptoethanol, 200 μCi of α-^{32}P-dATP (5000 Ci/mmol), 100 μg/ml bovine serum albumin (nuclease free), and 20 units of terminal transferase. Purification of the tailed probe is by oligo(dT)-cellulose chromatography. Using this approach, oligonucleotide probes of specific activity up to approximately 3×10^{10} dpm/μg should be achievable[71] (Table 2).

Lewis et al.[72] have used the 3′-tailing reaction with terminal transferase and ^{125}I-dCTP to prepare oligonucleotide probes labeled to high specific activity with ^{125}I for *in situ* hybridization histochemistry. They also used the primer extension method with the Klenow fragment of DNA polymerse I (Section IV.B) to prepare ^{125}I-labeled oligonucleotide probes with similar specific activity.

V. RADIOIODINATION OF NUCLEIC ACIDS

A. Introduction

The description by Commerford[73] in 1971 of a simple chemical method for the iodination of DNA and RNA led to its immediate widespread use for the preparation of nucleic acids of high specific activity. At that time there were really no suitable methods for the routine labeling of nucleic acids so it is not surprising that the method found immediate acceptance. The now widely used nick translation procedure for labeling double-stranded DNA (Section II.A) was discovered in 1970,[9] but it took several years to become widely used. Part of the reason for this was the limited commercial availability of α-^{32}P-labeled nucleotides and their relative low specific activity at that time. ^{3}H- and ^{14}C-labeled nucleosides and nucleotides were more commonly used but their low specific activity and high cost precluded their extensive use for labeling nucleic acids.

For the above reasons, radioiodinated nucleic acids became widely used in the 1970s but their use has since declined considerably with the advent of numerous methods(described in this chapter) for preparing labeled nucleotides using high specific activity ^{32}P- labeled nucleotides. In addition, the potential hazard associated with the volatilization of radioactive iodine that may be formed during various handling procedures and the accumulation of any adsorbed iodine in the thyroid gland requires stringent laboratory protection.

B. Properties of Radioactive Iodine

Either ^{125}I or ^{131}I can be used to label nucleic acids to a high specific activity. ^{125}I is the much preferred isotope because of its longer half-life and lower energies of emission (Table 1).

^{125}I decays with a half-life of 60 d and is supplied as the nonvolatile sodium iodide. It emits a range of soft gamma rays with a maximum energy of emission of 35 keV, as well as a number of β particles (electrons) with a maximum energy of about 35 keV.[74] By comparison, the E_{max} of the β particle emissions of ^{32}P, ^{14}C and ^{35}S are 1.71 meV, 0.16 meV, and 0.17 meV. The low energies of the gamma ray emissions (and of the β particles) allow ready protection by 1 mm thicknesses of lead. However, the low energy emissions do not allow efficient monitoring by the normal Geiger-Muller tubes found in laboratory radiation monitors, so that special detection equipment is required.

^{131}I decays with a half-life of 8.6 d; 79% of the gamma radiation is at the maximum level of 0.37 meV, while about 90% of the β-radiation is in the form of a 0.608 meV E_{max} particle.[75] Although ^{131}I has practical disadvantages because of its short half-life and high energy of radiation emissions, it does have the potential advantage of an extremely high specific activity at 100% isotope enrichment of 16,100 Ci/mmol as compared with 2400 Ci/mmol for ^{125}I (Table 1).

C. Chemical Coupling of Radioiodine to Nucleic Acids

1. General Conditions

The original method described by Commerford,[73] has been used extensively with only minor modifications.[75] The reaction is carried out in two stages:

1. The nucleic acid (3 to 1000 μM as cytosine) is heated at 60°C for 5 to 20 min in the presence of 30 to 100 μM radioiodide, thallium trichloride ($TlCl_3$) at 4 to 10 times the iodide concentration, and 30 to 200 mM sodium acetate buffer, pH 4.7.
2. The pH of the mixture is then raised to 7 to 8 and heated again for 20 to 30 min at 60°C.

The first heating step results in the covalent bonding of iodine to the 5-position of the cytosine base and the formation of an unstable 5-iodo, 6-hydroxy-dihydropyrimidine. The latter compound is dehydrated on further heating at higher pH to give the stable 5-iodo-cytosine derivative of the nucleic acid. About 20 to 40% of iodine is bound to DNA or RNA when equimolar amounts of iodine and cytosine are used.

In the iodination reaction, $TlCl_3$ acts as an oxidizing agent and converts iodide to I_2 and IOH, and it is the latter that reacts with cytosine. The pH of the solution is critical to the iodination reaction; above pH 5.5 there is negligible iodination while pH values below about 4.5 should be avoided to minimize depurination, especially of DNA, at the high temperatures required.

In order to ensure an even distribution of ^{125}I in double-stranded DNA, Chan et al.[76] carried out the iodination reaction in concentrated $NaClO_4$ (5.8 M) at 60°C under which conditions the DNA is partially dissociated. Several other denaturing agents (e.g., dimethyl sulfoxide) interfered with the iodination whereas $NaClO_4$ did not. Initial experiments indicated that double-stranded RNA could be iodinated under the same conditions.

2. Iodination of RNA

RNA is more difficult to iodinate than DNA because of the lability of the internucleotide bond on heating and the ever present problem of contaminating RNases. In the first detailed method described for RNA, Heiniger et al.[77] included 100 μg/ml of sodium heparin in the iodination mixture as an RNase inhibitor. The iodination reaction was carried out at 60°C for 15 min, the reaction stopped by cooling on ice and the addition of a solution of 0.1 M Na_2SO_3 to a concentration of about 0.01 M. The second heating step in the original Commerford[73] procedure was not carried out and no mention was made of it, nor was it apparently required in a later paper by Bean et al.[78] describing the iodination of low amounts (4 μg) of influenza viral RNA. It was, however, used by Tereba and McCarthy [79] and in a rapid procedure described by Lee and Fowlks[80] for the ^{125}I-labeling of RNA for two-dimensional gel fingerprinting. These variations in techniques are surprising in view of the demonstration of the instability on heating above pH 7 of up to 70% of the ^{125}I incorporated into RNA and polynucleotides at pH 5.[81] Essentially all of the stable ^{125}I in RNA is present as 5-iodocytosine and only 3 to 5% as 5-iodouracil.[81]

According to Scherberg and Refetoff[82] and Altenburg et al.[83] the following sequence of events occurs during the iodination of RNA (Fig. 16). The IOH formed by the oxidation of radioiodide by $TlCl_3$ is incorporated into both pyrimidine groups to roughly the same extent. The resulting addition products, the 5-iodo, 6-hydroxy bases, vary in their stability and subsequent end-product. In the case of cytosine, the C5,6 saturated intermediate is unstable and is converted by the loss of water to the stable 5-iodo derivative. The saturated uracil derivative is more stable at pH 5 and is readily recovered as a component of the reaction mixture. However, when it decomposes, it does so by the loss of iodine to give the unsubstituted uracil. The rate of conversion of both bases is pH-dependent. At a pH above 7,

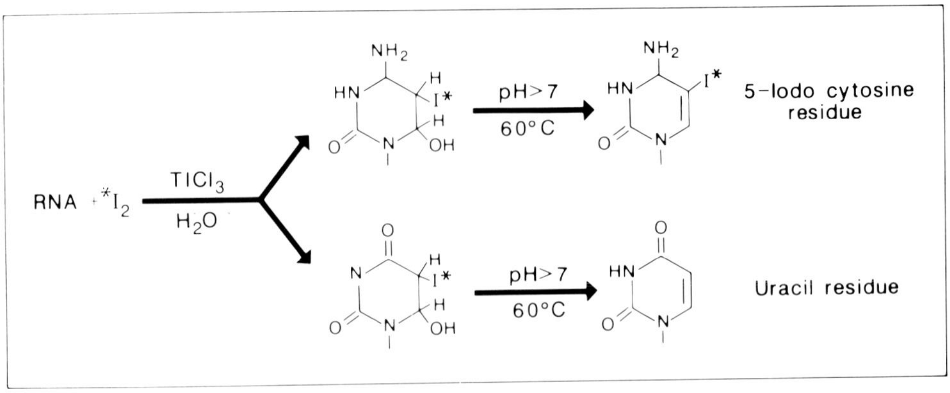

FIGURE 16. Proposed reaction mechanism[82,83] for the iodination of RNA.

exposure of RNA at 60°C for 10 min is sufficient treatment to take the conversion of both C and U bases to completion. Hence, iodinated RNA is primarily labeled only in the C residues.

D. Enzymatic Incorporation of Iodinated Nucleotides into Nucleic Acids

RNA polymerase. Oligonucleotides have been labeled by primer extension on a complementary DNA template using ^{125}I-d CTP and the Klenow fragment of *E. coli* ^{125}I- and ^{131}I-CTP and dCTP have been used as labeled precursors for the *in vitro* synthesis of RNA and DNA, respectively.[84,85] The iodinated nucleotides were prepared by the iodination of CTP and dCTP using $TlCl_3$ as the oxidizing agent followed by purification by ion-exchange chromatography. Iodinated dCTP was a substrate for *E. coli* DNA polymerase I, avian myeloblastosis virus reverse transcriptase, and rat liver DNA polymerase, while iodinated CTP was a substrate for *E. coli* RNA polymerase. Oligonucleotides have been labeled by primer extension on a complementary DNA template using ^{125}I-dCTP and the Klenow fragment of *E. coli.* DNA polymerase, as well as by the addition of 3′-tails using terminal deoxynucleotidyl transferase and ^{125}I-dCTP as the sole substrate.[72]

E. Uses of Radioiodinated Nucleic Acids

^{125}I- or ^{131}I-labeled DNA and RNA have been widely used wherever radioactive nucleic acids are required.[82,83] ^{125}I-labeled nucleic acids are particularly suitable for *in situ* hybridization because of their high specific activity and the low energy of emission of their β-particles (Table 1) which makes them equivalent to ^{3}H-labeled nucleic acids.[72,78,83] The simple radiodination of RNA allows the rapid comparison of two or more purified RNAs by two-dimensional fingerprint analysis.[86]

ACKNOWLEDGMENTS

The writing of this chapter was supported by the Centre of Gene Technology which was set up by a Commonwealth Government Grant in the Department of Biochemistry of Adelaide University. We thank Ruth Evans for preparing the diagrams, Jennifer Cassady and Tammy Edmonds for photography, and Ros Murrell for typing the manuscript.

REFERENCES

1. **Maniatis, T., Fritsch, E. F., and Sambrook, J.,** *Molecular Cloning. A Laboratory Manual,* Cold Spring Harbor Laboratory, Cold Spring Harbor, N.Y., 1982.
2. **Glover, D. M.,** Ed., *DNA Cloning. A Practical Approach,* Vol 1, IRL Press, Oxford, U.K., 1985.
3. **Glover, D. M.,** Ed., *DNA Cloning. A Practical Approach,* Vol 2, IRL Press, Oxford, U.K., 1985.
4. **Galt, M. J.,** Ed., *Oligonucleotide Synthesis. A Practical Approach,* IRL Press, Oxford, U.K., 1984.
5. **Caruthers, M. H.,** Gene synthesis machines: DNA chemistry and its uses, *Science,* 230, 281, 1985.
6. **Meinkoth, J. and Wahl, G.,** Hybridization of nucleic acids immobilized on solid supports. *Anal. Biochem.,* 138, 267, 1984.
7. **Hames, B. D. and Higgins, S. J.,** Eds., *Nucleic Acid Hybridization. A Practical Approach,* IRL Press, Oxford, U.K., 1985.
8. **Ambinder, R. F., Charache, P., Staal, S., Wright, P., Forman, M., Hayward, S. D., and Hayward, G. S.,** The vector homology problem in diagnostic nucleic acid hybridization of clinical specimens. *J. Clin. Microbiol.,* 24, 16, 1986.
9. **Kelly, R. B., Cozzarelli, N. R., Deutscher, M. P., Lehman, I. R., and Kornberg, A.,** Enzymatic synthesis of deoxyribonucleic acid: XXXII. Replication of duplex deoxyribonucleic acid by polymerase at a single strand break, *J. Biol. Chem.,* 245, 39, 1970.
10. **Schachat F. H. and Hogness, D. S.,** Repetitive sequences in isolated Thomas circles from Drosophila melanogaster, *Cold Spring Harbor Symp. Quant. Biol.,* 38, 371, 1974.
11. **Rigby, P. W. J., Diekmann, M., Rhodes, C., and Berg, P.,** Labeling deoxyribonucleic acid to high specific activity *in vitro* by nick translation with DNA polymerase I, *J. Mol. Biol.,* 113, 237, 1977.
12. **Koch, J., Kϕlvraa, S., and Bolund, L.,** An improved method for labelling of DNA probes by nick translation, *Nucleic Acids Res.,* 14, 7132, 1986.
13. **Feinberg, A. P. and Vogelstein, B..,** A technique for radiolabeling DNA restriction endonuclease fragments to high specific activity, *Anal. Biochem.,* 132, 6, 1983.
14. **Feinberg, A. P. and Vogelstein, B.,** A technique for radiolabeling DNA restriction endonuclease fragments to high specific activity: addendum, *Anal. Biochem.,* 137, 266, 1984.
15. **Klenow, H., Overgaard-Hansen, K., and Patkar, S. A.,** Proteolytic cleavage of native DNA polymerase into two different catalytic fragments, *Eur. J. Biochem.,* 22, 371, 1971.
16. **Taylor, J. M., Illmensee, R., and Summers, J.,** Efficient transcription of RNA into DNA by avian sarcoma virus polymerase, *Biochim. Biophys. Acta.,* 442, 324, 1976.
17. **Hodgson, C. P. and Fisk, R. Z.,** Hybridization probe size control: optimized 'oligolabelling', *Nucleic Acids Res.,* 15, 6295, 1987.
18. **Jovin, T. M., Englund, P. T., and Kornberg, A.,** Enzymatic synthesis of deoxyribonucleic acid: XXVII. Chemical modifications of deoxyribonucleic acid polymerase, *J. Biol. Chem.,* 244 3009, 1969.
19. **Hu, N. and Messing, J.,** The making of strand-specific M13 probes, *Gene,* 17, 271, 1982.
20. **Brown, D. M., Frampton, J., Goelet, P., and Karn, J.,** Sensitive detection of RNA using strand-specific M13 probes, *Gene* , 20, 139, 1982.
21. **Messing, J.,** New M13 vectors for cloning, in *Methods in Enzymology,* Vol. 101, Wu, R., Grossman, L., and Moldave, K., Eds., Academic Press, New York, 1983, 20.
22. **Barker, J. M., McInnes, J. L., Murphy, P. J., and Symons, R. H.,** Dot-blot procedure with [^{32}P]DNA probes for the sensitive detection of avocado sunblotch and other viroids in plants, *J. Virol. Methods,* 10, 87, 1985.
23. **Bruening, G., Gould, A. R., Murphy, P. J., and Symons, R. H.,** Oligomers of avocado sunblotch viroid are found in infected avocado leaves, *FEBS Lett.,* 148, 71, 1982.
24. **Boyer, P. D.,** A sensitive technique for detection of RNA with single-stranded probes, *Nucleic Acids Res.,* 14, 7505, 1986.
25. **Morris, C. F., Hama-Inaba, H., Mace, D., Sinha, N. K., and Alberts, B.,** Purification of the gene 43,44,45, and 62 proteins of the bacteriophage T4 DNA replication apparatus, *J. Biol. Chem.,* 245, 6787, 1979.
26. **O'Farrell, P.,** Replacement synthesis method of labelling DNA fragments, *FOCUS (Bethesda Research Laboratories/Life Technologies Inc.),* 3(3), 1, 1981.
27. **Richardson, C. C.,** Phosphorylation of nucleic acid by an enzyme from T4 bacteriophage-infected *Escherichia coli, Proc. Natl. Acad. Sci. U.S.A.,* 54, 158, 1965.
28. **Berkner, K. L. and Folk, W. R.,** Polynucleotide kinase exchange reaction: quantitative assay for restriction endonuclease-generated 5′-phosphoryl termini in DNAs, *J. Biol. Chem.,* 252, 3176, 1977.
29. **Maxam, A. M. and Gilbert, W.,** Sequencing end-labelled DNA with base-specific chemical cleavages, in *Methods in Enzymology,* Vol. 65, Grossman, L., and Moldave, K., Eds., Academic Press, New York, 1980, 499.
30. **Gaastra, W. and Josephsen, J.,** Radiolabeling of DNA using polynucleotide kinase, in *Methods in Molecular Biology,* Vol. 2, Walker, J. M., Ed., Humana Press, Clifton, N. J., 1984, chap. 39.

31. **Harrison, B. and Zimmerman, S. B.**, Stabilization of T4 polynucleotide kinase by macromolecular crowding, *Nucleic Acids Res.*, 14, 1863, 1986.
32. **Bollum, F. J.**, Terminal deoxynucleotidyl transferase, in *The Enzymes*, Vol. 10, 3rd. ed., Boyer, P. D., Ed., Academic Press, New York, 1974, chap. 5.
33. **Roychoudhury, R., Jay, E., and Wu, R.**, Terminal labeling and addition of homopolymer tracts to duplex DNA fragments by terminal deoxynucleotidyl transferase, *Nucleic Acids Res.*, 3, 863, 1976.
34. **Tu, C-P. D. and Cohen, S. N.**, 3'-End labeling of DNA with [α-^{32}P] cordycepin-5'-triphosphate, *Gene* 10, 177, 1980.
35. **Gaastra, W. and Klemm, P.**, Radiolabeling of DNA with 3' terminal transferase, in *Methods in Molecular Biology*, Vol. 2, Walker, J. M., Ed., Humana Press, Clifton, NJ, 1984, chap. 40.
36. **Chirala, S. S. and Wakil, S. J.**, Radio-labeling of the *Pst*I restriction fragments and improvement in the sequencing procedure, *Gene* , 47, 297, 1986.
37. **Gaastra, W. and Josephsen, J.**, Radio-labeling of DNA with the Klenow fragment of DNA polymerase, in *Methods in Molecular Biology*, Vol. 2, Walker, J. M., Ed., Humana Press, Clifton, N.J., 1984, chap. 41.
38. **Rogers, S. G. and Weiss, B.**, Exonuclease III of *Escherichia coli* K-12, an AP endonuclease, in *Methods in Enzymology*, Vol. 65, Grossman, L. and Moldave, K., Eds., Academic Press, New York, 1980, 201.
39. **Donelsen, J. E. and Wu, R.**, Nucleotide sequence analysis of deoxyribonucleic acid: VII. Characterization of *Escherichia coli* exonuclease III activity for possible use in terminal nucleotide sequence analysis of duplex deoxyribonucleic acid, *J. Biol. Chem.*, 247, 4661, 1972.
40. **Verma, I. M.**, Reverse transcriptase, in *The Enzymes*, Vol. 14, 3rd. ed., Boyer, P. D., Ed., Academic Press, New York, 1981, chap. 6.
41. **Ashley, P. L. and MacDonald, R. J.**, Synthesis of single-stranded hybridization probes from reusable DNA templates bound to solid support, *Anal. Biochem.*, 140, 95, 1984.
42. **Melton, D. A., Krieg, P. A., Rebagliati, M. R., Maniatis, T., Zinn, K., and Green, M. R.**, Efficient *in vitro* synthesis of biologically active RNA and RNA hybridization probes from plasmids containing a bacteriophage SP6 promoter, *Nucleic Acids Res.*, 12, 7035, 1984.
43. **Krieg, P. A. and Melton, D. A.**, *In vitro* RNA synthesis with SP6 RNA polymerase, in *Methods in Enzymology*, Vol. 155, Wu, R., Ed., Academic Press, New York, 1987, in press.
44. **Butler, E. T. and Chamberlin, M. J.**, Bacteriophage SP6-specific RNA polymerase. I. Isolation and characterization of the enzyme, *J. Biol. Chem.*, 257, 5772, 1982.
45. **Kassavetis, G. A., Butler, E. T., Roulland, D., and Chamberlin, M. J.**, Bacteriophage SP6-specific RNA polymerase. I. Mapping of SP6 DNA and selective *in vitro* transcription, *J. Biol. Chem.*, 257, 5779, 1982.
46. **Kotani, H., Ishizaki, Y., Hiraoka, N., and Obayashi, A.**, Nucleotide sequence and expression of the cloned gene of bacteriophage SP6 RNA polymerase, *Nucleic Acids Res.*, 15, 2653, 1987.
47. **Birnstiel, M. L., Sells, B. H., and Purdom, I. F.**, Kinetic complexity of RNA molecules, *J. Mol. Biol.*, 63, 21, 1972.
48. **Casey, J. and Davidson, N.**, Rates of formation and thermal stabilities of RNA:DNA and DNA:DNA duplexes at high concentrations of formamide, *Nucleic Acids Res.*, 4, 1539, 1977.
49. **Gray, D. M., Liu, J-J., Ratliff, R. L., and Allen, F. S.**, Sequence dependence of the circular dichroism of synthetic double-stranded RNAs, *Biopolymers* , 20, 1337, 1981.
50. **Dunn. J. J. and Studier, F. W.**, Complete nucleotide sequence of bacteriophageT7 DNA and the locations of T7 genetic elements, *J. Mol. Biol.*, 166, 477, 1983.
51. **Davanloo, P., Rosenberg, A. H., Dunn, J. J., and Studier, F. W.**, Cloning and expression of the gene for bacteriophage T7 RNA polymerase, *Proc. Natl. Acad. Sci. U.S.A.*, 81, 2035, 1984.
52. **Tabor, S. and Richardson, C. C.**, A bacteriophage T7 RNA polymerase/promoter system for controlled exclusive expression of specific genes, *Proc. Natl. Acad. Sci. U.S.A.*, 82, 1074, 1985.
53. **Eperon, I. C.**, M13 vectors with T7 polymerase promoters: transcription limited by oligonucleotides, *Nucleic Acids Res.*, 14, 2830, 1986.
54. **Krupp, G. and Söll, D.**, Simplified *in vitro* synthesis of mutated RNA molecules: An oligonucleotide promoter determines the initiation site of T7 RNA polymerase on ssM13 phage DNA, *FEBS Lett.*, 212, 271, 1987.
55. **Chamberlin, M., Kingston, R., Gilman, M., Wiggs, J., and DeVera, A.**, Isolation of bacterial and bacteriophage RNA polymerases and their use in synthesis of RNA *in vitro*, in *Methods in Enzymology*, Vol. 101, Wu, R., Grossman, L., and Moldave, K., Eds., Academic Press, New York, 1983, 540.
56. *FOCUS (Bethesda Research Laboratories/Life Technologies Inc.)*, 9(1), 11, 1987.
57. **Wölfl, S., Quaas, R., Hahn, U., and Wittig, B.**, Synthesis of highly radioactively labelled RNA hybridization probes from synthetic single stranded DNA oligonucleotides, *Nucleic Acids Res.*, 15, 858, 1987.
58. **Walker, G. C., Uhlenbeck, O. C., Bedows, E., and Gumport, R. I.**, T4-Induced RNA ligase joins single-stranded oligoribonucleotides, *Proc. Natl. Acad. Sci. U.S.A.*, 72, 122, 1975.

59. **Uhlenbeck, O. C. and Gumport, R. I.,** T4 RNA ligase, in *The Enzymes,* Vol. 15, 3rd. ed., Boyer, P. D., Ed., Academic Press, New York, 1982, chap. 2.
60. **England, T. E. and Uhlenbeck, O. C.,** Enzymatic oligoribonucleotide synthesis withT4 RNA ligase, *Biochemistry,* 17, 2069, 1978.
61. **England, T. E. and Uhlenbeck, O. C.,** 3′-Terminal labelling of RNA with T4 RNA ligase, *Nature (London),* 275, 560, 1978.
62. **England, T. E., Bruce, A. G., and Uhlenbeck, O. C.,** Specific labeling of 3′-termini of RNA with T4 RNA ligase, in *Methods in Enzymology,* Vol. 65, Grossman, L. and Moldave, K., Eds., Academic Press, New York, 1980, 65.
63. **Wallace, R. B., Shaffer, J., Murphy, R. F., Bonner, J., Hirose, T., and Itakura, K.,** Hybridization of synthetic oligodeoxyribonucleotides to ΦX 174 DNA: the effect of single base pair mismatch, *Nucleic Acids Res.,* 6, 3543, 1979.
64. **Szostak, J. W., Stiles, J. I., Tye, B.-K., Chiu, P., Sherman, F., and Wu, R.,** Hybridization with synthetic oligonucleotides, in *Methods in Enzymology,* Vol. 68, Wu. R., Ed., Academic Press, New York, 1979 419.
65. **Conner, B. J., Reyes, A. A., Morin, C., Itakura, K., Teplitz, R. L., and Wallace, R. B.,** Detection of sickle cell β^S-globin allele by hybridization with synthetic oligonucleotides, *Proc. Natl. Acad. Sci. U.S.A.,* 80, 278, 1983.
66. **Itakura, K., Rossi, J. J., and Wallace, R. B.,** Synthesis and use of synthetic oligonucleotides, *Ann. Rev. Biochem.,* 53, 323, 1984.
67. **Glover, D. M.,** *Gene Cloning: The Mechanics of DNA Manipulation,* Chapman and Hall, London, 1984.
68. **Berent, S. L., Mahmoudi, M., Torczynski, R. M., Bragg, P. W., and Bollon, A. P.,** Comparison of oligonucleotide and long DNA fragments as probes in DNA and RNA dot, southern, northern, colony and plaque hybridizations, *BioTechniques,* 3, 208, 1985.
69. **Studencki, A. B. and Wallace, R. B.,** Allele-specific hybridization using oligonucleotide probes of very high specific activity: discrimination of the human β^A and β^S-globin genes, DNA, 3, 7, 1984.
70. **Dattagupta, N., Rabin, D., Michaud, G., and Rae. P. M. M.,** A simple method for generation of high specific activity oligonucleotide probes, *BioTechniques,* 55, 38, 1987.
71. **Collins, M. L. and Hunsaker, W. R.,** Improved hybridization assays employing tailed oligonucleotide probes: a direct comparison with 5′-end-labelled oligonucleotide probes and nick-translated plasmid probes, *Anal. Biochem,* 151, 211, 1985.
72. **Lewis, M. E., Arentzen, R., and Baldino Jr., F.,** Rapid, high-resolution *in situ* hybridization histochemistry with radioiodinated synthetic oligonucleotides, *J. Neurosci Res.,* 16, 117, 1986.
73. **Commerford, S. L.,** Iodination of nucleic acids *in vitro, Biochemistry* , 10,l 1993, 1971.
74. **Ertl, H. H., Feinendegen, L. E., and Heiniger, H. J.,** Iodine-125, a tracer in cell biology: Physical properties and biological aspects, *Phys. Med. Biol.,* 15, 447, 1970.
75. **Prensky, W.,** The radioiodination of RNA and DNA to high specific activities, in *Methods in Cell Biology,* Vol. 13, Prescott, D. M., Ed., Academic Press, New York, 1976, chap. 7.
76. **Chan. H-C., Ruyechan, W. T., and Wetmur, J. G.,** *In vitro* iodination of low complexity nucleic acids without chainscission, *Biochemistry,* 15, 5487, 1976.
77. **Heiniger, H. J., Chen, H. W., and Commerford, S. L.,** Iodination of ribosomal RNA *in vitro, Int. J. Appl. Radiat. Isotop.,* 24, 425, 1973.
78. **Bean, Jr., W. J., Sriram, G., and Webster, R. G.,** Electrophoretic analysis of iodine-labeled influenza virus RNA segments, *Anal. Biochem.,* 102, 228, 1980.
79. **Tereba, A. and McCarthy, B. J.,** Hybridization of ^{125}I-labeled ribonucleic acid, *Biochemistry,* 12, 4675, 1973.
80. **Lee, Y. F. and Fowlks, E. R.,** Rapid *in vitro* labeling procedures for two-dimensional gel fingerprinting, *Anal. Biochem.,* 119, 224, 1982.
81. **Scherberg, N. H. and Refetoff, S.,** The radioiodination of ribopolymers for use in hybridizational and molecular analyses, *J. Biol. Chem.,* 249, 2143, 1974.
82. **Scherberg, N. H. and Refetoff, S.,** Radioiodine labeling of ribopolymers for special applications in biology, in *Methods in Cell Biology,* Vol. 10, Prescott, D. M., Ed., Academic Press, New York, 1975, chap. 19.
83. **Altenburg, L. C., Getz, M. J., and Saunders, G. F.,** ^{125}I in molecular hybridization experiments, in *Methods in Cell Biology,* Vol. 10, Prescott, D. M., Ed., Academic Press, New York, 1975, chap. 18.
84. **Scherberg, N. H. and Refetoff, S.,** The preparation of carrier-free iodine isotope-substituted cytosine nucleotides, *Biochim. Biophys. Acta.,* 340, 446, 1974.
85. **Bhalla, R. B., Geraci, D., Modak, M. J., Prensky, W., and Marcus, S. L.,** Preparation of (^{125}I)-dCTP and its use as a substrate for RNA- and DNA-directed DNA synthesis, *Biochem. Biophys. Res. Commun.,* 72, 513, 1976.
86. **Dickson, E., Prensky, W., and Robertson, H. D.,** Comparative studies of two viroids: analysis of potato spindle tuber and citrus exocortis viroids by RNA fingerprinting and polyacrylamide-gel electrophoresis, *Virology,* 68, 309, 1975.

Chapter 2

PREPARATION AND DETECTION OF NONRADIOACTIVE NUCLEIC ACID AND OLIGONUCLEOTIDE PROBES

J. L. McInnes and R. H. Symons

TABLE OF CONTENTS

I. INTRODUCTION

There is increasing interest worldwide in the development of nucleic acid probes which are detected by nonradioactive means. In the research laboratory, the use of ^{32}P for detection is undoubtedly the method of choice and is likely to remain so for the forseeable future, in spite of the half life of only 14 days for ^{32}P. In the diagnostic laboratory on the other hand, the use of nonradioactive probes has many potential advantages. Perhaps the major one is that nonradioactive probes are stable for at least 6 to 12 months, and probably much longer if properly stored, thus leading to a substantial reduction in cost by obviating the need to prepare them every 2 to 3 weeks. In addition, there is no radiation exposure from routine daily use and there are no storage and disposal problems.

The general field of nonradioactive probes is in a first generation stage where many different approaches are being investigated in the preparation of these probes and in the development of suitable detection methods. Very few systems using nonradioactive probes have reached the market place and there is unlikely to be a dramatic increase until the 1990s. There are a number of reasons for this. The well established ELISA procedures have a significant advantage in greater simplicity and their application to automated, quantitated print-out of data. They have found widespread application in diagnostic laboratories world-wide so that the use of nonradioactive probes must offer unique advantages to allow a competitive alternative. Perhaps the most important aspect is one of education, by demonstrating that probes can provide significant improvements in sensitivity and flexibility of design and application.

Numerous methods are described in this chapter for the preparation by enzymatic and chemical techniques of nonradioactive nucleic acid and oligonucleotide probes. In many cases, the resulting probes have yet to be fully tested under hybridization conditions. In others, initial results look very promising since some nonradioactive probes can provide a sensitivity of detection of target sequences similar to that provided by ^{32}P-labeled probes.

II. ENZYMATIC PREPARATION TECHNIQUES

A. Preparation of Nonradioactive, Labeled DNA

1. Incorporation of Biotinylated Nucleotides into DNA

a. Nick Translation Reaction

A landmark in the development of methods for labeling nucleic acid probes by nonisotopic means was the publication of Langer et al.[1] describing the synthesis of biotinylated nucleotides and their enzymatic incorporation into DNA and RNA. The biotin was attached via a linker arm to the 5-position of the pyrimidine ring of dUTP (Figure 1) or UTP. Such nucleotides could be incorporated by the respective polymerases into DNA or RNA. The length of the linker arm between the biotin and the pyrimidine ring was found to be important in the subsequent detection of biotinylated DNA when used as a probe; linker arms of either 11 or 16 atoms were better than those with only 4 atoms (Figure 1) for both *in situ*[2] and dot-blot[3] hybridizations.

Biotinylation of double-stranded DNA is readily achieved in a standard nick translation reaction (see Chapter 1, Section II.A) catalyzed by *E. coli* DNA polymerase I. For example, bio-11-dUTP is incorporated in place of dTTP into the DNA to the same extent as dTTP but at a slightly slower rate.[1] Using standard nick translation procedures, substitution of between 20 to 70% of the available deoxythymidine residues with biotinylated dUTP can be achieved. Reassociation kinetics of denatured normal and biotinylated double-stranded DNAs were identical,[1] indicating that the biotinylated DNA exists as an unperturbed double-helix. Hence, established hybridization procedures can be used with biotinylated DNA probes prepared in this way.

In addition to the biotinylated dUTP analogues (Figure 1), a series of biotinylated dATP and dCTP analogues have been prepared which were incorporated into DNA probes by nick translation.[4] Bio-7-dATP and bio-7-dCTP (Figure 2) gave the highest incorporation in the series of analogues where N (the number of atoms between the carbonyl group of the biotin moiety and the amino group of adenine or cytosine) varied from 3 to 17. All the results obtained with DNA probes labeled with either bio-7-dATP or bio-7-dCTP were very similar to those obtained with bio-11-dUTP.[2,3] Hence, none of these biotinylated probes offers any advantages over the latter.

b. Replacement Synthesis Using T4 DNA Polymerase

As considered in Chapter 1 (Section II.D), T4 DNA polymerase can be used to generate high specific activity hybridization probes from double-stranded DNA. The same reaction can be used to incorporate bio-11-dUTP into DNA fragments with almost the same efficiency as the incorporation of dTTP.[5] The resulting biotinylated DNA probe was successfully used in colony hybridization.

c. Use of Other DNA Polymerases

Bio-dUTP is not a suitable substrate for the avian myeloblastosis virus (AMV) reverse transcriptase[1,6] in the preparation of long cDNA copies of RNA. However, the recently cloned form of the reverse transcriptase of Moloney murine leukemia virus[7,8] does incorporate bio-11-dUTP into full length cDNA, although somewhat less efficiently (35%) as compared with dTTP.[6]

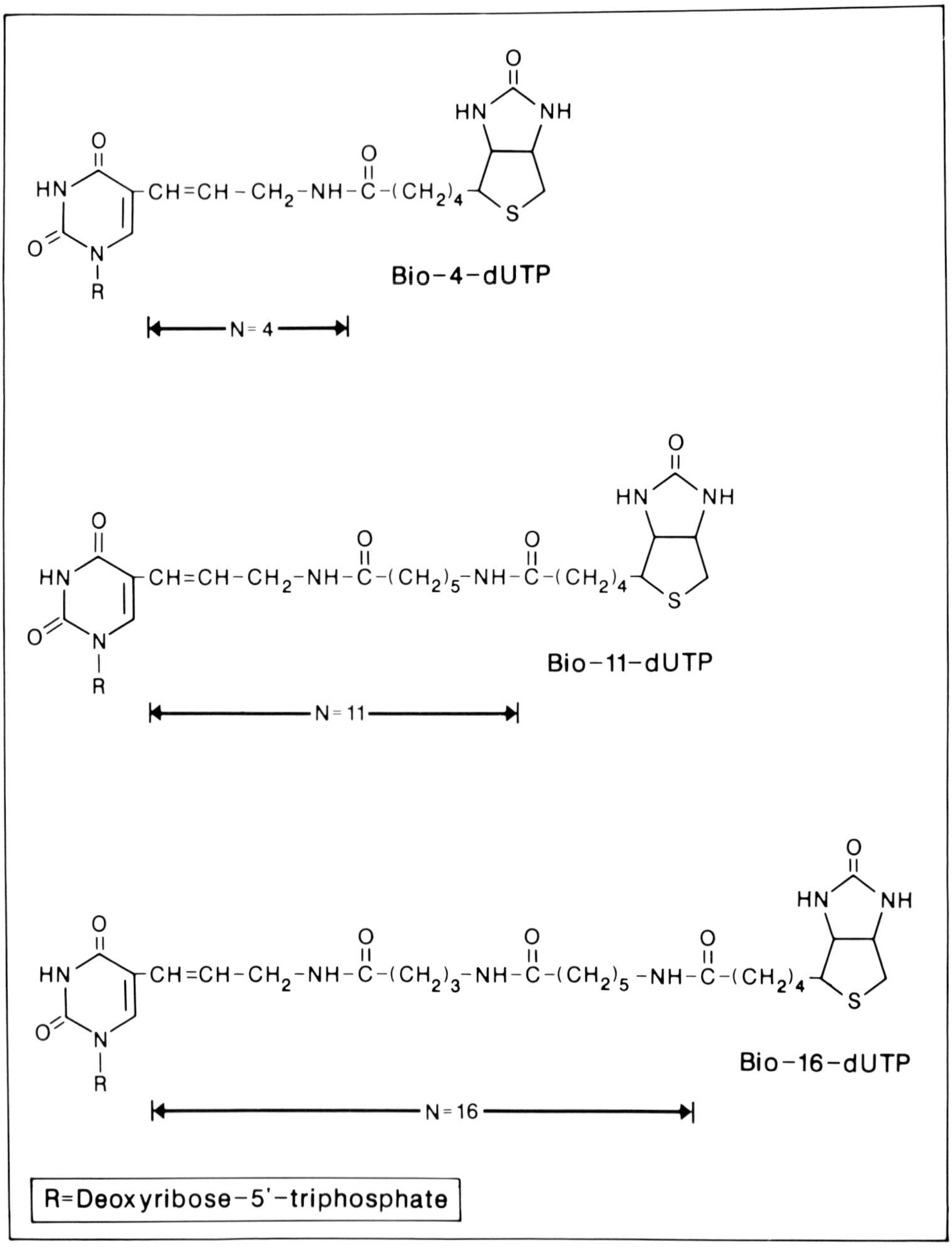

Bio-4-dUTP

N = 4

Bio-11-dUTP

N = 11

Bio-16-dUTP

N = 16

R=Deoxyribose-5'-triphosphate

FIGURE 1. Structures of biotinylated analogues of deoxyuridine triphosphate (dUTP); Bio-4-dUTP, Bio-11-dUTP and Bio-16-dUTP.[1-3] Biotin is coupled by a linear chain of 4, 11, or 16 atoms to C5 of dUTP.

2. Terminal Labeling of DNA

Bio-11-dUTP can also serve as a substrate for terminal deoxynucleotidyl transferase (terminal transferase; see Chapter 1, Section II.F) in a 3′-hydroxy terminal addition reaction with double-stranded DNA.[9] Briefly, the method involves the initial treatment of the DNA with DNase I to produce 3′-hydroxyl termini at random sites throughout the molecule followed by the synthesis of single-stranded DNA tails at these termini with terminal transferase and bio-11-dUTP. dCTP and/or dTTP are added to the reaction mixture to promote the incorporation of bio-11-dUTP. Biotinylated probes prepared in this way were successfully used in Southern hybridizations.[9]

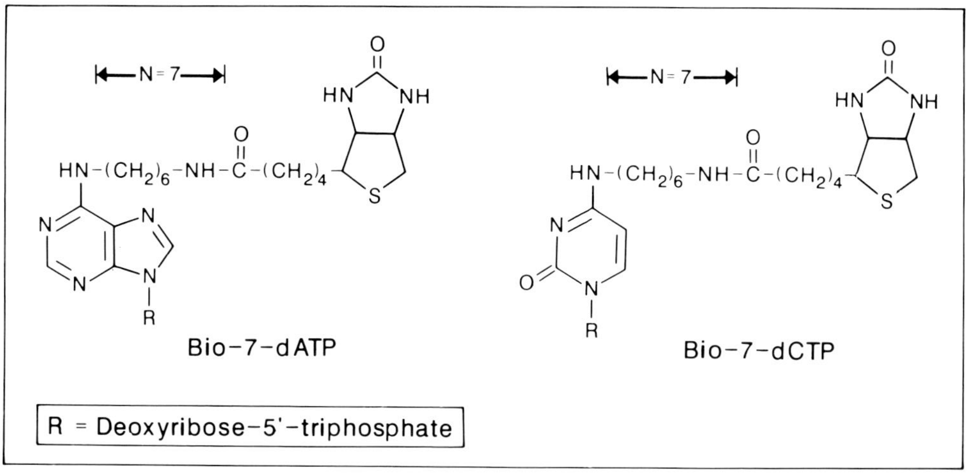

FIGURE 2. Structures of biotinylated analogues[4] of deoxyadenosine triphosphate (dATP) and deoxycytidine triphosphate (dCTP); Bio-7-dATP and Bio-7-dCTP.

In an interesting variation of this procedure, the commercial firm Enzo Biochem Inc. has described[10] the following procedure. Double-stranded DNA was treated with DNase I as above or cleaved with a restriction enzyme to generate 3′-hydroxyl termini. Single-stranded dT polymer tails were generated at the termini using terminal transferase and dTTP (Figure 3). The resulting dT-tailed probe was denatured and hybridized to the target nucleic acid on a filter. The hybridized probe was detected, after thorough washing of the filter, by hybridization to a biotinylated, single-stranded poly(dA) chain (this was synthesized by the terminal transferase addition of bio-dATP to a short oligo(dA) primer). In this way, the bio-poly(dA) can be used as a general signal-generating polymer for any poly(dT) labeled DNA probe (Figure 3). A high signal to background ratio was reported for this system.

3. Applications of Enzymatically Labeled DNA Probes

The success of biotin-labeled probes prepared by enzymatic procedures[1,3] is shown by an ever increasing number of reports in the literature on their use. Such probes have been used, for example, in the detection of a variety of viral infections,[11] of infectious bacteria,[12] of the gene responsible for sickle cell anemia,[13] and for *in situ* hybridization studies.[2,14-20] One disadvantage of the enzymatic labeling approach is that different modified nucleotides and enzymes are required depending on whether DNA or RNA (see Section II.B) is to be nonisotopically labeled. Also, if large quantities of probe are needed, this procedure can prove costly in terms of enzymes and substrates.

B. Preparation of Nonradioactive, Labeled RNA

1. Several Enzymes Can Be Used

Biotinylation of RNA can be achieved in a standard phage SP6 RNA polymerase transcription system (see Chapter 1, Section III.A) using the biotinylated ribonucleotide, bio-11-UTP, which is the ribose analogue of bio-11-dUTP (Figure 1). To prepare single-stranded, biotinylated RNA as a probe, the appropriate DNA sequence is first cloned into an appropriate vector which contains the SP6 promoter upstream from the polylinker region (vectors pSP64 and pSP65).[21,22] After linearization of the DNA clone downstream from the cloned insert, the RNA transcript of defined length is produced by the SP6 RNA polymerase using ATP, CTP,GTP and bio-11-UTP as substrates.[23]

In addition to the above system, the phage T7 and T3 RNA polymerases can be used in

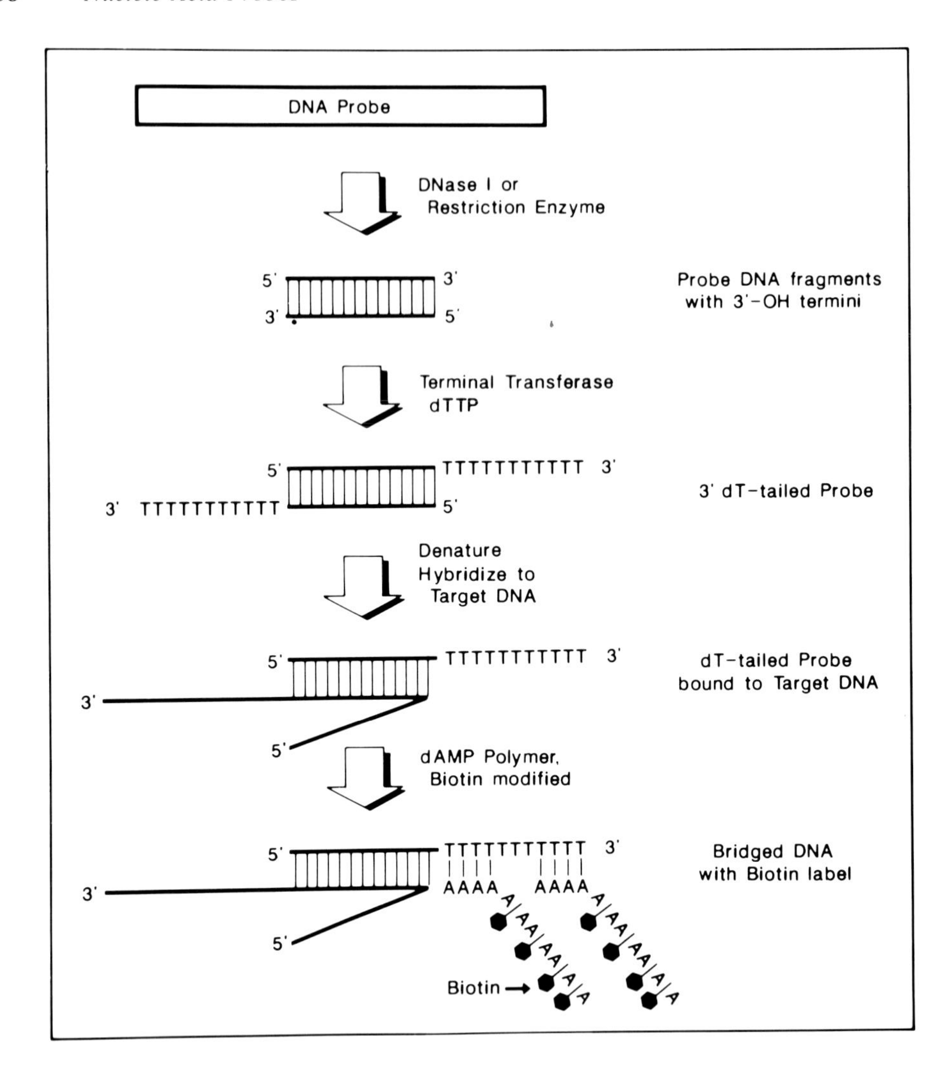

FIGURE 3. Indirect labeling procedure for DNA via a *biotin-bridge* system (ENZO Biochem, Inc.). A poly dT-tailed DNA probe, after hybridization to target DNA, is detected by hybridization to a biotin-labeled oligo (dA) prepared by primer extension of short oligo (dA) with bio-dATP using terminal transferase.

a similar way (see also Chapter 1, Sections III.B,C,D,). While the three-phage RNA polymerases (SP6, T7, T3) perform similarly when conventional nucleotides are used as subtrates, they do differ in their ability to incorporate bio-11-UTP instead of UTP into RNA.[24] The T3 enzyme incorporates bio-11-UTP the most efficiently, synthesizing approximately three times as much RNA as the SP6 enzyme, and twice as much as the T7 enzyme.[24]

As an alternative to the above procedure, allylamine-UTP (a UTP analogue with an allylamine group attached at the 5 position of the uracil base) can be first incorporated into RNA by any of the phage RNA polymerase transcription systems.[23,24] The resulting RNA is then biotinylated by reaction with the *N*-hydroxysuccinimide ester of biotinyl-6-aminohexanoic acid[25] (Figure 4) to give a product which is essentially the same as RNA prepared by the incorporation of bio-11-UTP.

A final approach to the enzymatic labeling of RNA makes use of the bacteriophage enzyme, T4 RNA ligase (refer to Chapter 1, Section III.E). This versatile enzyme catalyzes a variety of reactions; the intra- and inter-molecular joining of RNA or DNA, the addition of mono-

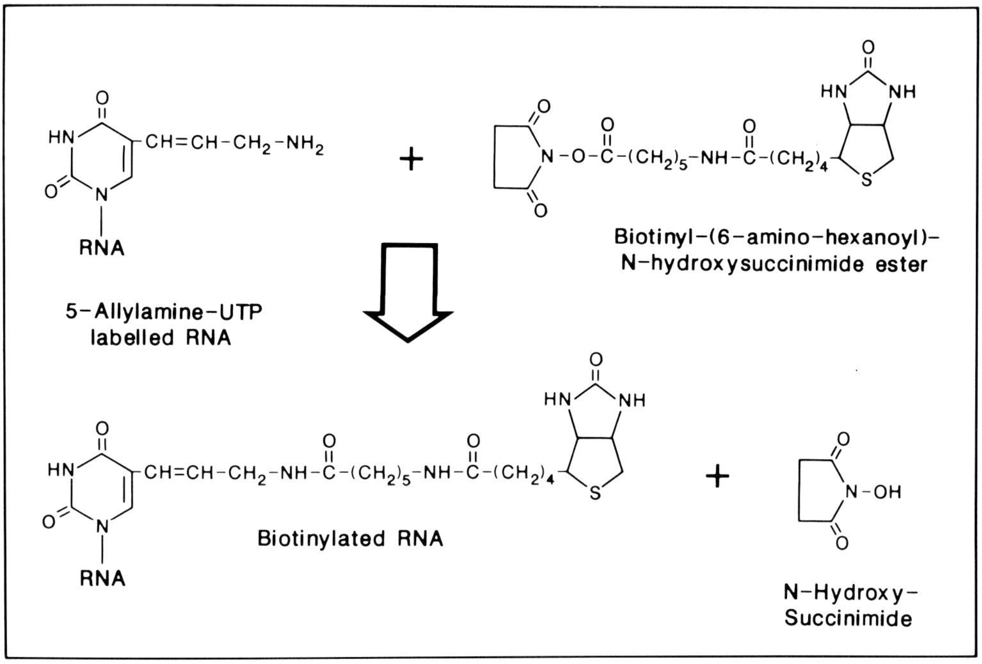

FIGURE 4. Preparation of biotinylated RNA probes after the enzymatic incorporation of the UTP analogue, 5-allylamine-UTP, into RNA.[25]

nucleotides to RNA or DNA, and the addition of nonnucleotide compounds to the 3′-hydroxyl of RNA in an ATP-independent reaction.[26] In the case of the latter reaction, England et al.[27] have shown that a number of β-substituted ADP derivatives of the form adenosine-5′-P-P-X will transfer P-X to the 3′-hydroxyl of an RNA acceptor to form a 3′-phosphodiester bond and the release of 5′-AMP. Richardson and Gumport[28] exploited this reaction and prepared adenosine-5′-P-P-(6-aminohexan-l-yl) to which they coupled biotin, fluorescein, and tetramethylrhodamine; all three derivatives were successfully coupled to oligomers, tRNA and 5S RNA, in high yield. Preliminary experiments indicated that such products could be used in Southern hybridization analysis.[28]

C. Nonradioactive Labeling of Oligonucleotides

Two methods are briefly described for the enzymatic addition of bio-dUTP to the 3′-terminus of synthetic oligodeoxynucleotides. Terminal labeling should avoid any effect that internal biotinylated residues may have on the stability of the hybridized probe to the target sequence.

In the first of these methods (Figure 5), a single biotinylated residue was added by a primer extension reaction.[29] An excess of a short oligodeoxynucleotide template complementary to the 3′-end of the probe (primer) but containing an extra 5′-A residue was hybridized to the primer. In the presence of the Klenow fragment of *E. coli* DNA polymerase and bio-dUTP, a single bio-dUTP residue was added to the primer. The product was then purified by polyacrylamide gel electrophoresis under denaturing conditions.

In the second technique,[30] terminal deoxynucleotidyl transferase was used to add bio-dUTP residues to the 3′-terminus of oligodeoxynucleotides. For example, in a 4-h reaction at pH 7.5 and 35°C, four to six bio-dUTP residues were added. Optimal addition of biotinylated residues was achieved using cobalt as the divalent metal ion.

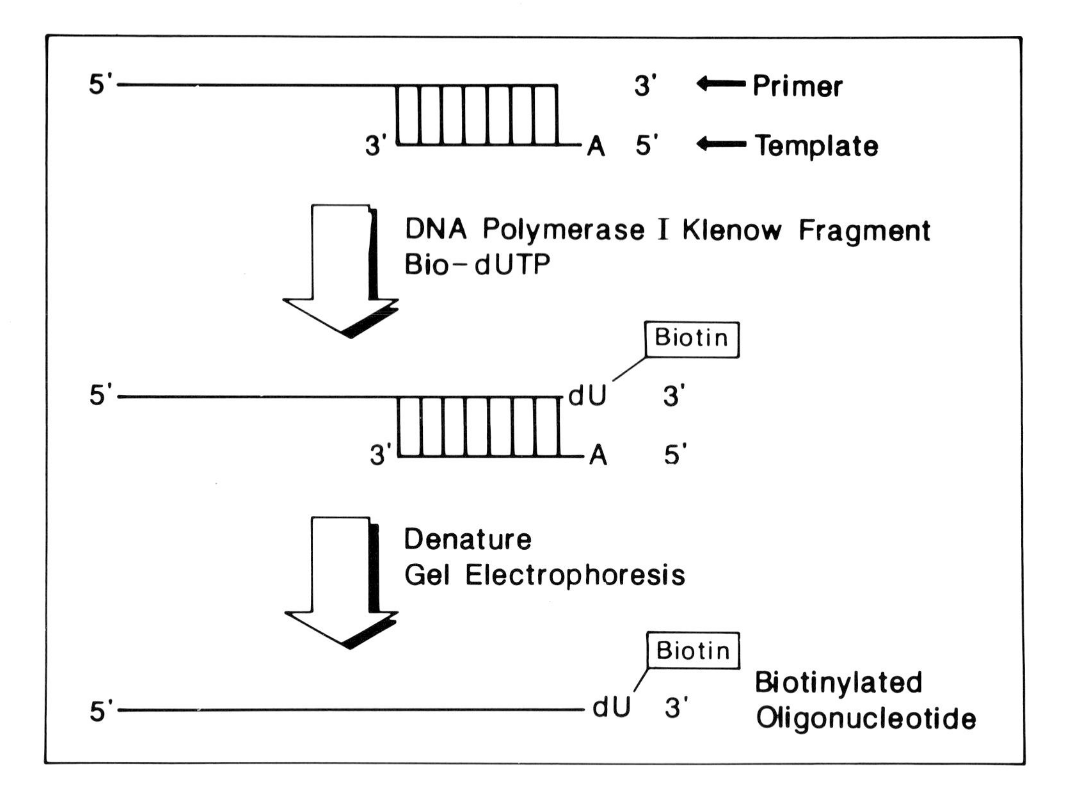

FIGURE 5. General approach for the synthesis of olinonucleotide probes with a single biotin residue at the 3′ terminus.[29] Only one nucleotide, bio-dUTP, is added to the primer by the Klenow fragment of DNA polymerase I. (From Murasugi, A. and Wallace, R. B., *DNA*, 3, 269, 1984. With permission.)

III. CHEMICAL PREPARATION TECHNIQUES

A. Reagents for the Chemical Modification of Nucleic Acids

1. N-Acetoxy-N-2-Acetylaminofluorene

This reagent has been used by a number of workers to label both DNA and RNA with a hapten which can then be detected by immunological methods.[31-33] At neutral pH, DNA and RNA react readily *in vitro* with *N*-acetoxy-*N*-2-acetylaminofluorene in a simple reaction (37°C, 30 min); covalent coupling of 2-acetylaminofluorene groups takes place, mainly at the C8 position of guanine residues[34,35] (Figure 6). The probe can be detected after hybridization by the sequential addition of antibodies specific for the modified bases and then enzyme-conjugated second antibodies (see Section V).

Nucleic acids modified in this way have been used widely for *in situ,*[31] dot-blot,[32,36] Southern, colony and plaque hybridization.[36] The abililty to label both double-stranded as well as single-stranded nucleic acids extends the applicability of the technique. Probes prepared with this reagent are stable for at least 6 months.[33] The major drawback with this approach is that *N*-acetoxy-*N*-2-acetylaminofluorene is a carcinogen; hence, safety and disposal problems are associated with this technique.

2. Photobiotin®

We have developed[37] a simple chemical procedure for labeling nucleic acids using a photoactivatable analogue of biotin, called Photobiotin® (Figure 7). The reagent consists of biotin attached by a linker arm to a photoactivatable (azido) group. When a mixture of nucleic acid and Photobiotin® is exposed to strong visible light for 10 to 20 min under

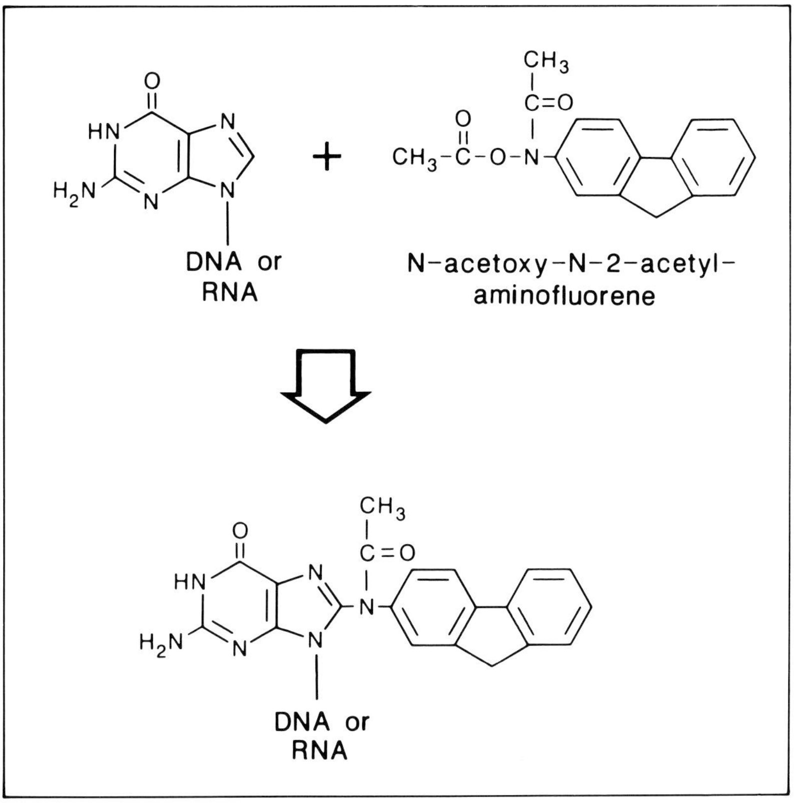

FIGURE 6. Covalent coupling of DNA or RNA with 2-acetylaminofluorene groups.[31-33] Reaction takes place mainly at the C8 position of guanine residues.[34,35]

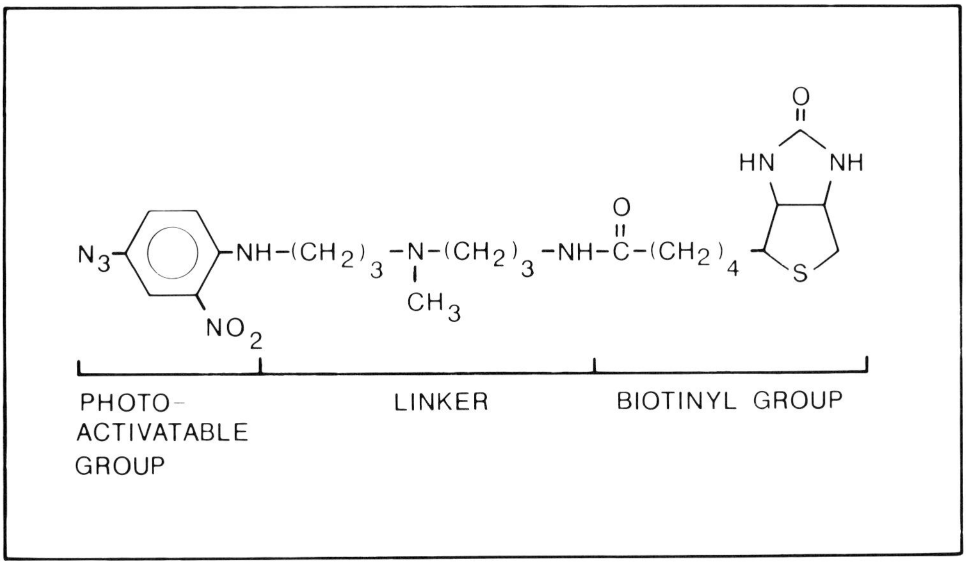

FIGURE 7. Structure of Photobiotin®.

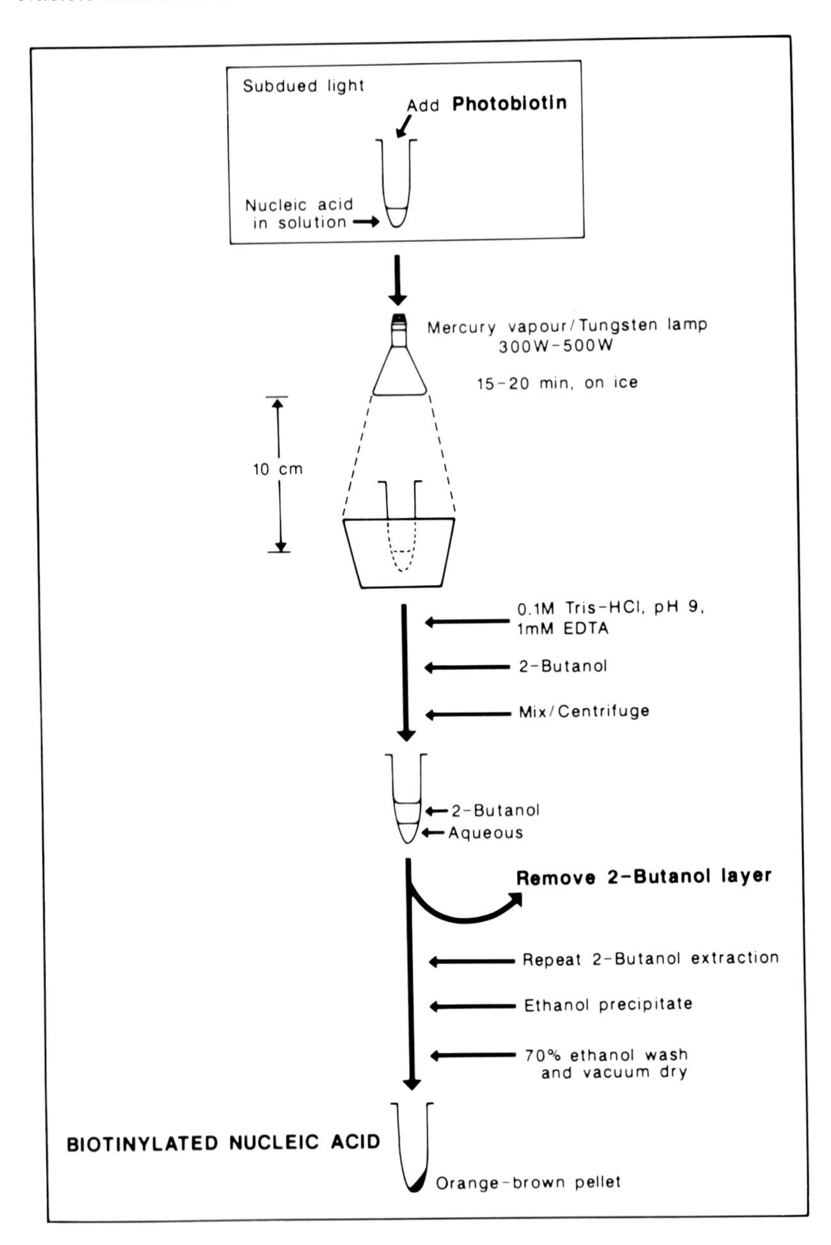

FIGURE 8. Schematic diagram showing the chemical biotinylation procedure for labeling nucleic acids with Photobiotin®. (From McInnes, J. L., Vize, P. D., Habili, N., and Symons, R. H., *Focus (Bethesda Research Laboratories/Life Technologies Inc.)*, 9 (4), 1, 1987. With permission.)

defined conditions, the azido group is converted to an extremely reactive nitrene which allows the formation of stable linkages to the nucleic acid. Excess photolyzed Photobiotin® can be removed by 2-butanol extraction. An outline of the labeling procedure is given in Figure 8.

Both single-stranded DNA and RNA and double-stranded DNA (double-stranded RNA has not been tested) are readily labeled with Photobiotin® with the incorporation of one biotin per 100 to 150 residues. The site(s) of coupling to the nucleic acid have not been determined. Following hybridization, the presence of the biotinylated probe can be detected by a variety of techniques (see Section V). Probes, once prepared, can be stored for long periods without detectable loss of sensitivity in hybridization reactions. Photobiotin®-labeled single-stranded M13 DNA probes appear stable for at least 12 months when stored at −20°C. No breakdown of Photobiotin®-labeled double-stranded DNA probes occurred after 8 months storage at −20°C. We have also successfully used Photobiotin®-labeled single-stranded RNA probes after storage at −20°C for 12 months. An important advantage of this reagent is that large (100 μg to milligram amounts) of biotinylated probe can be rapidly and economically prepared.

During the last 2 years, Photobiotin®-labeled DNA (and to a lesser extent, RNA) probes have been rigorously tested in our laboratory in different hybridization analyses. Various studies have shown that Photobiotin® can replace the use of radioactivity in preparing gene probes for Southern, Northern, and dot-blot analyses.[38,39] In each case, the detection of target nucleic acids was at least as sensitive as with probes labeled with ^{32}P. In particular, Photobiotin®-labeled DNA probes could detect, in Southern transfer analysis, the unique genes metallothionein IIA[38] and histone H5[39] in human and chicken genomal DNA, respectively. Low abundance chicken histone H5 mRNA was also detected in total cytoplasmic RNA isolated from an erythroleukemic cell line by Northern transfer.[38] Recently, we have utilized a Photobiotin®-labeled DNA probe to detect the RNA of a serious plant pathogen, barley yellow dwarf virus, in plant extracts by dot-blot hybridization[40] (see also Chapter 4).

Finally, Photobiotin® is proving to be an attractive reagent to other researchers. Milde and Löning[41] have reported the successful use of Photobiotin® to label a DNA probe for the detection of human papilloma virus DNA by *in situ* hybridization in oral papillomas and carcinomas. Flavivirus RNA has been detected in infected tissue-culture cells by a spot-hybridization assay using Photobiotin®-labeled probes.[42] Photobiotin® was employed to label one of two probes for use in a rapid sandwich hybridization protocol[43] and also to prepare a biotinylated RNA probe for its selective enrichment of cDNA sequences in a subtractive hybridization procedure involving avidin and copper-chelate agarose.[44] The use of Photobiotin® as a sensitive probe for protein labeling has also been reported.[45]

3. Biotinylated Psoralen

Sheldon et al.[46] have reported a novel method for the nonisotopic labeling of DNA probes. It involves the single-stranded circular DNA of the bacteriophage M13[47] and a unique labeling reagent, a biotinylated psoralen derivative: *N*-biotinyl-*N*′-(4′-methylene trioxsalen)-3,6,9-trioxaundecane-1,11-diamine[48] (Figure 9A).

A "gapped-circle" DNA probe was first prepared by hybridizing the single-stranded circular M13 DNA containing the desired cDNA insert in the BamHI site to BamHI-linearized M13 DNA replicative form, essentially as described by Courage-Tebbe and Kemper[49] (Figure 9B). The gapped-circle DNA was combined with biotinylated psoralen at 100 μg/ml and 60 μ*M*, respectively, in 10 m*M* Tris-HC1, 1m*M* EDTA, pH 8.0, and irradiated with 360 nm ultraviolet light at 30 mW/cm^2 for 10 min. The resulting probe (Figure 9B), containing covalently bound biotinylated psoralen, was ethanol precipitated and resuspended in the same buffer. Following hybridization, the presence of biotinylated probe bound to target nucleic acid was detected with a streptavidin-horseradish peroxidase conjugate which converted a colorless substrate to a visible blue precipitate in less than 1 h (see Section V).

DNA probes prepared as described above were labeled with 5 to 10 biotinylated psoralen moieties per 100 base pairs of double-stranded DNA.[46] Such probes have been successfully used for the nonisotopic detection of the patterns of HLA class II restriction fragment length (RFLP) polymorphism.[46] These probes, now marketed as See-Quence® HLA-D Gene Probes (Cetus), can detect single-copy genes in less than 0.5 μg of total human DNA on Southern blots and generate the same specific RFLP patterns as do probes labeled with ^{32}P by nick translation. Biotinylated psoralen M13 DNA probes have also provided nonradioactive detection of the sickle cell and normal alleles of the β-globin locus.[50]

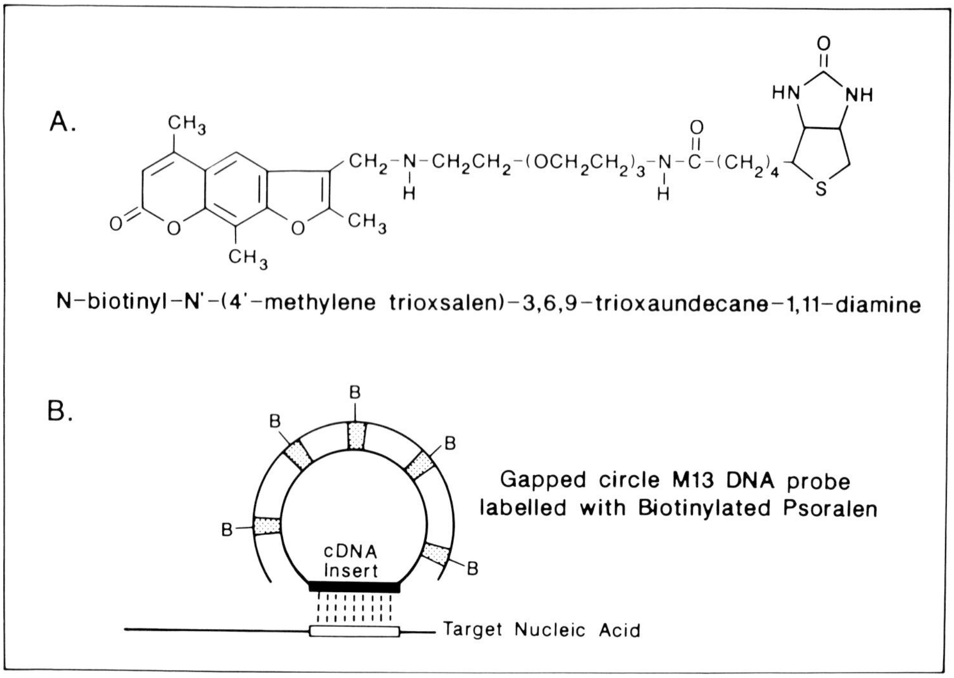

FIGURE 9. Nonisotopic labeling of DNA probes with a biotinylated psoralen.[46,48] (A) Structure of the biotinylated psoralen labeling reagent.[48] (B) Schematic representation of the biotinylated psoralen probe hybridized to target nucleic acid.[46] (From Sheldon, E. L., Kellogg, D. E., Watson, R., Levenson, C. H., and Erlich, H. A., *Proc. Natl. Acad. Sci. U.S.A.*, 83, 9085, 1986. With permission.)

4. Cytosine Modification Protocols

Viscidi et al.[51] have prepared biotin-labeled nucleic acids by the method outlined in Figure 10. The procedure involves the transamination of the 4-amino group of unpaired cytosine residues of nucleic acids with a bifunctional amine, ethylene diamine, in the presence of sodium bisulfite. The primary amino groups on the cytosine derivatives are then coupled with biotinyl-(6-amino-hexanoyl-*N*-hydroxysuccinimide ester to yield the biotinylated nucleic acid. Following hybridization, the presence of the biotinylated probe was detected colorimetrically (see Section V.)

Denatured lambda bacteriophage DNA modified by the recommended protocol of Viscidi et al.[51] contained 3 to 4% of their cytosine residues biotinylated. Probes of 0.5 to 50kb in length functioned efficiently in dot-blot and Southern hybridizations and detected 1 to 2 pg of nitrocellulose filter-bound DNA, a result comparable to that obtained with probes prepared by the enzymatic incorporation of biotinylated nucleotides by nick translation (Section II.A 1.a).

Two important features of this method should be noted. First, the bisulfite-catalyzed transamination reaction is *single-strand specific*. Hence, it is essential to denature double-stranded DNA before the modification reaction. This works satisfactorily for linear DNA but not for covalently closed circular DNA. In the latter case, heat denaturation and snap cooling of the DNA presumably provide single-stranded regions which reanneal to reform the duplex structure under the conditions used for the transamination reaction (2 *M* $NaHSO_3$,pH 7 to 8, 3 h at 42°C). We have found negligible transamination using covalently closed circular plasmid DNA.[52] Second, sodium bisulfite also catalyzes the deamination of cytosine residues to uracil; this reaction has an acid pH optimum and can be minimized by working at neutral pH.[53]

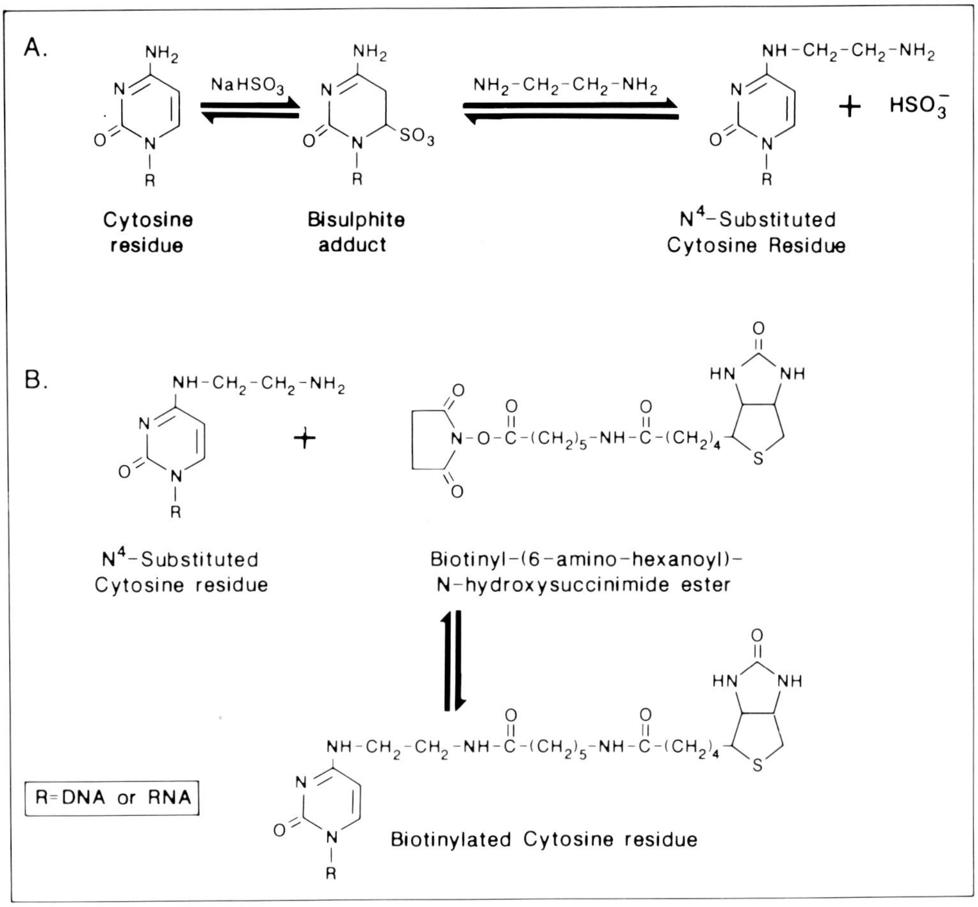

FIGURE 10. Biotin-labeling of DNA or RNA via cytosine modification.[51] (A) Bisulphite-catalyzed transmination reaction with ethylene diamine. (B) Biotinylation reaction. (From Viscidi, R. P., Connelly, C. J., and Yolken, R. H., *J. Clin. Microbiol.*, 23, 311, 1986. With permission.)

Another approach describing cytosine modification has been recently reported by Reisfeld et al.[54] Their method involves a transamination reaction between biotin hydrazide and unpaired cytosine residues of the sample DNA (Figure 11). The reaction, catalyzed by sodium bisulfite at pH 4.5, required a 24 h incubation at 37°C, conditions known to give significant deamination of cytosine to uracil.[53] Probes prepared in this way were able to detect about 1 pg of nitrocellulose-bound target sequence using a streptavidin-alkaline phosphatase conjugate for detection.[54]

5. Cross-Linking Protocols

The ability of basic proteins to bind to nucleic acids has been exploited to prepare nonradioactive probes. Biotinylated basic proteins (e.g., cytochrome *c*, histone H1) were bound to DNA or RNA and then cross-linked to the nucleic acid by formaldehyde or glutaraldehyde.[55-57] Such probes proved to be unstable under certain hybridization conditions[58] and detection of target DNA with these probes in conjunction with an avidin-peroxidase conjugate (see Section V) proved less sensitive than radioactive techniques.[57]

Syvänen et al.[59] varied this approach by using the *E. coli* single-strand binding protein to bind to single-stranded bacteriophage M13 DNA containing a cloned insert for use as probe. After cross-linking the protein with glutaraldehyde it was biotinylated by reaction

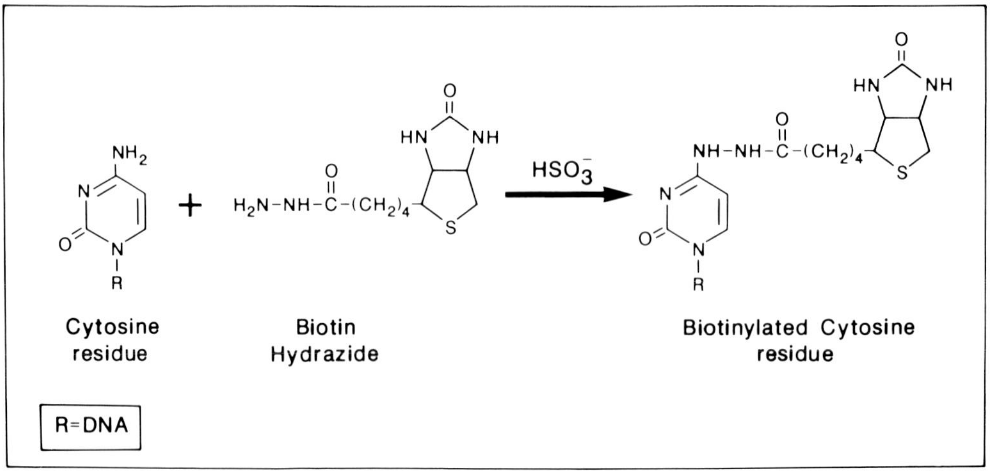

FIGURE 11. Bisulphite-catalyzed transamination reaction between DNA and biotin hydrazide.[54] (From Reisfeld, A., Rothenberg, J. M., Bayer, E. A., and Wilchek, M., *Biochem. Biophys. Res. Commun.*, 142, 519, 1987. With permission.)

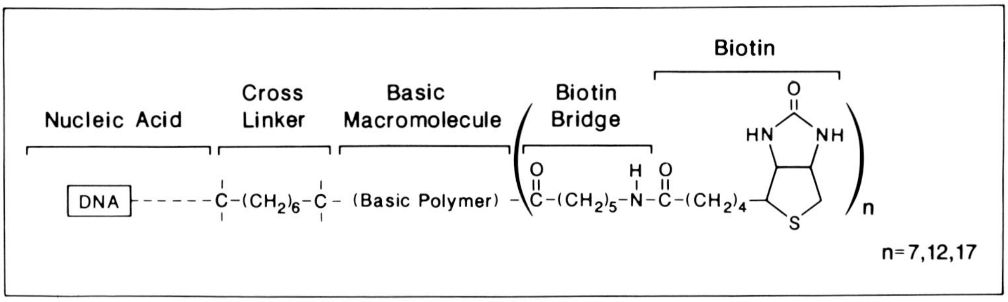

FIGURE 12. General structure of the nonradioactive *reporter* group.[60] (From Al-Hakim, A. H. and Hull, R, *Nucleic Acids Res.*, 14, 9965, 1986. With permission).

with biotin-*N*-hydroxysuccinimide ester. These M13 DNA-protein complexes proved disappointing as probes,[59] presumably because the bound protein interfered with the hybridization of probe to target.

In a further variation of this general approach, Al-Hakim and Hull[60] cross-linked biotin via a basic macromolecule to single-stranded phage M13 DNA (Figure 12). Basic macromolecules tested included histone H1, cytochome *c* and two amine polymers — polyethyleneimine of average mol wt 60,000 and 1,400. Cross-linking agents included formaldehyde, glutaraldehyde, 1,2,7,8-diepoxyoctane, *bis*(succinimidyl)suberate and its sulfosuccinimidyl derivative. The results obtained on hybridization and detection of bound biotin byavidin-peroxidase or avidin-alkaline phosphatase indicated that the sensitivity of detection of target sequences was dependent on the nature of all components used and the length of the biotin bridge (Figure 12). For example, the longer the biotin bridge and the cross-linking reagent, the more sensitive was the detection. In the case of the size of the basic macromolecule, polyethyleneimine (1,400) gave the best sensitivity followed by the two basic proteins and then polyethyleneimine (60,000). The sensitivity of detection of target sequences was down to 0.01 to 0.05 pg for the best probe, levels significantly better than those achieved by previous workers using this cross-linking approach.[57]

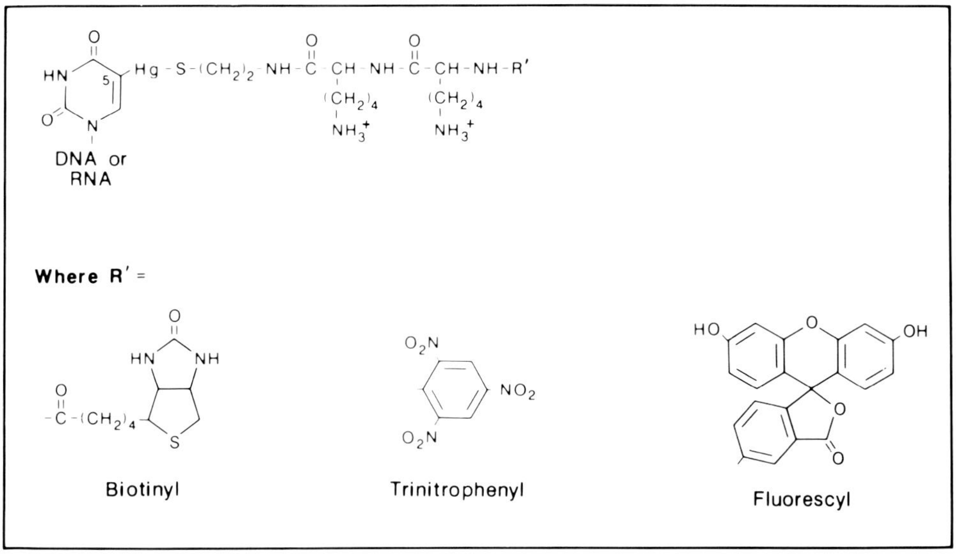

FIGURE 13. General structure of a mercurated nucleic acid probe *reported* group.[62,63] The mercurated cytosine residue of the *in situ* hybridized probe requires reaction with the sulfhydryl group of a lysine containing ligand-hapten for detection purposes. Hapten groups shown are biotinyl, 2,4,6-trinitrophenyl and fluorescyl. (Adapted from Hopman, A. H. N., Wiegant, J., Tesser, G. I., and Van Duijn, P., *Nucleic Acids Res.*, 14, 6471, 1986.)

6. Sulfonation

Incubation of 1 *M* $NaHSO_3$ and 1 *M* methylhydroxylamine with single-stranded DNA or RNA results in the conversion of cytosine residues to the relatively stable N^4 methoxy-5,6-dihydrocytosine-6-sulfonate derivatives.[61] This reaction forms the basis of a nonradioactive DNA labeling system of Orgenics Ltd., Israel (the Chemiprobe system). After hybridization, the sulfonated probe is detected by immunological methods.

7. Mercuration

A nonradioactive labeling technique based on the introduction of mercury at the C5 position of pyrimidine bases of nucleic acid molecules has been described by Hopman et al.[62,63] (Figure 13). Contrary to the other nonradioactive protocols described here, the ligand or hapten required for detection is introduced *after* the hybridization step. Mercurated probes in the presence of cyanide ions are stable under hybridization conditions.

The DNA or RNA to be used as probe was initially mercurated in an overnight reaction at 50°C of the nucleic acid in 5 m*M* mercuric acetate at pH 5.5 to 6.0. After purification by gel filtration, the probes were used as such for both *in situ* and dot-blot hybridization.[62] The target-bound probes were then coupled to the ligand or hapten by compounds having the general structure:

sulfhydryl -- positively charged linker -- ligand (hapten)

The sulfhydryl group was necessary for reaction with the mercury atoms while the positively charged amino groups of the linker (two lysine residues) contributed to the electrostatic binding of the compound to the negatively charged phosphate groups of the nucleic acid. The haptens used were trinitrophenyl and fluorescyl while the ligand was biotinyl. The sensitivity of detection of target sequences was similar to that reported for other nonradioactive probes.[1,3,32,37]

B. Direct Coupling of Enzymes to Nucleic Acid Probes

As reported in Section III A.1.5, biotinylated proteins or other basic macromolecules have been chemically cross-linked to nucleic acids to provide nonradioactive probes. After hybridization, the probe is detected by conventional cytochemical techniques.

An alternative strategy, which circumvents the biotin/avidin or streptavidin system, has been developed by Renz and Kurz.[64] Here, peroxidase or alkaline phosphatase was cross-linked to polyethyleneimine with *p*-benzoquinone, and the resultant conjugates were then covalently cross-linked to single-stranded DNA using glutaraldehyde. Probes constructed in this way had a large protein to mass ratio (approximately 30) and proved efficient in Southern blot hybridizations. The limit of detection of target DNA on nitrocellulose filters was in the range of 1 to 5 pg, which is similar to that found with biotinylated probes.

The method also allows a flexible approach to target detection. By using two differently labeled probes (one labeled with peroxidase and the other with alkaline phosphatase), different target DNAs could be detected in a one-step hybridization reaction followed by the sequential addition of the two different enzyme substrates.

While this approach appears attractive, two aspects need to be considered. First, the hybridization conditions are limited by the stability of the enzymes linked to the DNA probes. For example, Renz and Kurz[64] performed their hybridizations at 38°C for 2 to 16 h, presumably because of the instability of the coupled enzymes for similar time periods at higher temperatures. Second, as Renz[57] has suggested, extensive labeling of the probe nucleic acid with a large enzyme may sterically interfere with the efficiency of the hybridization reaction with target sequences bound to a solid support.

IV. PREPARATION OF OLIGODEOXYNUCLEOTIDE PROBES

There is an increasingly intense interest in the development of methods for the nonradioactive labeling of synthetic oligodeoxynucleotides for use as probes. Apart from obviating the need for radioactivity, these probes allow enormous flexibility in probe design. Some important aspects which illustrate the powerful nature of this appraoch are as follows:

1. The rapidly expanding sequence data banks allow the precise selection of target sequences for probes.
2. The commercial availability of machines for the rapid and reliable synthesis of milligram quantities of synthetic oligodeoxynucleotides of defined sequence. Although most oligonucleotides used so far for probes are 20 to 30 residues long, these machines are capable of producing oligonucleotides 30 to 100 residues long, and even longer if necessary.
3. The chemical methods for the modification and derivatization of oligonucleotides with ligands, haptens, and enzymes are at a stage of rapid development.
4. In general, oligonucleotide probes obviate the need to prepare and characterize probes by cloning in suitable vectors, procedures which can be laborious and time-consuming. In addition, they eliminate what is sometimes an important practical problem, namely, background hybridization of vector sequences to nucleic acids in the sample under test.
5. Oligonucleotide probes, because of their small size, hybridize much faster than longer probes, thus reducing hybridization time. For example, hybridization can be reduced to 1 h as compared to the usual overnight hybridization with cloned probes.[65-67]

In spite of all these obvious advantages, it must be remembered that oligonucleotide probes only hybridize to what could be only a small fraction of the total target molecule. In this case, sensitivity is much reduced as compared with a much longer cloned probe, although

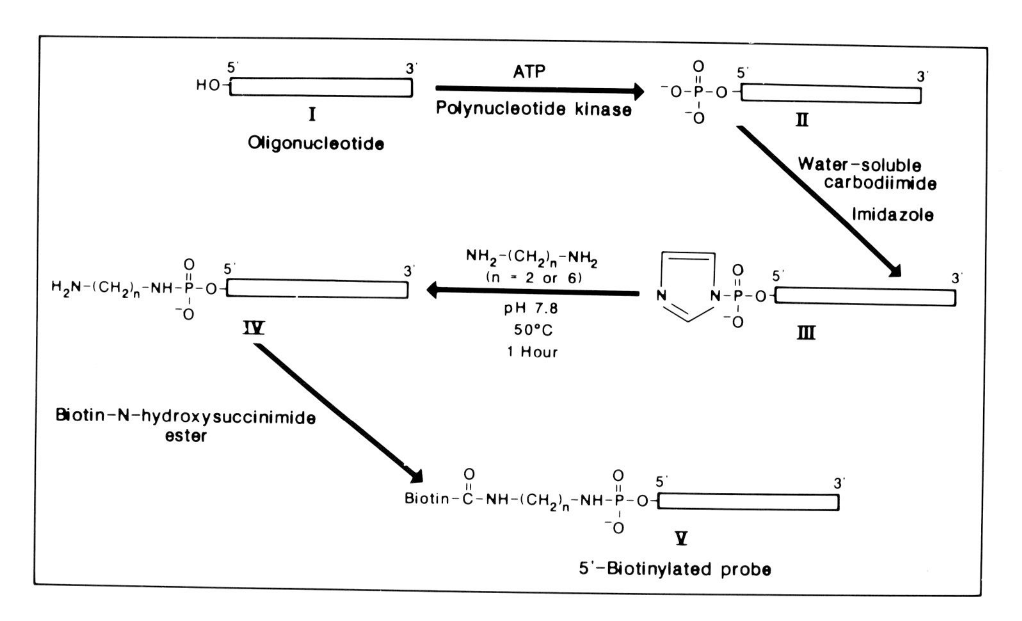

FIGURE 14. Biotin labeling procedure under aqueous conditions at the 5′-end of an unprotected oligonucleotide.[68-70]

the greater accessibility of a small probe to target molecules bound to a solid support such as a nitrocellulose or nylon membrane will partly alleviate this disadvantage. In addition, it is possible to have a number of oligonucleotide probes, rather than just one, that will hybridize to the target sequence although this will lead to a significant increase in cost of the probe.

A. Chemical Labeling of Oligonucleotides with Ligand or Hapten

1. Labeling at the 5′-End of Unprotected Oligonucleotide

As described below in section IV.A.2, most synthetic oligonucleotides are derivatized while still fully protected and bound to the solid support of the DNA synthesizer. Although this approach is preferable, purified, deprotected oligonucleotides can be derivatized via the free 5′-hydroxyl by the method outlined in Figure 14.[68-70]

In this method, the oligonucleotide I was first phosphorylated using ATP and polynucleotide kinase followed by conversion of II to the phosphorimidazolidate III by reaction at pH 6 in the presence of imidazole and a water-soluble carbodiimide. Incubation with excess 1,2-diaminoethane or 1,6-diaminohexane for 1 h at 50°C and pH 7.8 provided the phosphoramidate of the diamine (IV) which was then biotinylated by reaction with biotin-*N*-hydroxysuccinimide ester. Probes V prepared in this way were successfully used in dot-blot and Southern hybridizations.[68,69]

Care needs to be taken in using probes containing a phosphoramidate linkage. First, this linkage is labile below pH 7 so that all reactions, including the detection, must be carried out under alkaline conditions. Second, Chu and Orgel[69] reported a significant loss of the diaminohexane-linked adduct and a larger loss of the diaminoethane adduct during the washing procedure used in the avidin-alkaline phosphatase detection system. This was presumably due to hydrolysis of the phosphoroamidate linkage or a displacement reaction at the same activated linkage by one or more components in the washing solutions.

2. Derivatization of Oligonucleotide at 5′-End during Solid Phase Synthesis

The most practical route for the derivatization of synthetic oligonucleotides is while they are still bound to the solid phase in the now common solid phase step-wise synthesis. At the end of each monomeric addition, all reactive groups of the oligonucleotide contain suitable protecting groups. Synthesis proceeds in the 3′- to 5′-direction and at the end of each step the 5′-hydroxyl is protected by a dimethoxytrityl group. This is removed by brief acid treatment prior to the addition of the next activated nucleotide monomer, all other protecting groups are stable to this treatment. At the end of the synthesis of the oligonucleotide the 5′-hydroxyl is the site of activation and conversion to a form to which the ligand or hapten can be added, preferably while the oligonucleotide is still bound to the solid support. A variety of chemical methods have been developed to achieve this and the key ones are described here.

a. Activation of 5′-Hydroxyl with Carbonyl Diimidazole

The route for the preparation of 5′-biotinylated oligonucleotide by the method of Wachter et al.[71] is outlined in Figure 15. The 5′-terminal hydroxyl of the support coupled, protected oligonucleotide I was reacted with carbonyl diimidazole in dioxane under completely anhydrous conditions for 30 to 45 min to produce the 5′-imidazolyl derivative II. The activated derivative was then reacted with 1,6-diaminohexane in aqueous dioxane for 20 min under which conditions the displacement reaction went to completion to form the *N*-alkyl carbamate linkage in III. This linkage is stable to treatment with both thiophenol for 30 min to remove the phosphate protecting groups and conc. NH_4OH over 5 h at 55°C to remove all the protecting groups from the oligonucleotide and to cleave it from the solid support. The crude product IV was then converted to the 5′-biotinyl derivative by reaction with biotin-*N*-hydroxysuccinimide ester at alkaline pH and the product purified by gel electrophoresis. Its efficiency as a hybridization probe was not reported.

b. Cyclic Phosphoramidite Provides 5′-Aminoethyl Phosphate

Applied Biosystems Inc. of California have recently introduced a cyclic phosphoramidite (called Aminolink 1, see Figure 16) for the single step addition of an amino group on the terminal 5′-hydroxyl group of an oligonucleotide as the final step of the automated synthesis prior to deprotection with ammonia. Aminolink 1 is a liquid and is diluted with 0.1 *M* dimethylaminopyridine in anhydrous acetonitrile before use. Reaction of this compound with the 5′-hydroxyl group of a support bound, protected oligonucleotide I produces product II. Oxidation by iodine converts the phosphite of II to phosphate; thiophenol treatment removes the methyl group from the phosphate and NH_4OH treatment cleaves the oligonucleotide from the support and hydrolyzes all protecting groups to produce the 5′-amino-ethylphosphoryl oligonucleotide III. This oligonucleotide can then be derivatized on the reactive amino group with the appropriate ligand or hapten.

Although Aminolink I is reported to have high coupling yields, it is more sensitive to moisture than the standard nucleoside phosphoramidites. This may be an important factor in determining its widespread use.

c. Versatile Phosphoramidite Provides 5′-Aminophenylethyl Phosphate

We have developed[72] a reliable procedure for the introduction of a reactive amino group via a phosphoramidite I (Figure 17), prepared as described by Uhlmann and Engels.[73] When reacted with a synthetic oligonucleotide (II) after removal of the final dimethoxytrityl group on an automated DNA synthesizer, product III was produced in high yield. After the standard iodine, thiophenol, and ammonia treatments to remove protecting groups and release of the oligonucleotide from the solid support, compound IV was purified from failure sequences by reverse phase HPLC. At this stage, Uhlmann and Engels[73] removed the *p*-nitrophenylethyl group by treatment with a strong hindered base to give the 5′-phosphorylated oligonucleotide VI, the end point of their approach. However, treatment of IV with $(NH_4)_2S$ (Zinin reduction) gave product V with a reactive amino group which was converted to the biotinyl derivative

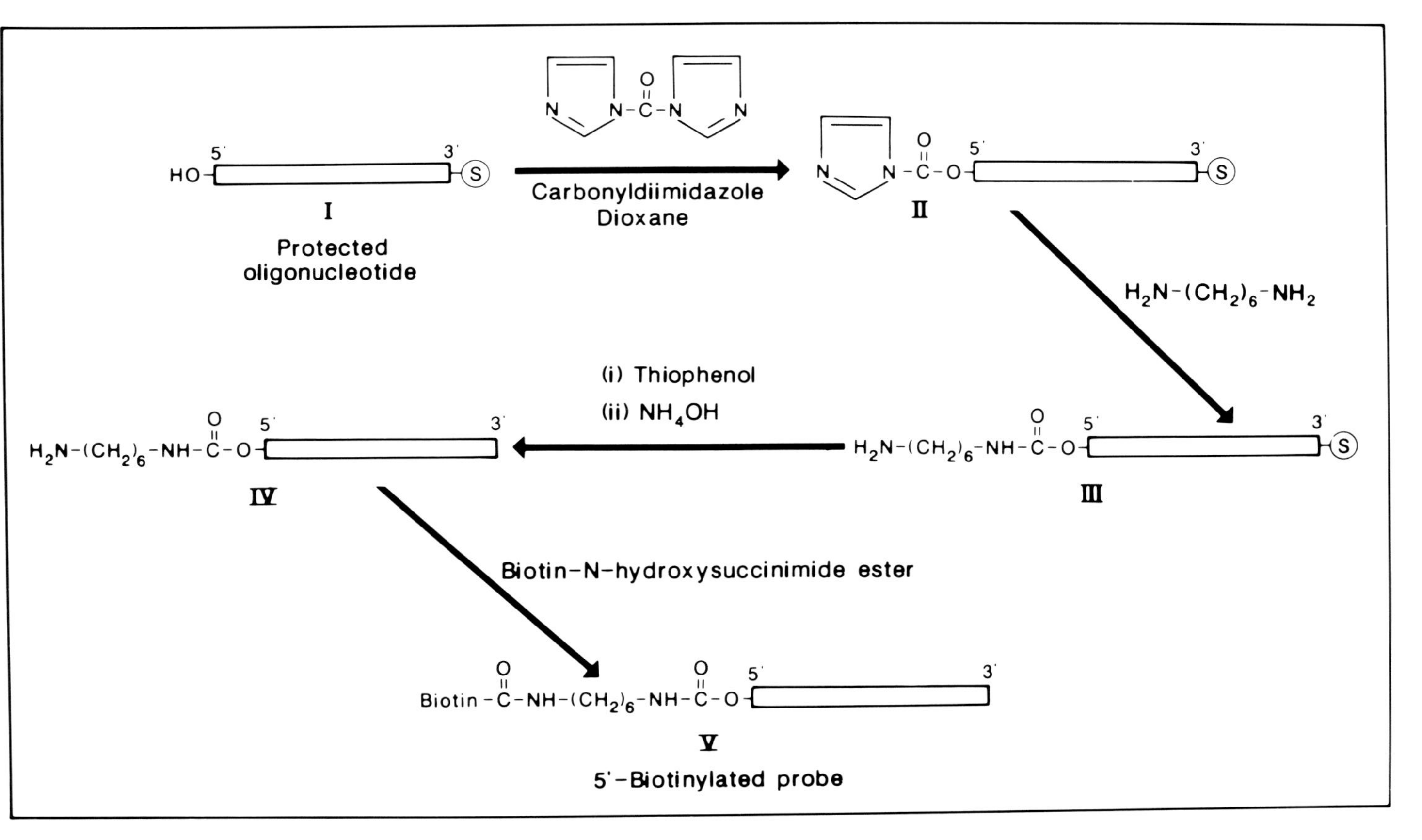

FIGURE 15. Route for the preparation of a 5′-biotinylated oligonucleotide probe.[71] Compound II was prepared from I under anhydrous conditions, all other reactions were under aqueous conditions. S, solid support used for step-wise chemical synthesis of oligonucleotide.

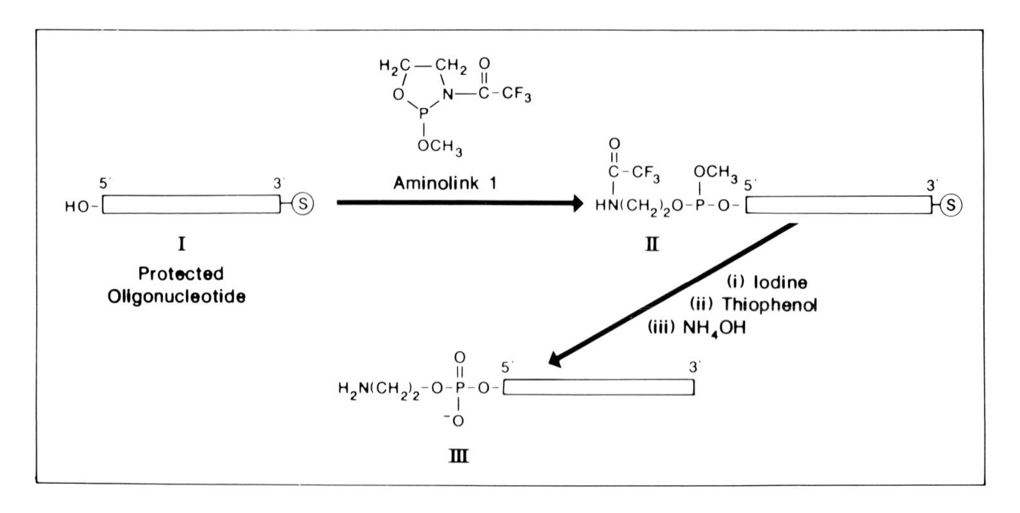

FIGURE 16. Route for the preparation of a 5′-aminoethyl-phosphoryl derivative of an oligonucleotide during solid phase synthesis using Aminolink 1 (Applied Biosystems Inc.). S, solid support used for step-wise synthesis of oligonucleotide.

by reaction with biotinyl-*N*-hydroxysuccinimide ester. This overall procedure has considerable potential for the preparation of a range of 5′-derivatives labeled with various ligands and haptens. Their use as hybridization probes is being investigated.

d. N-Protected, Aminoalkyl Phosphoramidites

A similar approach to that used in c has been used by three separate groups to prepare 5′-aminoethyl- or 5′-aminopropyl-phosphoryl derivatives of oligonucleotides for subsequent derivatization with a ligand or hapten.[74-76] The three analogous reagents prepared for the phosphorylation of the terminal 5′-hydroxyl group of a support bound, protected oligonucleotide are given in Figure 18. All compounds are diisopropylaminophosphoramidites with either cyanoethyl (I,II) or methyl (III) protecting groups on the phosphite plus an aminoalkyl group (aminoethyl or aminopropyl) protected by trifluoracetyl (I), 9-fluorenylmethoxycarbonyl (FMoc) (II) or methoxytrityl (III) groups. Efficient phosphorylation was obtained with all compounds as the final step of the assembly of oligonucleotides on an automated DNA synthesizer. The trifluoracetyl and FMoc groups were removed during the standard ammonia treatment for the removal of protecting groups while the methoxytrityl group was removed by treatment with 80% acetic acid.[76]

The 5′-aminoalkyl oligonucleotides were further derivatized (1) with biotin using biotin-*N*-hydroxysuccinimide ester[74,75] or biotin-*p*-nitrophenol ester,[76] (2) with the fluorescent dansyl group using dansyl chloride,[76] and (3) with three other fluorescent groups—fluorescein and tetramethylrhodamine via their isothiocyanates, and 4-fluoro-7-nitrobenzofurazan via its fluoride.[75]

e. N-Protected 5′-Amino-5′-Deoxy-3′-Phosphoramidites of Deoxythymidine

An alternative approach to the labeling of the 5′-terminus of synthetic oligonucleotides has been reported by Smith et al.[77] Like reagent III in Figure 18, they similarly used a diisopropylaminophosphoramidite with a methyl protecting group on the phosphite, but the essential group was a *N*-protected 5′-amino-5′-deoxy-2′-deoxy-thymidine (Figure 19). The 5′-amino protecting group (R, Figure 19) was either FMoc or trifluoroacetyl.

These nucleoside phosphoramidites gave efficient phosphorylation of the 5′-hydroxyl of support bound oligonucleotides (Figure 20). After removal of all protecting groups and cleavage from the support, the 5′-amino group was coupled to four fluorescent compounds;

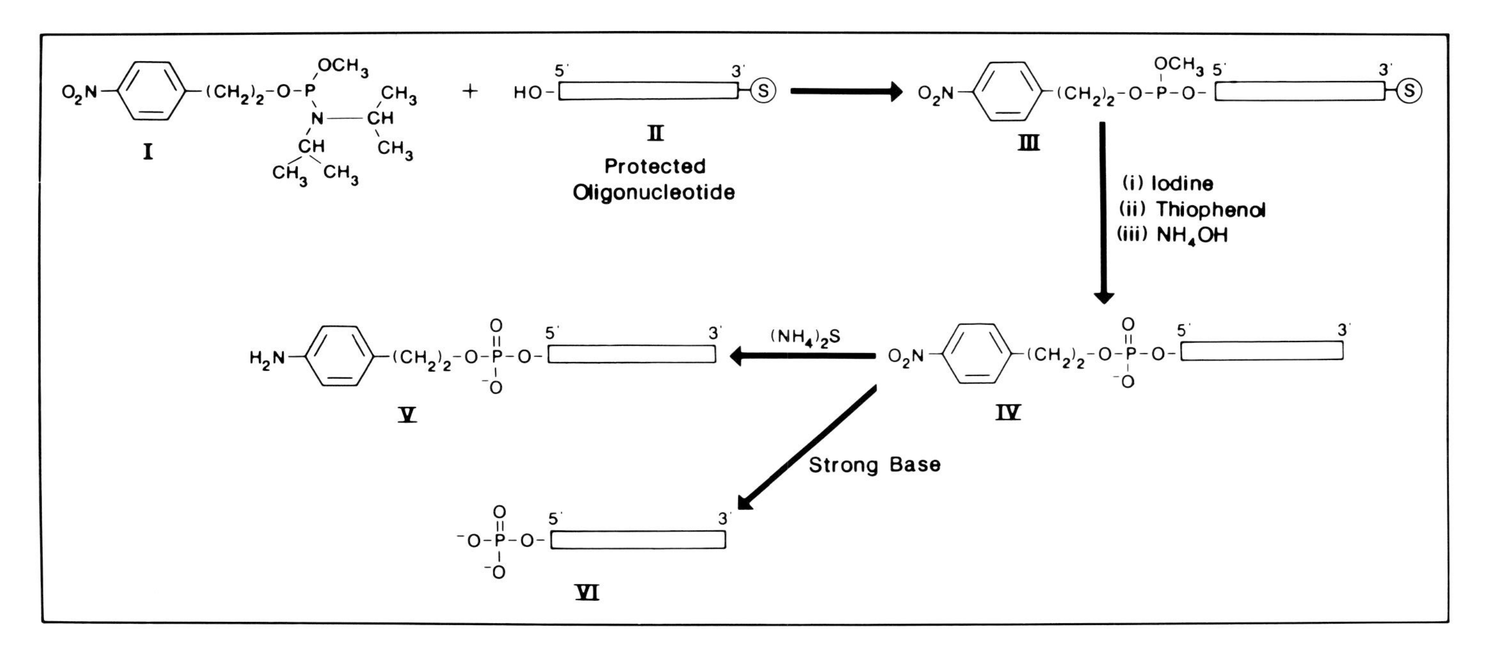

FIGURE 17. Route for the preparation of a 5′-aminophenylethyl-phosphoryl derivative of an oligonucleotide during solid phase synthesis using a versatile phosphoramidite.[72] S, solid support used for step-wise synthesis of oligonucleotide.

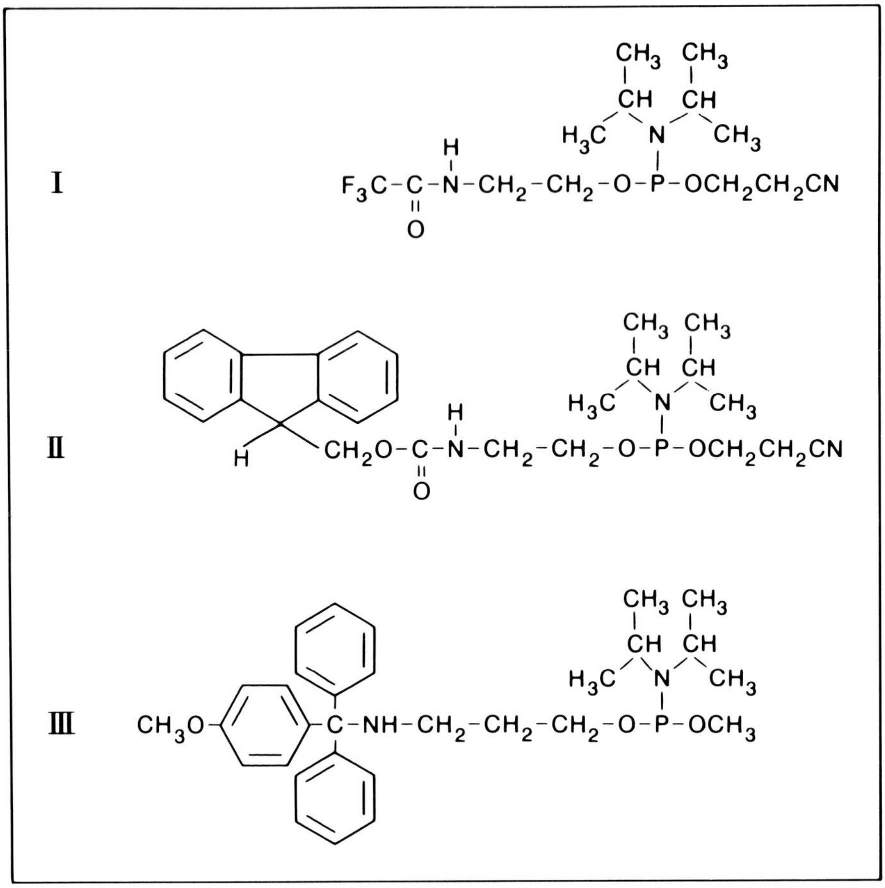

FIGURE 18. Structures of *N*-protected, aminoalkyl phosphoramidites used by 3 groups[74-76] to label the 5′-terminal hydroxyl group of oligonucleotides with a reactive amino group as the final synthetic step on an automated DNA synthesizer. Amino group protection was provided by trifluoroacetyl (I), FMoc (II), and methoxytrityl (III).

fluorescein isothiocyanate, Texas Red, tetramethyl rhodamine isothiocyanate, and 4-fluoro-7-nitrobenzofurazan (Figure 20). The products V were then purified by reverse phase HPLC for use as nonradioactively labeled DNA primers for sequencing[78] by the Sanger dideoxy-nucleotide technique.[79]

A similar approach to that taken by Smith et al.[77,78] (Figure 20) was described by Sproat et al.[80] where 3′-0-cyanoethyl phosphoramidites were prepared but the 5′-protected amino group was replaced by a (*S*-triphenylmethyl) mercapto group. After synthesis of the oligonucleotide and removal of the protecting groups, the 5′-mercapto-oligonucleotide was derivatized with fluorescent labels for use in automated DNA sequencing.[81] The same group[82] also prepared 5′-amino oligonucleotides using cyanoethyl phosphoramidites (Figure 19) and derivatized them with biotin, fluorophores, and metal cluster derivatives for use as probes.

Hence, the general approach described here lends itself to the preparation of oligonucleotides 5′-terminally labeled with a variety of ligands and haptens. However, their potential for use as diagnostic probes has yet to be fully explored. In addition, the widespread use of the methods described will depend on the commercial availability of the essential phosphoramidites of the type shown in Figure 19.

f. C-5 Nucleoside Derivatized 3′-Phosphoramidites

In a further variation on modified deoxynucleoside phosphoramidites,[83] deoxyuridine was

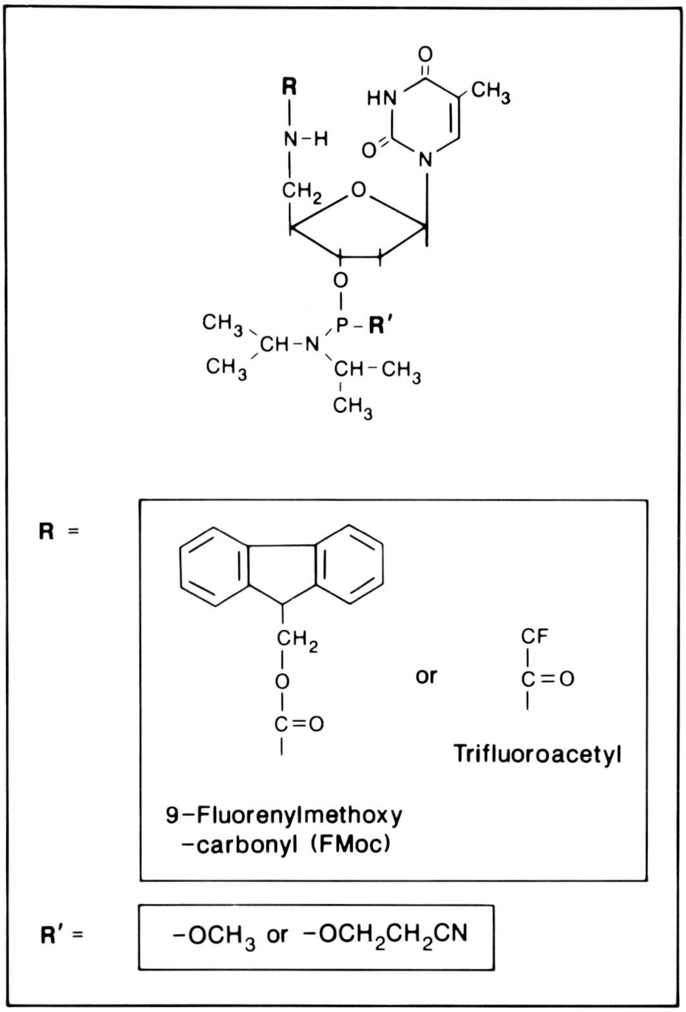

FIGURE 19. Structures of *N*-protected amino-derivatized 2′-deoxythymidine phosphoramidites used by 2 groups[77,82] to label the 5′-terminal hydroxyl group of oligonucleotides with a reactive amino group as the final synthetic step on an automated DNA synthesizer.

modified at C-5 with a 10 atom linker arm terminated with a protected amino group, somewhat analogous to the C-5 derivatization in bio- 11-dUTP (Figure 1). Oligonucleotides were then synthesized and extended at the 5′-end with the addition of one or more residues of this phosphoramidite. After deprotection, the oligonucleotide was coupled to fluorescein and 5′-^{32}P-labeled; the efficiency of hybridization was monitored by radioactivity. However, no results on the sensitivity of target detection by fluorescence were reported.

One advantage of the approach described here as compared to those described in *a* to *e* is the potential to add several extra residues, and hence also fluorescent residues, to the 5′-end of the oligonucleotide probe. This may provide enhanced sensitivity. The approach also has the potential to add residues at the 3′-end of the oligonucleotide before the required oligonucleotide is synthesized.

3. Labeling at 3′-End via Chemically Modified Oligonucleotide

Perhaps the simplest procedure for the preparation of terminally modified oligonucleotides for subsequent derivatization has been described by Li et al.[67] The starting material for

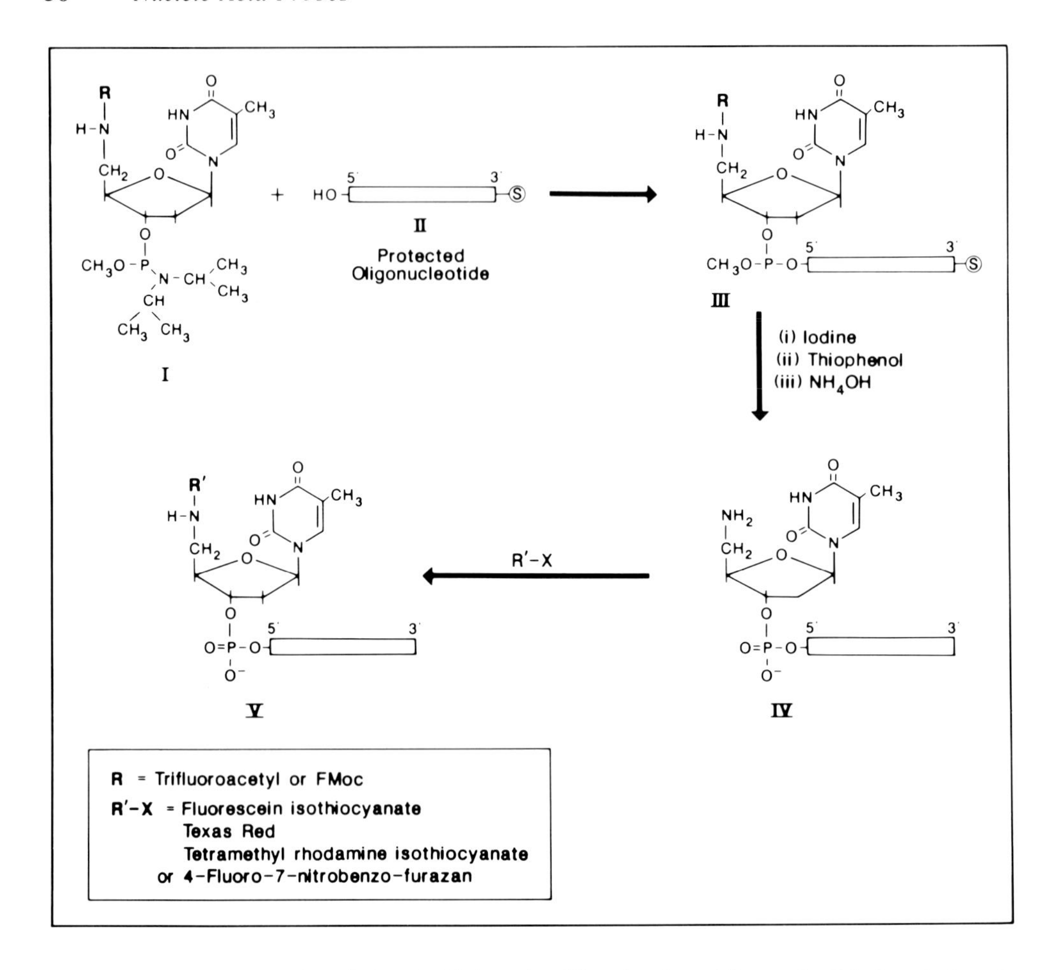

FIGURE 20. Coupling reaction of 5′-amino-2′-deoxythymidine phosphoramidites to the 5′-hydroxyl of an oligonucleotide and their subsequent attachment to fluorescent dye molecules. Method shown, as reported by Smith et al.[77] S, solid support used for stepwise synthesis of oligonucleotides.

oligonucleotide syntheses was a silica support containing a deoxycytidine residue modified at C4 with FMoc-protected diaminoethane (I, Figure 21). After the addition of three residues (e.g., AAA) as a spacer arm by standard automated phosphoramidite procedures, the specific probe sequence was synthesized to give the support bound product II. After the standard thiophenol and NH_4OH treatments to remove protecting groups and hydrolyze the oligonucleotide from the support, product III was obtained with a 3′-terminal deoxycytidine residue with a reactive amino group. Product III was then converted to the 3′-biotinylated probe (IV) by reaction with biotin-*N*-hydroxysuccinimide ester. Probes prepared in this way were effective in dot-blot analyses.[84]

The major advantage of the approach described here is that an extra step is not required with every oligonucleotide synthesis to provide an active amino group for subsequent derivatization. One batch preparation of the support I is sufficient for numerous syntheses. Further, when this method is coupled to one of the procedures in Section IV.A.2, oligonucleotides can be readily prepared with a reactive amino group at both the 5′-and 3′- ends. In this way, oligonucleotide probes would contain two ligands or haptens which may provide enhanced sensitivity of detection, especially in the case of fluorescent ligands. In other cases, the large size of a ligand or hapten may prevent their addition at both ends.

In a variation of this theme, Zuckermann et al.[85] introduced a thiol into the 3′-nucleoside

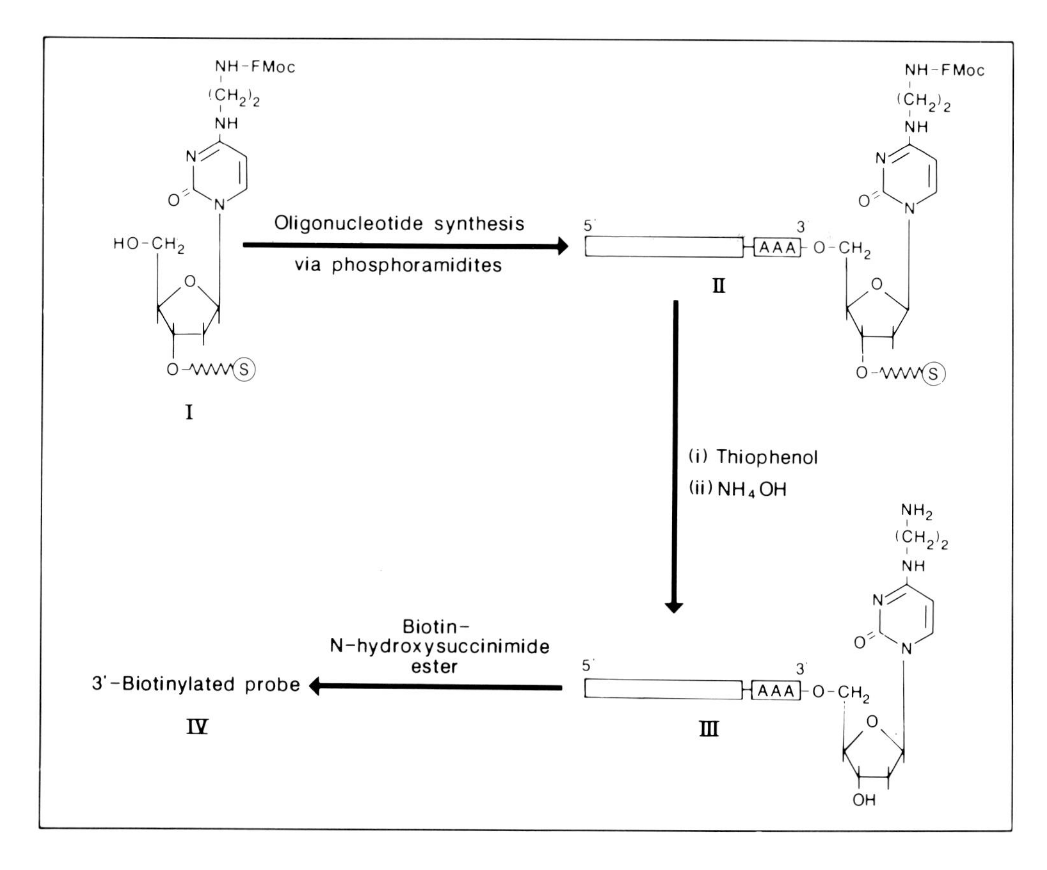

FIGURE 21. Scheme for the preparation of a 3′-biotinylated oligonucleotide probe during solid phase synthesis by the method of Li et al.[67] S, solid support used for stepwise synthesis of oligonucleotides.

support linkage prior to oligonucleotide synthesis. The purified oligonucleotide was then coupled to fluorescent probes but no hybridization studies were reported.

B. Labeling of Oligonucleotide Probes with Enzymes as the Detector Ligand

Biotin is the most widely used ligand in nonradioactive nucleic acid and olignucleotide probes. Its detection is normally via an avidin-enzyme or streptavidin-enzyme complex (See Section V). A practical problem associated with this method of detection is nonspecific background reaction due to binding of avidin (and less importantly streptavidin), not only to the biotinylated probe, but also to endogenous biotinylated proteins and other glycoproteins commonly found associated with clinical and other specimens.[86] Hence, serious background problems may result and this could account for the limited use of biotinylated probes in DNA colony hybridization.[86,87]

One way to bypass this problem is to couple the detecting enzyme directly to the nucleic acid probe, as already considered in Section III.B. The same approach has recently been reported for oligonucleotide probes by Jablonski et al.[88] and Li et al.[67]

In the method of Jablonski et al.[88] the desired oligonucleotide, 21 to 26 residues long, was chemically synthesized with a 5′-deoxyuridine residue derivatized on C5 with an aminoheptyl-3-acrylamido arm (I, Figure 22). Reaction with an excess of suberyl-di-*N*-hydroxysuccinimide (II) gave III which was purified by gel filtration and concentrated by lyophilization. The direct 5′-enzyme-labeled oligonucleotide probe (IV) was then obtained by reaction of III with a two fold molar excess of alkaline phosphatase overnight at pH 8.25 and room temperature followed by removal of the excess enzyme by ion-exchange chro-

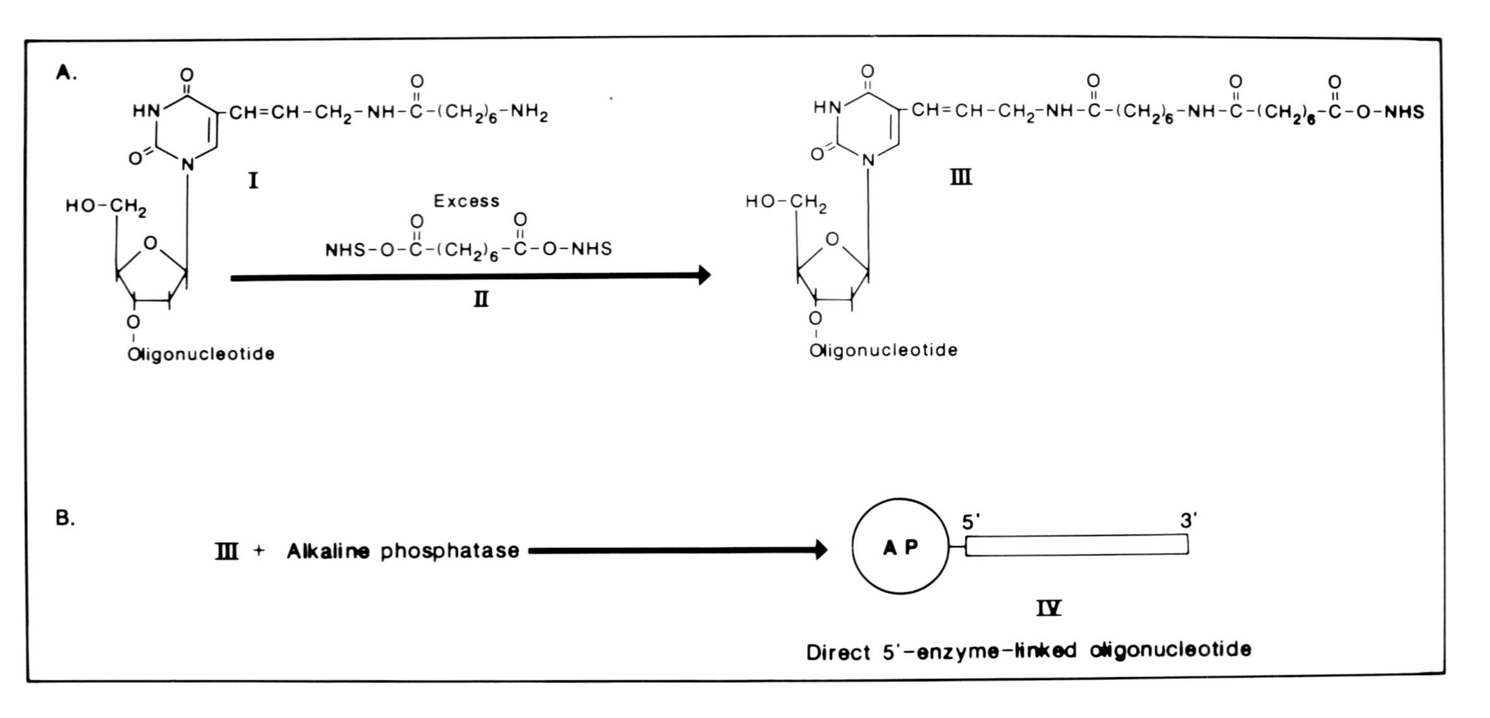

FIGURE 22. Scheme for the preparation of a direct 5′-enzyme-linked oligonucleotide probe by the method of Jablonski et al.[88] AP, calf intestine alkaline phosphatase. Note that the size of the spherical AP molecule is not in proportion to the length of the oligonucleotide probe.

matography. The enzyme was therefore separated from the oligonucleotide by a spacer arm of 20 atoms. The mass ratio of protein to DNA of probe IV is approximately 20.

Probes prepared in this way were used successfully in dot-blot and Southern hybridization. Surprisingly, the efficiency of hybridization appeared unaltered from that found with unmodified 5'-^{32}P-labeled oligomer and filter background was nonexistent.[88] Hybridization protocols were brief and simple; 15 to 30 min at 50°C in simple buffer mixtures with probe concentrations equivalent to 50 ng/ml of oligomer. Over a 4 h detection time for the hybridized oligomer, using nitro blue tetrazolium and 5-bromo-4-chloro-3-indolyl phosphate (see Section V), the sensitivity obtained was approximately 10^6 molecules of target DNA (2×10^{-18} mol) and was five fold greater than that obtained using a 5'-^{32}P-labeled oligomer and overnight autoradiography.

Two examples of the use of direct 5'-enzyme-conjugated oligonucleotide probes, based on the methodology of Jablonski et al.[88] have been reported. McLaughlin et al.[89] synthesized an alkaline phosphatase linked 21-base oligonucleotide probe and used it to detect purified *Plasmodium falciparum* (malaria) DNA by dot-blot hybridization. Likewise, alkaline phosphatase-conjugated 26-base oligonucleotide probes have been used to identify enterotoxigenic *E. coli* strains by colony and stool blot hybridization.[90]

Li et al.[67] have taken a different approach to prepare direct enzyme-linked oligonucleotide probes, as outlined in Figure 23. The starting 3'-aliphatic amine substituted oligonucleotide (I) was prepared as described in Section IVA.3 (Figure 21, compound III). This was derivatized by reaction with an excess of dithio-*bis*-propionyl-NHS (II) to give a mixture of III and IV which, on reduction by dithiothreitol (DTT), were converted to the oligonucleotide (V) with a 3'-terminal reactive sulfhydryl group. Alkaline phosphatase (VI) was derivatized by brief (5 min at 0°C) incubation with bromoacetyl-NHS (VII) and purified from excess reagent by rapid gel filtration. The sulfhydryl-oligonucleotide (V) was then conjugated overnight at 4°C with a small excess of the bromoacetylated enzyme (VIII) and the product (IX) purified by ion-exchange chromatography.

The method outlined has proven reliable for the routine preparation of direct 3'-enzyme-labeled oligonucleotide probes which are stable for many months when stored dry or in solution at -20°C. Such probes have been successfully used in the diagnosis of strains of enterotoxigenic *E. coli* in clinical specimens.[67] After immobilization of the bacterial nucleic acids on the membrane filter, the total time taken for hybridization (30 min at 50°C), washing and color development was as short as 2.5 h. However, color development was often allowed to proceed overnight to ensure maximum color intensity for weakly positive samples.

For this work on enterotoxigenic *E. coli*, the sequences of the oligonucleotide probes were determined by the published sequences of the toxin genes in the plasmids containing these genes.[67] For the heat-stable enterotoxin gene, four unique oligonucleotide sequences, covering approximately 30% of the gene, were synthesized and coupled to alkaline phosphatase. It was considered important to have several oligonucleotide probes covering various parts of the gene in order to cater for the possibility of silent mutations, which would not alter the protein sequence of the toxin gene product but would decrease the efficiency of hybridization of one or more of the probes. The sensitivity of detection was approximately 3-4 $\times$ 10^6 molecules (6×10^{-18} mol) of target DNA and was similar to that obtained with the same probes 5'-^{32}P-terminally labeled.

V. DETECTION OF LIGANDS AND HAPTENS

Hybridized nonradioactive probes can be detected by a variety of methods; however in the majority of cases an enzyme, a fluorochrome, or an electron-dense label is normally detected. Detection systems can be classified into four main categories:

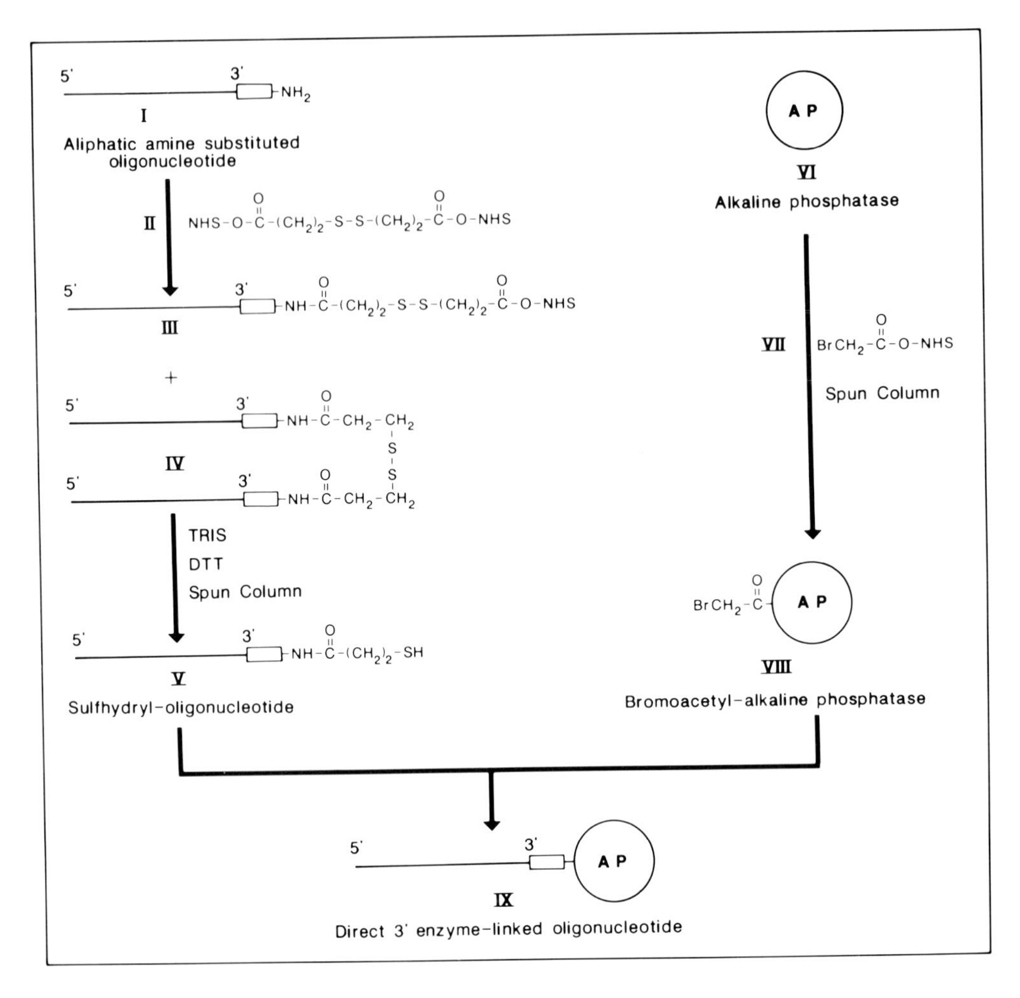

FIGURE 23. Scheme for the preparation of a direct 3′ enzyme-linked oligonucleotide probe.[67] Open box connected to the 3′ end of the oligonucleotide represents the spacer arm. Dithio-oligonucleotide intermediates are not named and only the most reactive bromoacetyl group in the derivatized alkaline phosphatase is shown. AP, calf intestine alkaline phosphatase. See also Figure 22. (From Li, P., Medon, P. P., Skingle, D. C., Lanser, J. A., and Symons, R. H., *Nucleic Acids Res.*, 15, 5275, 1987. With permission.)

1. Direct assay
2. Immunoassay; antibody involvement
3. Affinity assay; biotin/avidin (or biotin/streptavidin) involvement
4. Immunoaffinity assay; combination of (2) and (3)

This section describes some of the more commonly utilized detection systems; for further information, readers are referred to reviews by Pereira[11] and Syvänen.[91]

A. Enzymatic Detection

A schematic diagram summarizing the detection of hybridized nonradioactive probes by various enzymatic means is presented in Figure 24. Although many enzymes have been suggested as labels, most assays in current use employ either alkaline phosphatase or horseradish peroxidase; acid phosphatase and β-*D*-galactosidase are utilized to a lesser extent. Probes labeled directly with enzymes[64,67,88-90] are detected by direct enzymatic assay (Figure 24A). Antigenic probes[31-33] are normally detected by indirect immuno-enzymatic assay (Fig-

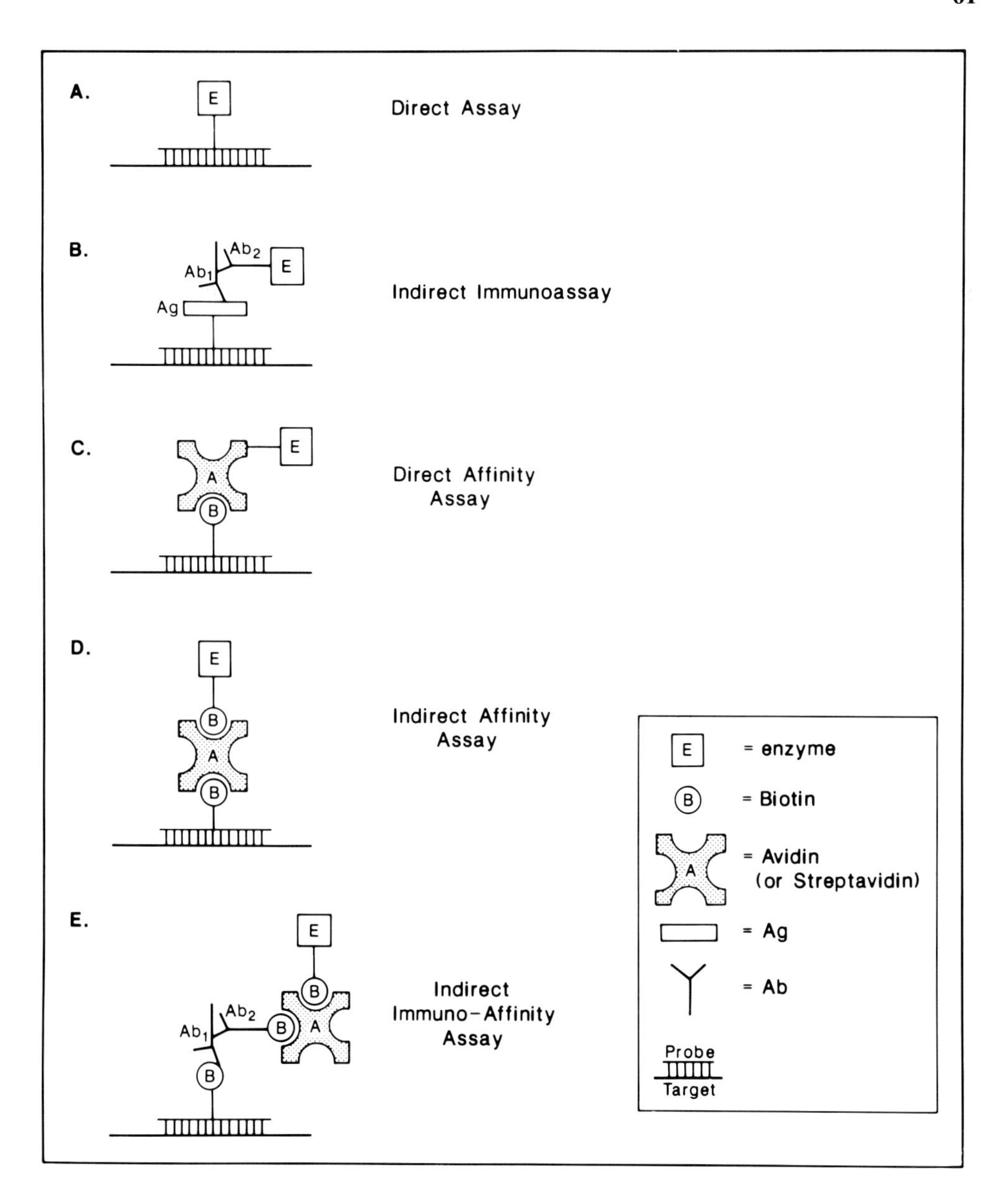

FIGURE 24. A schematic summary of the nonradioactive hybridization assays, based on enzymatic detection, as described in the text. (Adapted from Pereira, H. G., *BioEssays*, 4, 110, 1986. With permission.)

ure 24B) in which a first antibody (Ab_1), reacting with hapten (Ag) incorporated into the probe, is subsequently combined with an anti-antibody (Ab_2) labeled with an enzyme.

Biotinylated probes may be detected by either an affinity (direct or indirect) or an indirect immunoaffinity enzymatic assay. In the direct affinity approach (Figure 24C), biotin molecules linked to the probe are detected by a reaction with an avidin (or streptavidin) enzyme conjugate. The indirect affinity approach (Figure 24D) normally involves the use of an avidin (or streptavidin) - biotinylated enzyme complex which makes use of two of the four biotin-binding sites on avidin (or streptavidin). Detection by indirect immunoaffinity assay (Figure 24E) usually involves a primary reaction with antibody to biotin (Ab_1), followed by a secondary reaction with an anti-antibody (Ab_2) labeled with biotin which is subsequently reacted with an avidin (or streptavidin)-biotinylated enzyme complex.

The above mentioned multienzyme complexes (or polymers) have been designed to increase the number of enzyme moieties associated with hybridization and hence to provide an increase in sensitivity. Such complexes are commercially available in a preformed state, or they may be prepared, prior to use, by mixing avidin (or streptavidin) with biotinylated-enzyme in defined stoichiometry. Two recent approaches aimed at increasing the sensitivity of detection are discussed later (Section V.E).

A most important consideration in interpreting results using enzyme-coupled systems is the size of this system and the *steric hindrance* it offers in binding to its target on a solid surface. Given that a protein is spherical, the diameter, D, in angstroms is given by:

$$D = 3\sqrt{3.2 \times M_r \times \bar{v}}$$

where M_r is the relative molecular mass and $\bar{v}$ the partial specific volume.[92] For example, the M_r of dimeric calf intestine alkaline phosphatase is approximately 115,000.[93,94] Assuming $\bar{v} = 0.74$, a usual value for proteins,[92] then D = 65å. A protein of this size therefore has a diameter equivalent in length to an oligonucleotide of 20 to 30 nucleotides, depending on the internucleotide length in solution. When two alkaline phosphatase molecules are coupled to a single avidin molecule of M_r 68,000, the total M_r of the complex is approximately 300,000 and of sufficient size to cover roughly 50 to 100 residues of nucleic acid. On this basis, for biotin-labeled nucleic acid probes, increased sensitivity of detection is unlikely to be achieved by increasing the average density above 1 to 2 biotins per 100 residues, a conclusion supported by experimental data using avidin-alkaline phosphatase detection and Photobiotin®-labeled probes.[37]

1. Alkaline Phosphatase

Based on an original histochemical color reaction,[95] Leary et al.[3] introduced a sensitive alkaline phosphatase assay for the detection of nonradioactive probes on nitrocellulose membranes. The reaction, outlined in Figure 25, involves the formation and precipitation of two highly insoluble products, diformazan and indigo, at the site of hybrid formation. With the substrate, 5-bromo-4-chloro-3-indolyl phosphate (BCIP), the phosphate is cleaved by the enzyme and the hydroxyl group then undergoes tautomerization forming a ketone which oxidizes and dimerizes to produce the insoluble blue 5,5′-dibromo-4,4′-dichloro-indigo. In the process of dimerizing, hydrogen ions are released and these reduce the nitroblue tetrazolium (NBT) which then precipitates, forming an intense purple deposition of diformazan.

One distinct advantage of this technique over a previously described protocol[96] which uses Naphthol AS-MX phosphate and a diazo chromogenic compound, Fast Red TR salt, is the stability and solubility of the reactants in aqueous solutions. Thus, with increased incubation times, BCIP and NBT tend to be more stable, giving less nonspecific precipitation and less background staining of the nitrocellulose membrane.[97]

Numerous protocols involving alkaline phosphatase plus the substrates BCIP and NBT have indicated the popularity of this approach. The methods are sensitive and make use of all the assay procedures outlined in Figure 24. Indeed, many of the nonradioactive protocols mentioned in Sections II and III[29,30,37,51,54,60,64,67,88] have utilized one of these approaches. Commercial kits utilizing this alkaline phosphatase enzymatic approach are readily available.

We routinely employ the avidin-alkaline phosphatase system in our laboratory for colorimetric detection.[37-40] Maximum color development was normally obtained within a 3 to 4 h period; however, this time varied considerably according to the actual amount of biotin-labeled probe annealed to the target nucleic acid. The final detection signal was most evident on one side of the filter. It is advisable to carry out color development in the dark in order to reduce nonspecific background color.

The majority of our hybridization studies have involved the use of nitrocellulose filters;

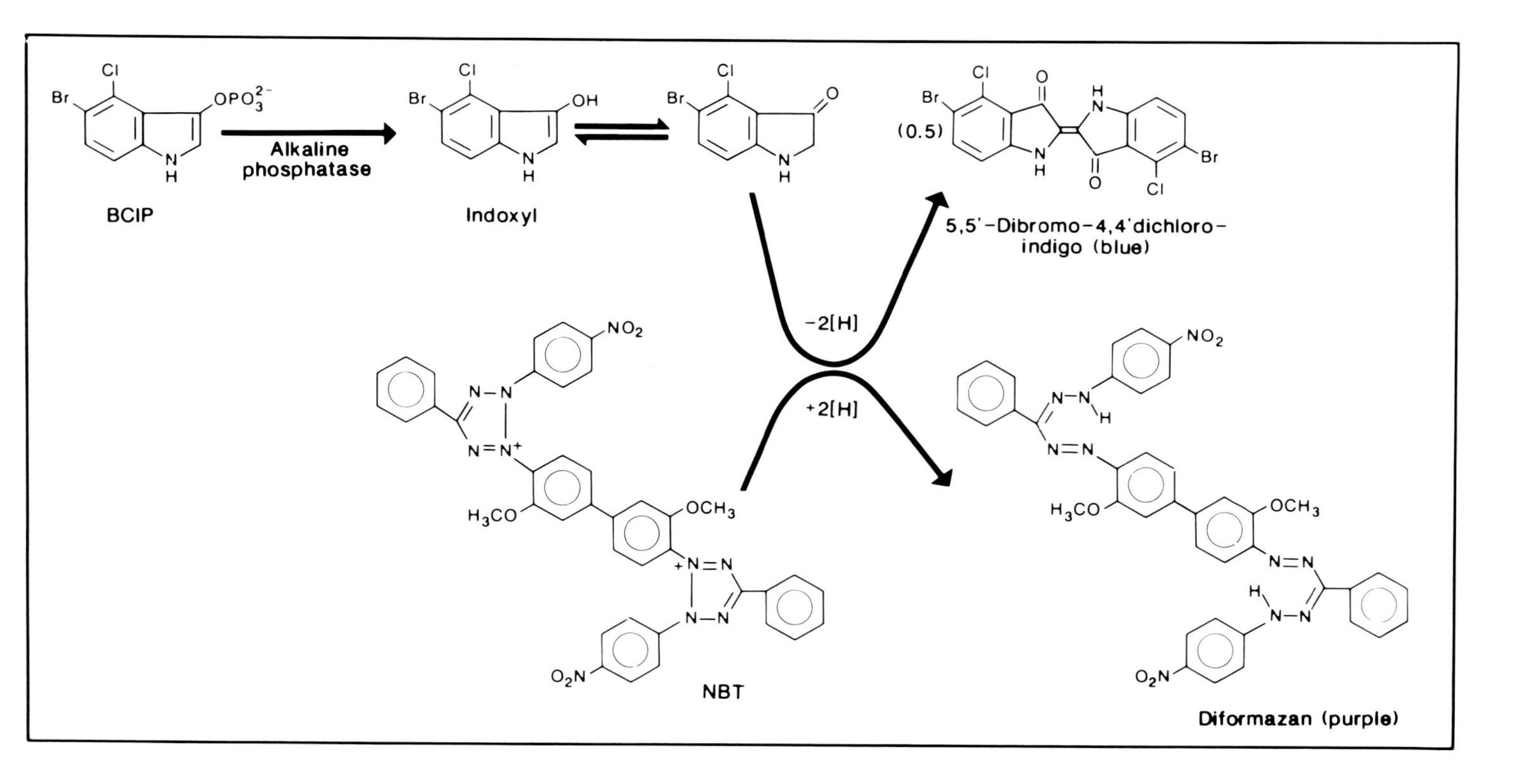

FIGURE 25. General mechanism of the alkaline phosphatase color reaction.[95] This reaction is the basis for the detection of nonradioactive probes on nitrocellulose membranes.[3]

we have routinely achieved lower backgrounds with nitrocellulose when compared to the use of nylon-based membranes. Acceptable backgrounds have been achieved with nylon-based membranes provided a stronger blocking protocol[98] is followed after hydridization. Unfortunately, nitrocellulose filters cannot be reused since the dyes, BCIP and NBT, appear to bind irreversibly. However, it is possible to remove the precipitated dyes from nylon membranes by heating in formamide, thus allowing them to be reprobed.[4]

2. *Horseradish Peroxidase*

Horseradish peroxidase (HRP) catalyzes the reaction:

$$AH_2 + H_2O_2 \rightarrow 2H_2O + A$$

where AH_2 is a hydrogen donor and A is its oxidized form. In the detection of HRP-conjugates (avidin/streptavidin) or HRP-bound antibodies on nitrocellulose membranes, a chromogenic hydrogen donor can be employed as substrate, which as a result of the peroxidase-catalyzed reaction, will form an insoluble colored oxidation product at the site of hybrid formation. These insoluble detection systems are also useful for *in situ* hybridization and electron microscopy studies.

HRP was one of the first enzymes to be used in combination with 3,3′-diaminobenzidine (DAB)[99] in cytochemical staining techniques. Since then a number of other histochemical procedures have been devised for visualization of HRP. Several chromogens currently employed for the detection of HRP-conjugates or HRP-bound antibodies, in addition to DAB, include 3,3′,5,5′-tetramethylbenzidine (TMB),[46,100,101] the Hanker-Yates reagent (phenylenediamine/pyrocatechol),[102] O-dianisidine (3,3′-dimethoxybenzidine),[103,104] and 4-chloro-1-naphthol.[105,106] The structures of these chromogens are given in Figure 26A, and the oxidation of TMB by H_2O_2 to an insoluble colored product is in Figure 26B.

DAB is still a widely employed reagent in histochemistry although one major disadvantage is its apparent carcinogenicity.[107] It is oxidized to a reddish-brown precipitate which deposits around the sites of the peroxidase-catalyzed reaction.[99] Heavy metal ions (Co^{2+}, Ni^{2+}) have been shown to intensify this DAB-based HRP reaction product.[108] This reaction product is also osmiophilic and so can be utilized in electron microscopy applications (see Section V.D).

In an endeavor to enhance the sensitivity of these DAB/peroxidase procedures for nitrocellulose membranes or *in situ* tissue sections, several protocols have been developed which involve a silver enhancement step. We have selected two such protocols for brief discussion (Figure 27). In the first of these,[109] small amounts of a biotinylated probe can be detected with an avidin-biotinylated peroxidase complex using hydrogen peroxide and polymerized nickel-complexed 3,3′-diaminobenzidine (Ni-DAB) as substrate. Silver enhancement, utilizing silver nitrate, is then carried out in a single step which intensifies the original DAB reaction deposit and results in a dark brown or black precipitate at the hybridization site (Figure 27A).

The second approach [110] initially involves production of a DAB/hydrogen peroxide signal by a streptavidin-enzyme or antibody-enzyme conjugate, followed by a silver amplification step. Here, amplification is based on the ability of DAB complexes to bind gold salts, e.g., sodium chloro-aurate,[108] which, when converted to gold sulfide (via sodium sulfide and HCl), precipitates metallic silver (via silver nitrate, formaldehyde)[111] (Figure 27B). The growing clump of silver grains makes the probe readily visible in tissue sections[112] and in scanning electron microscoy for the chromosome localization of genes.[113]

TMB, an apparent noncarcinogenic chromogenic substrate[114,115] is gaining popularity for HRP/H_2O_2 oxidation reactions (e.g., Sheldon et al.[46]). Here, the colorless substrate, TMB, is converted to a blue reaction product (Figure 26B) which is claimed to be easier to detect

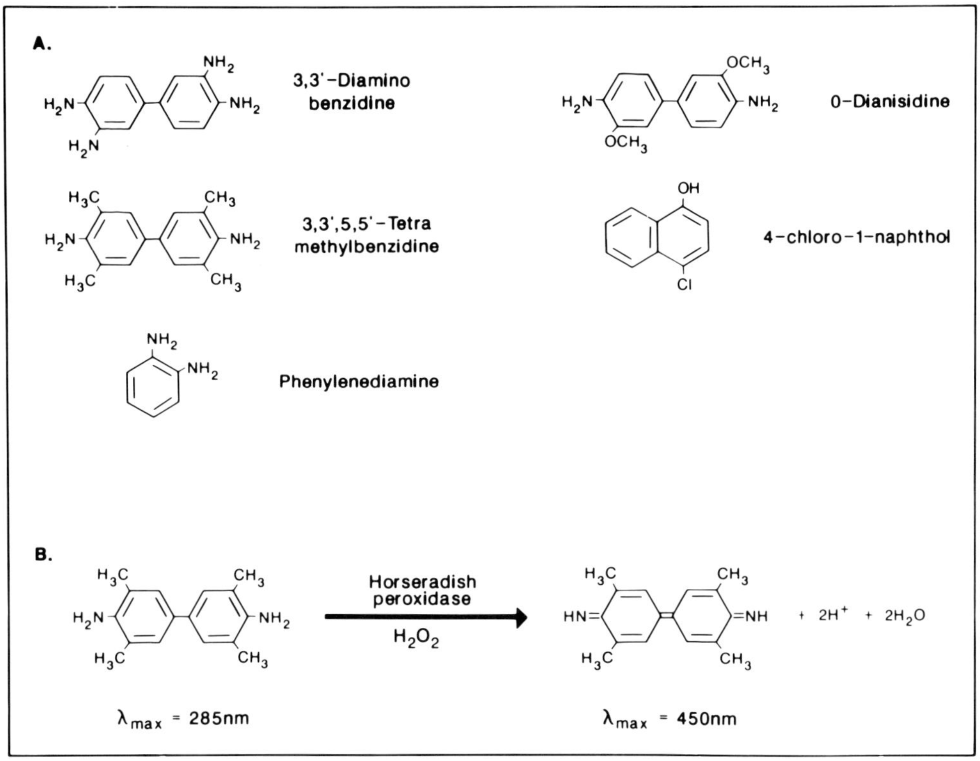

FIGURE 26. (A) Structures of several chromogens currently used to detect horseradish peroxidase (HRP)-conjugates or HRP-bound antibodies. (B) Reaction mechanism for the HRP-catalyzed oxidation of 3,3′,5,5′-tetramethylbenzidine.[101]

than the reddish brown product obtained with DAB, especially in bright-field microscopy.[100] Sheldon et al.[46] and Dykes[116] have utilized this TMB - HRP/H_2O_2 oxidation reaction to detect restriction fragment length polymorphisms (RFLPs) by the Southern blotting protocol; in both cases, excellent sensitivity was achieved.

Of the other chromogens, 4-chloro-1-naphthol[105,106] is also occasionally used, especially in immunoblotting. This reagent, however, is oxidized to a bluish-purple deposit around the sites of the enzyme reaction and tends to give smaller, less intense zones around these sites than DAB and TMB procedures. Sensitivity may, therefore, be a problem with this chromogen.[117] It is interesting to note that of the above mentioned chromogens (except 4-chloro-1-napthol), TMB appeared to be the most sensitive chromogen for the demonstration of neuronally transported HRP at the light microscopic level.[118,119]

3. Acid Phosphatase

An avidin (or streptavidin) conjugate with acid phosphatase can be used to detect biotinylated products (e.g., nucleic acid, antibodies) on nitrocellulose membrane matrices such as Southern transfers, immunoblots (Western transfers) or dot-blots. Using the recommended enzyme substrate/dye combination (naphthol AS-MX phosphate/Fast Violet B salt), the biotinylated nucleic acid or protein is visualized by the formation of a violet colored precipitate at the site of hybrid formation.

A commercially manufactured acid phosphatase signal generating system is available (DETEK 1-acp; Enzo Biochem, Inc.). Such a kit utilizes a streptavidin-biotinylated acid phosphatase complex in combination with the above substrate and dye. In our hands,[37] this

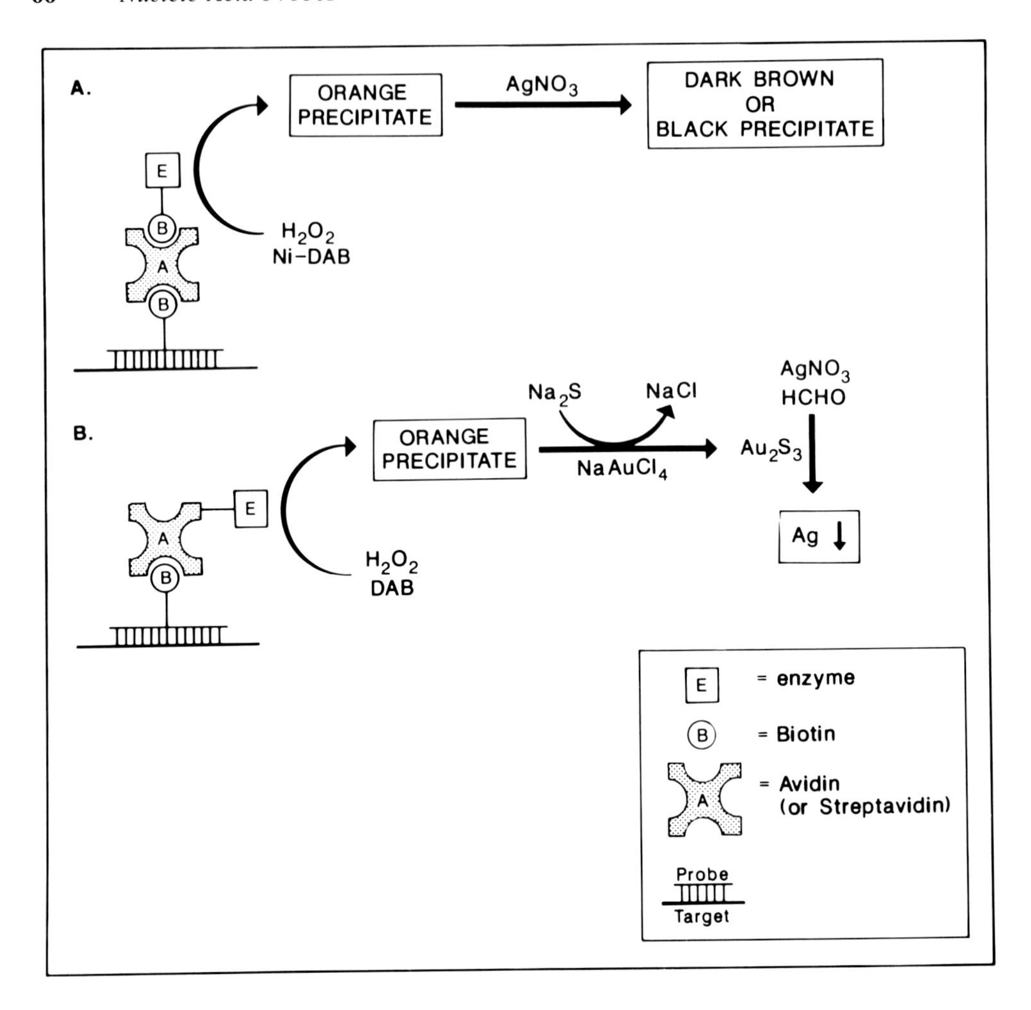

FIGURE 27. Silver enhancement protocols for the diaminobenzidine/horseradish peroxidase detection of biotinylated probes bound to target nucleic acid: (A) method of Przepiorka and Myerson;[109] (B) method of Burns et al.[110]

acid-phosphatase procedure required approximately a three fold longer detection time (e.g., normally overnight) to achieve the same level of sensitivity as that obtained with an alkaline phosphatase-based detection system. For *in situ* hybridizations and immunohistochemical studies, an alternative substrate/dye combination (viz., naphthol AS-BI phosphate/pararosanaline hydrochloride) is recommended (Enzo Biochem, Inc.).

4. β-Galactosidase

A relatively more recent addition to the range of marker enzymes used for nonradioactive probe detection is β-galactosidase. This enzyme, prepared from *Escherichia coli* and available as conjugates with either avidin (or streptavidin) or various antibodies, can be used to detect probe molecules bound to target species on solid matrices (e.g., nitrocellulose) or tissue sections. Because of its microbial origin, β-galactosidase can be used to label tissue samples without concern for endogenous enzyme activity.

Bondi et al.[120] have presented a protocol for the use of β-galactosidase as a tracer for immunocytochemical studies. Here, enzyme activity was revealed by use of the substrate 5-bromo-4-chloro-3-indolyl-β-D-galactopyranoside (BCIG, X-Gal) which resulted in the formation of a blue reaction product (5,5′-dibromo-4,4′-dichloro-indigo; Figure 25) which

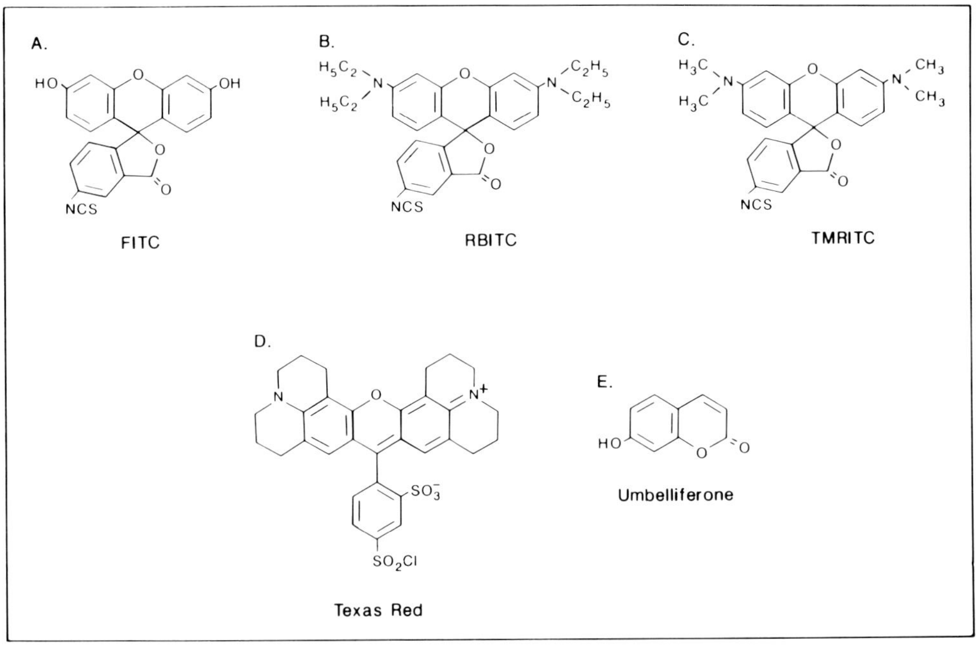

FIGURE 28. Structures of several fluorescent dyes currently employed in the detection of biotinylated nucleic acid hybrids *in situ*. FITC, fluorescein isothiocyanate; RBITC, rhodamine B isothiocyanate; TMRITC, tetra-methyl rhodamine isothiocyanate.

precipitated at the reaction site. Such a reaction could be utilized to detect hybrid formation where nucleic acid probes and targets are involved. The combination of β-naphthyl-β-D-galactopyranoside and Fast Blue B salt may also serve as substrate for β-galactosidase (Pharmacia).

B. Fluorescent Detection

1. Fluorescent Labels

Of the many fluorescent labels used in fluoroimmunoasays and immunofluorometric assays,[121] several have been successfully employed for the detection of biotinylated nucleic acid hybrids *in situ*. The structures of these fluorescent dyes are shown in Figure 28.

While fluorescein, as the isothiocyanate (FITC) (Figure 28A), is the most widely used fluorochrome due to its high fluorescent intensity, several rhodamines, e.g., rhodamine B isothiocyanate (RBITC) and tetramethyl rhodamine isothiocyanate (TMRITC) (Figure 28B and C) are also popular but the intensity of their fluorescence is less. The more recent fluorochrome, Texas Red (a product of Molecular Probes, Inc.) (Figure 28D), has also gained widespread recognition. Its excitation and emission spectra are well separated from those of fluorescein, making it a suitable alternative to rhodamine and ideal for double-labeling experiments; it gives a bright red signal and is resistant to fading.

Rhodymenia phycoerythrin (RPE),[122,123] an important fluorescent phycobiliprotein, is a constituent of the light-harvesting apparatus of red algae and some bacteria. Unlike fluorescein or rhodamine, RPE has a broad absorption spectrum and gives an extremely intense fluorescence, some 20 to 50 fold brighter than fluorescein on a molar basis.[124]

The simplest system of fluorescent detection of a biotinylated nucleic acid target *in situ* is to use a fluorochrome-avidin (or streptavidin) conjugate (e.g., FITC-avidin, TMRITC-avidin, Texas Red-avidin, RPE-avidin). Alternatively, biotinylated hybrids may be detected by a primary antibody directed against biotin, followed by a secondary antibody tagged with

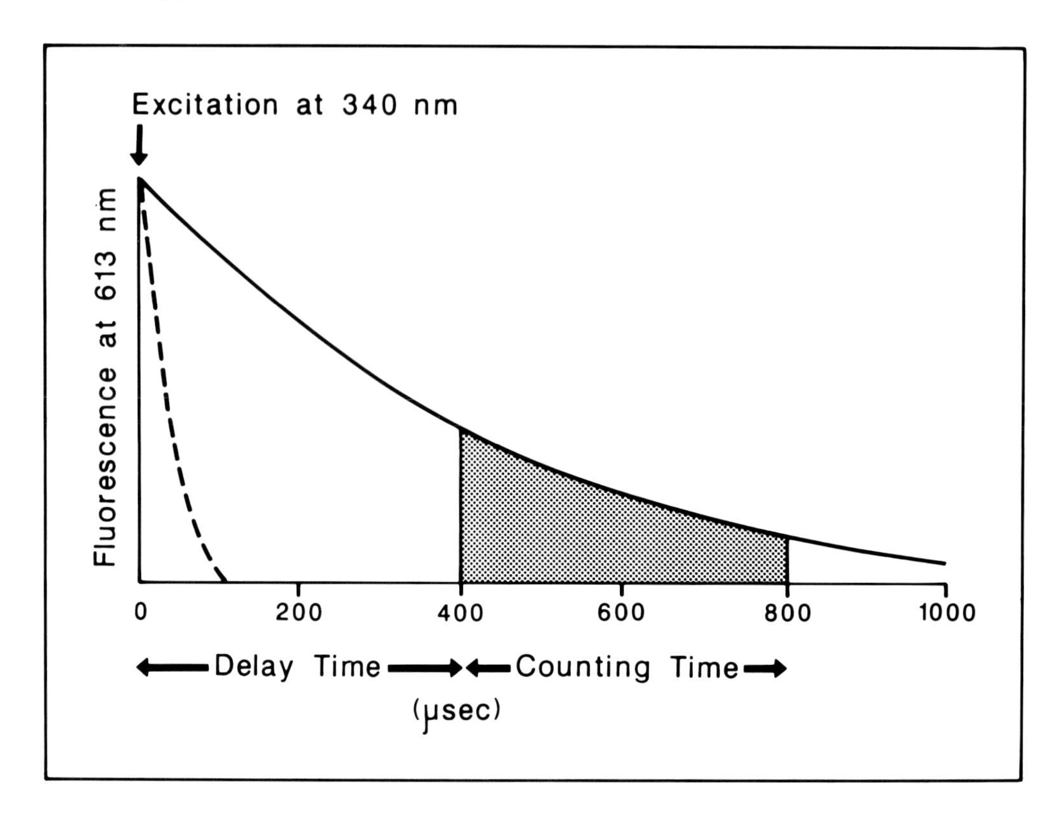

FIGURE 29. Principle of the time-resolved fluorometric measurement.[131] Long-lived fluorescent compounds are excited with a light pulse and the emitted fluorescence (—) is measured after a delay time, during which short-lived background fluorescence (---) disappears. The measuring cycle is repeated 1000 times giving a total measuring time of 1 sec per sample. (From Syvänen, A-C., Tchen, P., Ranki, M., and Söderlund, H., *Nucleic Acids Res.*, 14, 1017, 1986. With permission.)

the fluorescent molecule. Both approaches have found important application in cytogenetic analysis.[18,125,126]

Umbelliferone, a 7-hydroxy coumarin (Figure 28E) may also be used in flourescent probe detection of biotinylated nucleic acid hybrids. Nagata et al.[127] and Yokota et al.[128] have detected picogram levels of specific DNA immobilized in microtitre wells using the substrate, 4-methylumbelliferyl-β-D-galactoside. Briefly, biotinylate target nucleic acid hybrids were first complexed with an avidin-β-galactosidase conjugate; the substrate was then added and the resulting fluorescence emitted from the hydrolyzed product by β-galactosidase was measured.

2. *Time-Resolved Fluorescence*

When a fluorescent material is excited by a pulse of light, the intensity of the emitted light decreases exponentially at a rate governed by the speed of the energy transitions involved. For example, proteins usually display fluorescence with a lifetime of approximately 10^{-9} sec. In contrast, if competing avenues of energy loss are controlled, fluorescent groups characterized by slower electronic transitions may exhibit highly extended lifetimes. Certain β-diketonate chelates of lanthanides, e.g., europium, in an aqueous cocktail of detergents and synergistic agents, have lifetimes three to six orders of magnitude longer (10^{-6} to 10^{-3} sec) than typical background fluorescence. This exceptionally long fluorescent decay time of lanthanide chelates provides the basis for a technique named *time-resolved fluorometry* (Figure 29), in which pulsed light excitation is separated in time from the measurement of

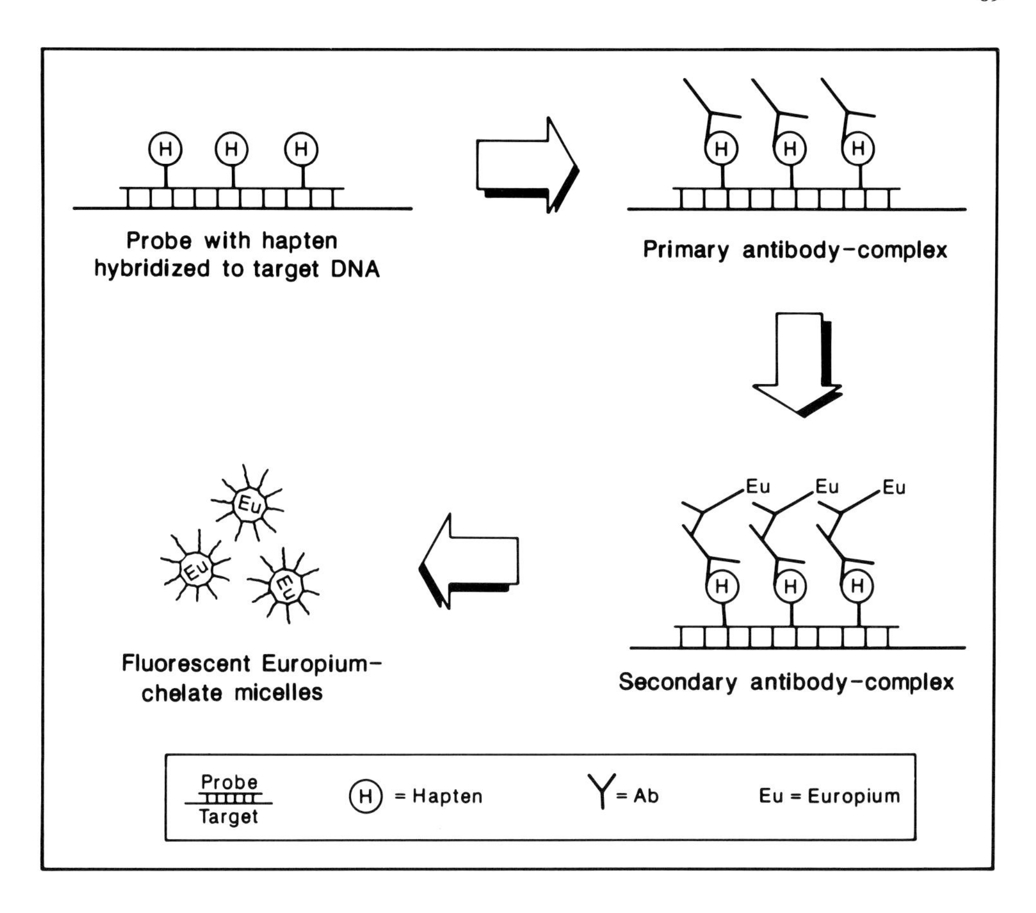

FIGURE 30. Nucleic acid hybrid detection by measurement of the time-resolved fluorescence of europium-diketone chelates. (Adapted from Syvänen, A.-C., Tchen, P., Ranki, M., and Söderland, H., *Nucleic Acids Res.*, 14, 1017, 1986.)

the fluorescence of the europium-chelate after the short-lived background fluorescence has decayed.[129-131]

Syvänen et al[131] have applied time-resolved fluorometry to detect nucleic acid hybrids. Their technique (Figure 30) involved an initial modification of the DNA probe with a hapten, either the immunogenic *N*-2-acetylaminofluorene[32] or sulfone[61] group (see Section III.A). After hybridization, the probe DNA was detected by a two-step immunological assay with the second antibody labeled with europium. Europium was released from the antibody complex with a fluorescent enhancement solution, in which the europium became chelated to diketones and trapped in detergent micelles. This method was found to be approximately ten fold more sensitive than enzyme-based assays carried out in parallel.

More recently, Oser et al.[132] have successfully prepared DNA probes labeled with europium chelates. Briefly, their technique involved labeling the probes with a specific chelate group-substituted psoralen followed by europium ion fluorescent detection. This method, however, would appear to have limited application since the europium ions need to be released from the DNA probe for time-resolved fluorometry measurements; hence, it is not suitable for Southern or Northern hybridizations or dot-blot hybridization unless each spot is treated separately. In both this report and that of Syvänen et al.,[131] the sensitivity of detection was in the low (0.3 to 5.0) picogram range.

In a further variation on this approach, Dahlén[133] detected bio-11-dUTP labeled DNA probes with streptavidin coupled with europium chelates. However, only a model hybridi-

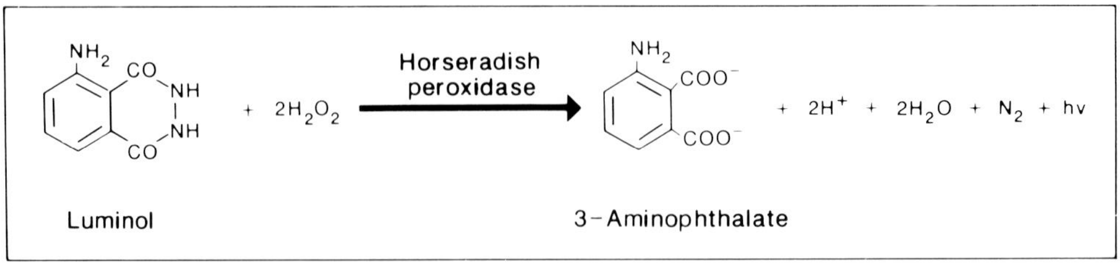

$$\text{Luminol} + 2H_2O_2 \xrightarrow{\text{Horseradish peroxidase}} \text{3-Aminophthalate} + 2H^+ + 2H_2O + N_2 + h\nu$$

FIGURE 31. Equation of the horseradish peroxidase-catalyzed chemiluminescent reaction involving luminol.[136-139]

zation system was used in which adenovirus DNA was bound to wells of a microtitre plate and then hybridized with the biotin-labeled probe at 100 ng/ml. After removal of excess probe and incubation with the streptavidin-europium chelate, the europium was solubilized into micelles and measured in a time-resolved fluorometer. The limit of detection was 10 pg of DNA per well. Presumably the sensitivity would be significantly less if the target nucleic acid in solution was detected by sandwich hybridization using two, non-overlapping probes, one of which was bound to the plate and the other in solution labeled with biotin. Other problems need to be resolved; e.g., enhancers of the hybridization rate, such as dextran sulfate and polyethylene glycol[134,135] could not be used because of elevated background levels. Further, there was a high loss of DNA bound to microtitre wells on hybridization at 65°C in 50% formamide although this was much less at 50°C.

C. Chemiluminescence

Chemiluminescence is a term used to describe the production of light by chemical reaction. Because of the potential for very high sensitivity, chemiluminescent catalysts and associated reagents are particularly attractive as a means of quantitating DNA hybrids bound to solid supports (e.g., nitrocellulose filters). Of particular interest here is the horseradish peroxidase-catalyzed chemiluminescent reaction involving luminol (3-aminophthalate hydrazide).[136-139] Briefly, peroxidase is responsible for the oxidation of luminol to 3-aminophthalate in the excited state with the loss of four electrons and light emission at 425 nm. For each molecule of luminol oxidized, two molecules of hydrogen peroxide are required (Figure 31).

Matthews et al.[140] have described an enhanced chemiluminescent method for the detection of DNA dot-blot hybridization assays. The method utilized DNA probes labeled with biotin, which were detected using a biotinylated streptavidin-horseradish peroxidase complex. The peroxidase enzyme then takes part in an enhanced chemiluminescent reaction with luminol, hydrogen peroxide, and an enhancer (e.g., *p*-hydroxycinnamic acid). The light emission is relatively intense and is stable for many min; it can be measured with a range of detectors such as photomultiplier tubes in luminometers, silicon photodiodes, or photagraphic film.

A detection limit of approximately 1 pg (plasmid DNA) was obtained by the above chemiluminescent protocol;[140] this is more or less equivalent to the other commonly used nonradioactive probing techniques. Previously we have seen that the insoluble colored dye products on nitrocellulose filters, e.g., alkaline phosphatase detection (Section V.A.1) cannot be removed and hence filters may only be probed once. A major advantage with this chemiluminescent technique is that the probe can be removed from the filter completely by several washes of the membrane in distilled water at 65°C over 90 min; filters may thus be reprobed, a valuable acquisition for research and clinical settings where filters need to be reused.

The technique does, however, have its disadvantages; e.g., a dot which hybridizes very strongly can interfere with neighboring samples when photographic detection is used. Since the light emitted from the probe decreases with time (17° of maximum after 30 min), better resolution of the individual dots is achieved for photographic quantitation by a 30 min delay before exposure to film. The use of a "mask" between the filter and the luminescent microtitre plate reader was found to reduce this "cross-talk" between neighboring samples.[140] Nylon membranes do not appear to be suitable for chemiluminescent detection of biotin-labeled DNA probes since large background labeling occurred during incubation with the horseradish peroxidase comples.[140]

Leong et al.[141] have further examined the luminescent detection method in a study involving Immunodot, Western, and Southern blots using a monoclonal antibody which contained binding sites for both biotin and horseradish peroxidase on the same molecule. Their results indicated that, although the method was suitable for Western blot development, it was less sensitive then autoradiographic methods using high-specific activity ^{32}P-labeled probes in southern blot development. The method as described was considered not to be sensitive enough for the detection of single-copy genes in genomic Southern blots.

In an interesting approach with the chemiluminescence system, Heller and Morrison[142] have used a different strategy; in their experimental system, two probes were used which bound to adjacent nucleic acid target sequences. One probe was labeled with a chemiluminescent complex, which could be induced to emit light of specific wavelength which then caused excitation of the label molecules on the other probe. This in turn produced an emission spectra that could be detected by a photomultiplier device. The novelties of this system, termed *nonradioactive energy transfer,* are twofold. First, the energy transfer will occur only if the two probes have hybridized correctly; false signals will be virtually eliminated. Second, the target nucleic acid does not need to be immobilized and no washing steps are necessary. This approach is ideally suited toward automation—a distinct advantage for large-scale testing. However, developmental work on this model system was at a stage where it was too early to determine its practical feasibility.

D. Detection via Electron-Dense Labels

In Section V.A.2, electron-dense detection was provided by the gold/silver enhancement of the colored product of the horseradish peroxidase reaction with diaminobenzidine. This approach has found widespread use for *in situ* hybridization with analysis at the light and electron microscope levels.

Other approaches avoid the requirement for an initial enzymatic reaction. For example, biotinylated probes were detected directly with avidin-derritin conjugates[143] or streptavidin-gold complexes.[144] Indirect antibody techniques have involved a primary incubation with anti-biotin antibodies which were then detected with a second antibody or protein A complexed with gold.[15,19,145] Similarly, acetylaminofluorene labeled probes were detected by the double antibody technique.[146] In most cases, the gold signal was enhanced by the use of silver lactate and hydroquinone.[147,148] All methods are applicable to hybridization on nitrocellulose filters or for *in situ* hybridization.

E. Protocols Aimed at Enhancing Sensitivity

1. Amplification of Signal by Enzymatic Cycling

Alkaline phosphatase and horseradish peroxidase are popular choices as enzyme indicators for nonradioactive nucleic acid probe detection (Sections V.A.1 and 2). The sensitivity of these systems, however, can be limited either in the detection of small numbers of indicator enzyme molecules or by background noise. Enzyme amplification is a process whereby the former of these two limitations may be improved. Briefly, the technique depends upon the use of the indicator enzyme to produce a catalytic activator for a secondary system, the activity of which is measured and used to quantify the indicator enzyme. The principle and use of this approach for the quantitation of enzymes and metabolites was originally reported in 1961 by Lowry et al.[149]

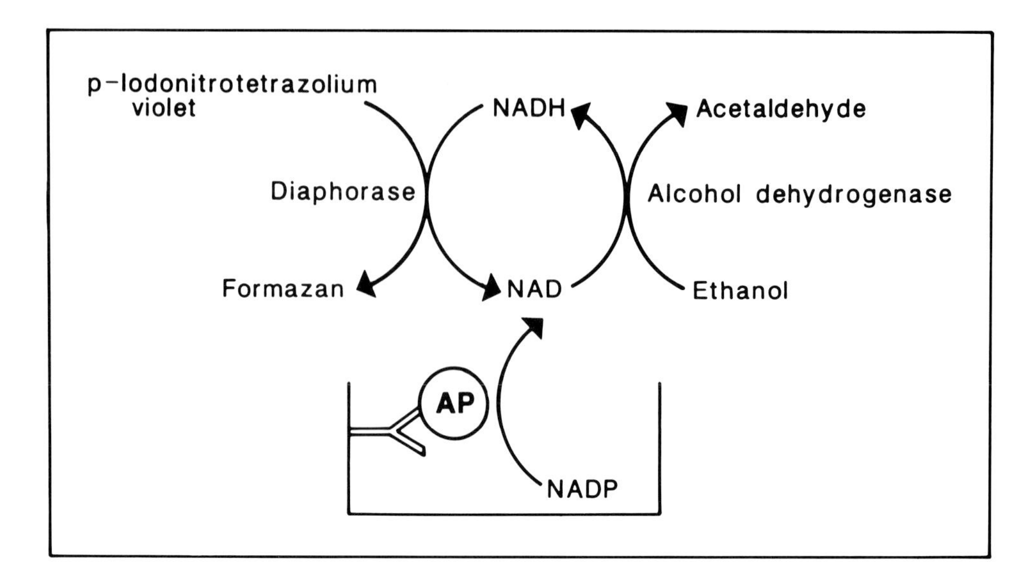

FIGURE 32. Example of the principle of signal amplification by enzymatic cycling.[150] See text for details. AP, alkaline phosphatase which is coupled to target-bound antibody.

Self[150] has described such an enzyme amplification system in which alkaline phosphatase was employed as the label, initiating two sequential catalytic reactions (Figure 32). The first catalytic step was the dephosphorylation by antibody-coupled alkaline phosphatase of nicotinamide adenine dinucleotide phosphate (NADP) to produce NAD. NAD, in turn, catalytically activated an NAD-specific redox cycle involving the enzymes alcohol dehydrogenase and diaphorase. In the cycle, NAD was reduced by ethanol and the NADH so formed reduced a tetrazolium salt (*p*-iodonitrotetrazolium violet) to regenerate NAD and to produce an intensely red-colored formazan dye. One molecule of formazan was produced with each turn of the cycle.

The rate of reduction of the tetrazolium salt is directly proportional to the concentration of NAD originally formed by the enzyme in the bound conjugate. NAD is not used up by the redox cycle and so the latter provides a method of catalytically amplifying the signal produced by alkaline phosphatase. The high specificity of the redox cycle for NAD prevents any interference by high NADP concentrations. However, the method requires reagents of very high purity since even trace contamination of the NADP or enzymes by NAD may lead to high backgrounds.

Stanley et al.[151] and Johannsson et al.[152] have both applied enzyme amplification to ELISA protocols involving thyroid-stimulating hormone. Both reports highlighted the extreme sensitivity shown by enzyme amplification; whereas the former reported a seventy fold increase in sensitivity when compared with the conventional ELISA, the latter showed that a detection level of 0.01 attomole (10^{-20} mole) was attainable. On a more recent note, amplified ELISAs have been reported showing considerable increases in sensitivity of detection for specific antibodies and immune complexes in human sera[153] and the plant virus, barley yellow dwarf virus, in the sap of oats and in individual vector aphids.[154] Commercial kits for enzyme amplification are now available (e.g., Ampak, IQ (Bio) Ltd; BRL; Wellcozyme HBsAg — the latter is an amplified enzyme immunoassay for the detection of Hepatitis B surface antigen in serum or plasma).

As can be seen from the above, enzyme amplification is particularly suited to liquid phase detection systems. Gatley[155] has reported limited success with enzyme amplification in the detection of nucleic acid probes on nitrocellulose filters. Briefly, biotinylated probe was

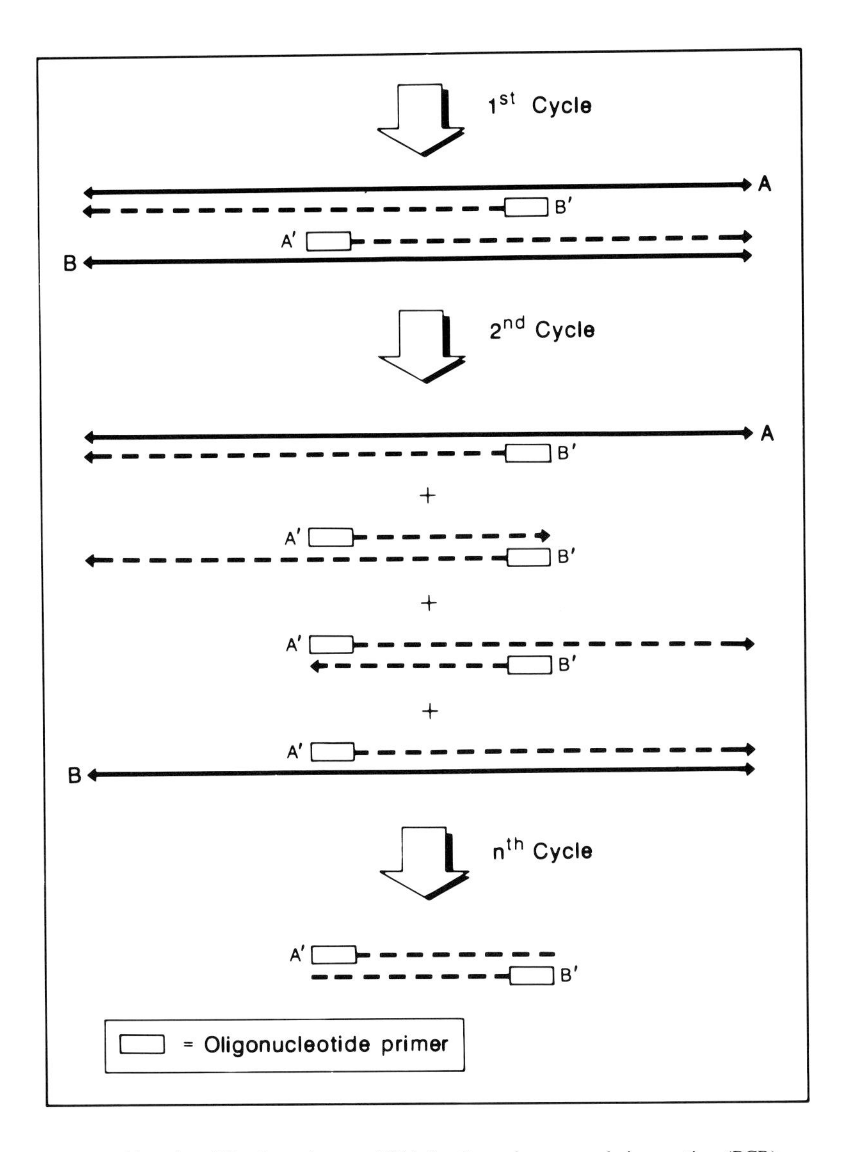

FIGURE 33. Amplification of target DNA by the polymerase chain reaction (PCR) procedure.[156-158] Original template DNA strands are labeled A and B, and the two oligonucleotide primers A′ and B′. The dotted line represents *in vitro* synthesized DNA.

fixed to nitrocellulose and exposed to an avidin-alkaline phosphatase conjugate. After thorough washing, the filter was cut up and individual target membrane pieces were placed into the wells of a 96 microtitre plate. Alkaline phosphatase determination was then performed using the amplification protocol of Figure 32; between 5 to 50 pg probe DNA only could be detected. Obviously, a substantial improvement in sensitivity will be required before enzyme amplification can be of routine use in nucleic acid probe/solid support systems.

2. Enzyme Amplification of Target DNA

Theoretically, sensitivity of detection can also be improved by increasing the number of target nucleic acid molecules in the original sample. Such an approach has been developed[156-158] whereby specific nucleic acid sequences are enzymatically amplified *in vitro* in an exponential fashion and with high fidelity. The technique, termed the polymerase chain reaction (PCR) procedure, is depicted in Figure 33.

The procedure is based on repeated cycles of denaturation, oligonucleotide primer annealing, and primer extension by the Klenow fragment of DNA polymerase. PCR has one prerequisite: that the sequences flanking the DNA region of interest be known. Given that, oligonucleotide primers (e.g., 20 bases) that flank the required region and complementary to opposite strands, are synthesized. These are added in large excess to the starting material. The first cycle gives a doubling of the region of interest but includes extension beyond it. As a result, these extension products overlap and, consequently, when denatured can serve as the template for the other primer (and vice versa) in the next cycle (Figure 33). This scenario, repeated in ensuing cycles of the reaction, produces a doubling of the amount of sequence defined by the two primers every cycle and thus overall an exponential increase in the concentration of the fragment. So far the longest stretch of DNA that has been amplified in this fashion is approximately 1 kb. However, in practice, amplification is usually restricted to sequences of 100 to 200 residues or less. A possible reason for this is that DNA chains at this size may start to assume tight secondary conformations which can block the enzyme during primer extension.

This PCR technique has enormous potential. In the original report, Saiki et al.[156] achieved approximately a 2×10^5 times amplification of specific β-globin target sequences in human genomic DNA after 20 PCR cycles (mean efficiency per cycle was approximately 85%). The technique has since become automated and can be performed directly on cell lysates, eliminating the need for DNA purification.[159] Amplifications of approximately 10^6-fold can now be achieved by automation in 3 h using a heat-stable species of the polymerase enzyme from the thermophilic bacterium, *Thermus aquaticus*.[160]

ACKNOWLEDGMENTS

The writing of this chapter was supported by the Centre of Gene Technology which was set up by a Commonwealth Government Grant in the Department of Biochemistry of Adelaide University. We thank Derek Skingle, Li Peng, and Peter Medon for discussions, Ruth Evans for preparing the diagrams, Jennifer Cassady and Tammy Edmonds for photography, and Ros Murrell for typing the manuscript.

REFERENCES

1. **Langer, P. R., Waldrop, A. A., and Ward, D. C.,** Enzymatic synthesis of biotin-labeled polynucleotides: novel nucleic acid affinity probes, *Proc. Natl. Acad. Sci. U.S.A.*, 78, 6633, 1981.
2. **Brigati, D. J., Myerson, D., Leary, J. J., Spalholz, B., Travis, S. Z., Fong, C. K. Y., Hsiung, G. D., and Ward, D. C.,** Detection of viral genomes in cultured cells and paraffin-embedded tissue sections using biotin-labeled hybridization probes, *Virology*, 126, 32, 1983.
3. **Leary, J. J., Brigati, D. J., and Ward, D. C.,** Rapid and sensitive colorimetric method for visualizing biotin-labeled DNA probes hybridized to DNA or RNA immobilized on nitrocellulose: bio-blots, *Proc. Natl. Acad. Sci. U.S.A.*, 80, 4045, 1983.
4. **Gebeyehu, G., Rao, P. Y., SooChan, P., Simms, D. A., and Klevan, L.,** Novel biotinylated nucleotide - analogs for labeling and colorimetric detection of DNA, *Nucleic Acids Res.*, 15, 4513, 1987.
5. *FOCUS (Bethesda Research Laboratories/Life Technologies Inc.)*, 7(4), 13, 1985.
6. *FOCUS (Bethesda Research Laboratories/Life Technologies Inc.)*, 7(1), 8, 1985.
7. **Kotewicz, M. L., D'Alessio, J. M., Driftmier, K. M., Blodgett, K. P., and Gerard, G. F.,** Cloning and overexpression of Moloney murine leukemia virus reverse transcriptase in *Escherichia coli*, *Gene*, 35, 249, 1985.
8. **Gerard, G. F., D'Alessio, J. M., Kotewicz, M. L., and Noon, M. C.,** Influence on stability in *Escherichia coli* of the carboxy-terminal structure of cloned Moloney murine leukemia virus reverse transcriptase, *DNA*, 5, 271, 1986.

9. **Brakel, C. L. and Engelhardt, D. L.,** Synthesis and detection of 3-OH terminal biotin-labeled DNA probes, in *Rapid Detection and Identification of Infectious Agents,* Kingsbury, D. T. and Falkow, S., Eds., Academic Press, New York, 1985, 235.
10. *ENZO Biochem, Inc.,* Catalog 1985.
11. **Pereira, H. G.,** Non-radioactive nucleic acid probes for the diagnosis of virus infections, *BioEssays,* 4, 110, 1986.
12. **Pollice, M. and Yang, H.-L.,** Use of nonradioactive DNA probes for the detection of infectious bacteria, *Clin. Lab. Med.,* 5, 463, 1985.
13. **Garbutt, G. J., Wilson, J. T., Schuster, G. S., Leary, J. J., and Ward, D. C.,** Use of biotinylated probes for detecting sickle cell anaemia, *Clin. Chem.,* 31, 1203, 1985.
14. **Langer-Safer, P. R., Levine, M., and Ward, D. C.,** Immunological method for mapping genes on *Drosophila* polytene chromosomes, *Proc. Natl. Acad. Sci. U.S.A.,* 79, 4381, 1982.
15. **Hutchison, N. J., Langer-Safer, P. R., Ward, D. C., and Hamkalo, B. A.,** *In situ* hybridization at the electron microscope level: hybrid detection by autoradiography and colloidal gold, *J. Cell Biol.,* 95, 609, 1982.
16. **Manuelidis, L., Langer-Safer, P. R., and Ward, D. C.,** High resolution mapping of satellite DNA using biotin-labeled DNA probes, *J. Cell Biol.,* 95, 619, 1982.
17. **Singer, R. H. and Ward, D. C.,** Actin gene expression visualized in chicken muscle tissue culture by using *in situ* hybridization with a biotinated nucleotide analog, *Proc. Natl. Acad. Sci. U.S.A.,* 79, 7331, 1982.
18. **Hopman, A. H. N., Wiegant, J., Raap, A. K., Landegent, J. E., van der Ploeg, M., and van Duijn, P.,** Bi-color detection of two target DNAs by non-radioactive *in situ* hybridization, *Histochemistry,* 85, 1, 1986.
19. **Binder, M., Tourmente, S., Roth, J., Renaud, M., and Gehring, W. J.,** *In situ* hybridization at the electron microscope level: localization of transcripts on ultrathin sections of Lowicryl K4M-embedded tissue using biotinylated probes and protein A-gold complexes, *J. Cell Biol.,* 102, 1646, 1986.
20. **Garson, J. A., van den Berghe, J. A., and Kemshead, J. T.,** Novel non-isotopic *in situ* hybridization technique detects small (1 Kb) unique sequences in routinely G-banded human chromosomes: fine mapping of N-myc and β-NGF genes, *Nucleic Acids Res.,* 15, 4761, 1987.
21. **Melton, D. A., Krieg, P. A., Rebagliati, M. R., Maniatis, T., Zinn, K., and Green, M. R.,** Efficient *in vitro* synthesis of biologically active RNA and RNA hybridization probes from plasmids containing a bacteriophage SP6 promoter, *Nucleic Acids Res.,* 12, 7035, 1984.
22. **Krieg, P. A. and Melton, D. A.,** *In vitro* RNA synthesis with SP6 RNA polymerase, in *Methods in Enzymology,* Vol. 155, Wu, R., Ed., Academic Press, New York, 1987, 397.
23. **McCracken, S.,** Preparation of RNA transcripts using SP6 RNA polymerase, *FOCUS (Bethesda Research Laboratories/Life Technologies Inc.),* 7 (2), 5, 1985.
24. **D'Alessio, J.,** Optimal synthesis of biotinylated RNA using phage polymerases, *FOCUS (Bethesda Research Laboratories/Life Technologies Inc.),* 7 (4), 13, 1985.
25. *FOCUS (Bethesda Research Laboratories/Life Technologies Inc.),* 7 (2), 11, 1985.
26. **Uhlenbeck, O. C. and Gumport, R. I.,** T4 RNA ligase, in *The Enzymes,* Vol. 15, 3rd ed., Boyer, P. D., Ed., Academic Press, New York, 1982, chap. 2.
27. **England, T. E., Gumport, R. I., and Uhlenbeck, O. C.,** Dinucleoside pyrophosphates are substrates for T4-induced RNA ligase, *Proc. Natl. Acad. Sci. U.S.A.,* 74, 4839, 1977.
28. **Richardson, R. W. and Gumport, R. I.,** Biotin and fluorescent labelling of RNA using T4 RNA ligase, *Nucleic Acids Res.,* 11, 6167, 1983.
29. **Murasugi, A. and Wallace, R. B.,** Biotin-labeled oligonucleotides: enzymatic synthesis and use as hybridization probes, *DNA,* 3, 269, 1984.
30. **Riley, L. K., Marshall, M. E., and Coleman, M. S.,** A method for biotinylating oligonucleotide probes for use in molecular hybridizations, *DNA,* 5, 333, 1986.
31. **Landegent, J. E., Jansen in de Wal, N., Baan, R. A., Hoeijmakers, J. H. J., and van der Ploeg, M.,** 2-Acetylaminofluorene-modified probes for the indirect hybridocytochemical detection of specific nucleic acid sequences, *Exp. Cell Res.,* 153, 61, 1984.
32. **Tchen, P., Fuchs, R. P. P., Sage, E., and Leng, M.,** Chemically modified nucleic acids as immunodetectable probes in hybridization experiments, *Proc. Natl. Acad. Sci. U.S.A.,* 81, 3466, 1984.
33. **Landegent, J. E., Jansen in de Wal, N., Van Duijn, P., and Van der Ploeg, M.,** Hybridocytochemistry with 2-acetylaminofluorene-modified probes, *Acta Histochem.,* Suppl.-Bd., 32, S.241, 1986.
34. **Miller, E. C., Juhl, U., and Miller, J. A.,** Nucleic acid guanine: reaction with the carcinogen *N*-acetoxy-2-acetylaminofluorene, *Science,* 153, 1125, 1966.
35. **Kriek, E., Miller, J. A., Juhl, U., and Miller, E. C.,** 8-(N-2-fluorenylacetamido) guanosine, an arylamidation reaction product of guanosine and the carcinogen *N*-acetoxy-*N*-2 fluorenylacetamide in neutral solution, *Biochemistry,* 6, 177, 1967.

36. **Masse, M. J. O., Meullen, P., Le Guern, A., and Kourilsky, P.,** Monoclonal antibody detection of 2-acetyl-aminofluorene-modified DNA probes for the specific detection of nucleic acids in hybridization procedures, *Ann. Inst. Pasteur/Immunol.*, 136D, 232, 1985.
37. **Forster, A. C., McInnes, J. L., Skingle, D. C., and Symons, R. H.,** Non-radioactive hybridization probes prepared by the chemical labelling of DNA and RNA with a novel reagent, photobiotin, *Nucleic Acids Res.*, 13, 745, 1985.
38. **McInnes, J. L., Dalton, S., Vize, P. D., and Robins, A. J.,** Non-radioactive photobiotin-labeled probes detect single copy genes and low abundance mRNA, *Bio/Technology*, 5, 269, 1987.
39. **McInnes, J. L., Forster, A. C., and Symons, R. H.,** Photobiotin-labelled DNA and RNA hybridization probes, in *Methods in Molecular Biology*, Vol. 4, Walker, J. M., Ed., Humana Press, Clifton, New Jersey, 1988, chap. 30.
40. **Habili, N., McInnes, J. L., and Symons, R. H.,** Nonradioactive, photobiotin-labelled DNA probes for the routine diagnosis of barley yellow dwarf virus, *J. Virol. Methods*, 16, 225, 1987.
41. **Milde, K. and Löning, T.,** Detection of papillomavirus DNA in oral papillomas and carcinomas: application of *in situ* hybridization with biotinylated HPV 16 probes, *J. Oral Pathol.*, 15, 292, 1986.
42. **Khan, A. M. and Wright, P. J.,** Detection of flavivirus RNA in infected cells using photobiotin-labelled hybridization probes, *J. Virol. Methods*, 15, 121, 1987.
43. **Syvänen, A-C., Laaksonen, M., and Söderlund, H.,** Fast quantification of nucleic acid hybrids by affinity-based hybrid collection, *Nucleic Acids Res.*, 14, 5037, 1986.
44. **Welcher, A. A., Torres, A. R., and Ward, D. C.,** Selective enrichment of specific DNA, cDNA and RNA sequences using biotinylated probes, avidin and copper-chelate agarose, *Nucleic Acids Res.*, 14, 10027, 1986.
45. **Lacey, E. and Grant, W. N.,** Photobiotin as a sensitive probe for protein labeling, *Anal. Biochem.*, 163, 151, 1987.
46. **Sheldon, E. L., Kellogg, D. E., Watson, R., Levenson, C. H., and Erlich, H. A.,** Use of nonisotopic M13 probes for genetic analysis: application to HLA class II loci, *Proc. Natl. Acad. Sci. U.S.A.*, 83, 9085, 1986.
47. **Messing, J.,** New M13 vectors for cloning, in *Methods in Enzymology*, Vol. 101, Wu, R., Grossman, L., and Moldave, K., Eds., Academic Press, New York, 1983, 20.
48. **Sheldon, E. L., Levenson, C. H., Mullis, K., and Rapoport, H.,** U.S. Patent 4,582,789, 1986.
49. **Courage-Tebbe, U. and Kemper, B.,** Construction of gapped circular DNA from phage M13 by *in vitro* hybridization, *Biochim. Biophys. Acta*, 697, 1, 1982.
50. **Sheldon, E., Kellogg, D. E., Levenson, C., Bloch, W., Aldwin, L., Birch, D., Goodson, R., Sheridan, P., Horn, G., Watson, R., and Erlich, H.,** Nonisotopic M13 probes for detecting the beta-globin gene: application to diagnosis of sickle cell anemia, *Clin. Chem.*, 33, 1368, 1987.
51. **Viscidi, R. P., Connelly, C. J., and Yolken, R. H.,** Novel chemical method for the preparation of nucleic acids for nonisotopic hybridization, *J. Clin. Microbiol.*, 23, 311, 1986.
52. **Li, P. and Symons, R. H.,** unpublished data, 1987.
53. **Draper, D. E.,** Attachment of reporter groups to specific, selected cytidine residues in RNA using a bisulfite-catalyzed transamination reaction, *Nucleic Acids Res.*, 12, 989, 1984.
54. **Reisfeld, A., Rothenberg, J. M., Bayer, E. A., and Wilchek, M.,** Nonradioactive hybridization probes prepared by the reaction of biotin hydrazide with DNA, *Biochem. Biophys. Res. Commun.*, 142, 519, 1987.
55. **Manning, J. E., Hershey, N. D., Broker, T. R., Pellegrini, M., Mitchell, H. K., and Davidson, N.,** A new method of *in situ* hybridization, *Chromosoma*, 53, 107, 1975.
56. **Sodja, A. and Davidson, N.,** Gene mapping and gene enrichment by the avidin-biotin interaction: use of cytochrome-c as a polyamine bridge, *Nucleic Acids Res.*, 5, 385, 1978.
57. **Renz, M.,** Polynucleotide-histone H1 complexes as probes for blot hybridization, *EMBO J.*, 2, 817, 1983.
58. **Pellegrini, M., Holmes, D. S., and Manning, J.,** Application of the avidin-biotin method of gene enrichment to the isolation of long double-stranded DNA containing specific gene sequences, *Nucleic Acids Res.*, 4, 2961, 1977.
59. **Syvänen, A-C., Alanen, M., and Söderlund, H.,** A complex of single-strand binding protein and M13 DNA as hybridization probe, *Nucleic Acids Res.*, 13, 2789, 1985.
60. **Al-Hakim, A. H. and Hull, R.,** Studies towards the development of chemically synthesized non-radioactive biotinylated nucleic acid hybridization probes, *Nucleic Acids Res.*, 14, 9965, 1986.
61. **Sverdlov, E. D., Monastyrskaya, G. S., Guskova, L. I., Levitan, T. L., Sheichenko, V. I., and Budowsky, E. I.,** Modification of cytidine residues with a bisulfite 0-methylhydroxylamine mixture, *Biochim. Biophys. Acta*. 340, 153, 1974.
62. **Hopman, A. H. N., Wiegant, J., Tesser, G. I., and Van Duijn, P.,** A non-radioactive *in situ* hybridization method based on mercurated nucleic acid probes and sulfhydryl-hapten ligands, *Nucleic Acids Res.*, 14, 6471, 1986.
63. **Hopmen, A. H. N., Wiegant, J., and Van Duijn, P.,** Mercurated nucleic acid probes, a new principle for non-radioactive *in situ* hybridization, *Exp. Cell. Res.*, 169, 357, 1987.

64. **Renz, M. and Kurz, C.**, A colorimetric method for DNA hybridization, *Nucleic Acids Res.*, 12, 3435, 1984.
65. **Meinkoth, J. and Wahl, G.**, Hybridization of nucleic acids immobilized on solid supports, *Anal. Biochem.*, 138, 267, 1984.
66. **Thein, S. L. and Wallace, R. B.**, The use of synthetic oligonucleotides as specific hybridization probes in the diagnosis of genetic disorders, in *Human Genetic Diseases: A Practical Approach.*, Davies, K. E., Ed., IRL Press, Oxford, U.K., 1986, chap. 3.
67. **Li, P., Medon, P. P., Skingle, D. C., Lanser, J. A., and Symons, R. H.**, Enzyme-linked synthetic oligonucleotide probes: non-radioactive detection of enterotoxigenic *Escherichia coli* in faecal specimens, *Nucleic Acids Res.*, 15, 5275, 1987.
68. **Chollet, A. and Kawashima, E. H.**, Biotin-labeled synthetic oligodeoxyribonucleotides: chemical synthesis and uses as hybridization probes, *Nucleic Acids Res.*, 13, 1529, 1985.
69. **Chu, B. C. F. and Orgel, L. E.**, Detection of specific DNA sequences with short biotin-labeled probes, *DNA*, 4, 327, 1985.
70. **Chu, B. C. F., Wahl, G. M., and Orgel, L. E.**, Derivatization of unprotected polynucleotides, *Nucleic Acids Res.*, 11, 6513, 1983.
71. **Wachter, L., Jablonski, J-A., and Ramachandran, K. L.**, A simple and efficient procedure for the synthesis of 5′-aminoalkyl oligodeoxynucleotides, *Nucleic Acids Res.*, 14, 7985, 1986.
72. **Skingle, D. C., Smyth, R., and Symons, R. H.**, unpublished data, 1987.
73. **Uhlmann, E. and Engels, J.**, Chemical 5′-phosphorylation of oligonucleotides valuable in automated DNA synthesis, *Tetrahedron Lett.*, 27, 1023, 1986.
74. **Coull, J. M., Weith, H. L., and Bischoff, R.**, A novel method for the introduction of an aliphatic primary amino group at the 5′ terminus of synthetic oligonucleotides, *Tetrahedron Lett.*, 27, 3991, 1986.
75. **Agrawal, S., Christodoulou, C., and Gait, M. J.**, Efficient methods for attaching non-radioactive labels to the 5′ ends of synthetic oligodeoxyribonucleotides, *Nucleic Acids Res.*, 14, 6227, 1986.
76. **Connolly, B. A.**, The synthesis of oligonucleotides containing a primary amino group at the 5′-terminus, *Nucleic Acids Res.*, 15, 3131, 1987.
77. **Smith, L. M., Fung, S., Hunkapiller, M. W., Hunkapiller, T. J., and Hood, L. E.**, The synthesis of oligonucleotides containing an aliphatic amino group at the 5′ terminus: synthesis of fluorescent DNA primers for use in DNA sequence analysis, *Nucleic Acids Res.*, 13, 2399, 1985.
78. **Smith, L. M., Sanders, J. Z., Kaiser, R. J., Hughes, P., Dodd, C., Connell, C. R., Heiner, C., Kent, S. B. H., and Hood, L. E.**, Fluorescence detectiopn in automated DNA sequence analysis, *Nature (London)*, 321, 674, 1986.
79. **Sanger, F., Coulson, A. R., Barrell, B. G., Smith, A. J. H., and Roe, B. A.**, Cloning in single-stranded bacteriophage as an aid to rapid DNA sequencing, *J. Mol. Biol.*, 143, 161, 1980.
80. **Sproat, B. S., Beijer, B., Rider, P., and Neuner, P.**, The synthesis of protected 5′-mercapto-2′, 5′-dideoxyribonucleoside-3′-O-phosphoramidites; uses of 5′-mercapto-oligodeoxyribonucleotides, *Nucleic Acids Res.*, 15, 4837, 1987.
81. **Ansorge, W., Sproat, B., Stegemann, J., Schwager, C., and Zenke, M.**, Automated DNA sequencing: ultrasensitive detection of fluorescent bands during electrophoresis, *Nucleic Acids Res.*, 15, 4593, 1987.
82. **Sproat, B. S., Beijer, B., and Rider, P.**, The synthesis of protected 5′-amino-2′, 5′-dideoxyribonucleoside-3′-O-phosphoramidites; applications of 5′-amino-oligodeoxyribonucleotides, *Nucleic Acids Res.*, 15, 6181, 1987.
83. **Haralambidis, J., Chai, M., and Tregear, G. W.**, Preparation of base-modified nucleosides suitable for non-radioactive label attachment and their incorporation into synthetic oligodeoxyribonucleotides, *Nucleic Acids Res.*, 15, 4857, 1987.
84. **Li, P. and Medon, P. P.**, unpublished data, 1987.
85. **Zuckermann, R., Corey, D., and Schultz, P.**, Efficient methods for attachment of thiol specific probes to the 3′-ends of synthetic oligodeoxyribonucleotides, *Nucleic Acids Res.*, 15, 5305, 1987.
86. **Sethabutr, O., Hanchalay, S., Echeverria, P., Taylor, D. N., and Leksomboon, U.**, A nonradioactive DNA probe to identify *Shigella* and enteroinvasive *Escherichia coli* in stools of children with diarrhea, *Lancet*, 2, 1095, 1985.
87. **Haas, M. J. and Fleming, D. J.**, Use of biotinylated DNA probes in colony hybridization, *Nucleic Acids Res.*, 14, 3976, 1986.
88. **Jablonski, E., Moomaw, E. W., Tullis, R. H., and Ruth, J. L.**, Preparation of oligodeoxynucleotide - alkaline phosphatase conjugates and their use as hybridization probes, *Nucleic Acids Res.*, 14, 6115, 1986.
89. **McLaughlin, G. L., Ruth, J. L., Jablonski, E., Steketee, R., and Campbell, G. H.**, Use of enzyme-linked synthetic DNA in diagnosis of *falciparum* malaria, *Lancet*, 1, 714, 1987.
90. **Seriwatana, J., Echeverria, P., Taylor, D. N., Sakuldaipeara, T., Changchawalit, S., and Chivoratanond, O.**, Identification of enterotoxigenic *Escherichia coli* with synthetic alkaline phosphatase-conjugated oligonucleotide DNA probes, *J. Clin. Microbiol.*, 25, 1438, 1987.

91. **Syvänen, A-C.,** Nucleic acid hybridization: from research tool to routine diagnostic method, *Med. Biol.*, 64, 313, 1986.
92. **Williams, V. R. and Williams, H. B.,** *Basic Physical Chemistry for the Life Sciences,* W. H. Freeman and Co., San Francisco, 1967, 279.
93. **Kam, W., Clauser, E., Kim, Y. S., Kan, Y. W., and Rutter, W. J.,** Cloning, sequencing, and chromosomal localization of human term placental alkaline phosphatase cDNA, *Proc. Natl. Acad. Sci. U.S.A.*, 82, 8715, 1985.
94. **Berger, J., Garattini, E., Hua, J-C., and Udenfriend, S.,** Cloning and sequencing of human intestinal alkaline phosphatase cDNA, *Proc. Natl. Acad. Sci. U.S.A.*, 84, 695, 1987.
95. **McGadey, J.,** A tetrazolium method for non-specific alkaline phosphatase, *Histochemie,* 23, 180, 1970.
96. **O'Connor, C. G. and Ashman, L. K.,** Application of the nitrocellulose transfer technique and alkaline phosphatase conjugated anti-immunoglobulin for determination of the specificity of monoclonal antibodies to protein mixtures, *J. Immunol. Methods,* 54, 267, 1982.
97. **Blake, M. S., Johnston, K. H., Russell-Jones, G. J., and Gotschlich, E. C.,** A rapid, sensitive method for detection of alkaline phosphatase-conjugated anti-antibody on Western blots, *Anal Biochem.*, 136, 175, 1984.
98. *FOCUS (Bethesda Research Laboratories/Life Technologies Inc.,)* 7 (2), 11, 1985.
99. **Graham, J., R. and Karnovsky, M. J.,** The early stages of absorption of injected horseradish peroxidase in the proximal tubules of mouse kidney: ultrastructural cytochemistry by a new technique, *J. Histochem. Cytochem.*, 14, 291, 1966.
100. **Hardy, H. and Heimer, L.,** A safer and more sensitive substitute for diamino-benzidine in the light microscopic demonstration of retrograde and anterograde axonal transport of HRP, *Neurosci. Lett.*, 5, 235, 1977.
101. **Josephy, P. D., Eling, T., and Mason, R. P.,** The horseradish peroxidase-catalyzed oxidation of 3,5,3′,5′-tetramethylbenzidine, *J. Biol. Chem.*, 257, 3669, 1982.
102. **Hanker, J. S., Yates, P. E., Metz, C. B., and Rustioni, A.,** A new specific, sensitive and non-carcinogenic reagent for the demonstration of horseradish peroxidase, *Histochem, J.*, 9, 789, 1977.
103. **Colman, D. R., Scalia, F., and Cabrales, E.,** Light and electron microscopic observations on the anterograde transport of horseradish peroxidase in the optic pathway in the mouse and rat, *Brain Res.*, 102, 156, 1976.
104. **de Olmos, J. S.,** An improved HRP method for the study of central nervous connections, *Exp. Brain Res.*, 29, 541, 1977.
105. **Nakane, P. K.,** Simultaneous localization of multiple tissue antigens using the peroxidase-labeled antibody method: a study on pituitary glands of the rat, *J. Histochem. Cytochem.*, 16, 557, 1968.
106. **Hawkes, R., Niday, E., and Gordon, J.,** A dot-immunobinding assay for monoclonal and other antibodies, *Anal. Biochem.*, 119, 142, 1982.
107. **Griswold, Jr., D. P., Casey, A. E., Weisburger, E. K., and Weisburger, J. H.,** The carcinogenicity of multiple intragastric doses of aromatic and heterocyclic nitro or amino derivatives in young female Sprague-Dawley rats, *Cancer Res.*, 28, 924, 1968.
108. **Adams, J. C.,** Heavy metal intensification of DAB-based HRP reaction product, *J. Histochem. Cytochem.*, 29, 775, 1981.
109. **Przeplorka, D. and Myerson, D.,** A single-step silver enhancement method permitting rapid diagnosis of cytomegalovirus infection in formalin-fixed, paraffin-embedded tissue sections by *in situ* hybridization and immunoperoxidase detection, *J. Histochem. Cytochem.*, 34, 1731, 1986.
110. **Burns, J., Chan, V. T. W., Jonasson, J. A., Fleming, K. A., Taylor, S., and McGee, J. O'D.,** Sensitive system for visualizing biotinylated DNA probes hybridized *in situ:* rapid sex determination of intact cells, *J. Clin. Pathol.*, 38, 1085, 1985.
111. **Timm, F.,** Zur histochemie der schwermetalle: Das sulfid-silberverfahren, *Z. Gesumie Gerichtliche Medizin (Berlin),* 46, 706, 1958.
112. **Burns, J., Redfern, D. R. M., Esiri, M. M., and McGee, J. O'D.,** Human and viral gene detection in routine paraffin embedded tissue by *in situ* hybridization with biotinylated probes: viral localization in herpes encephalitis, *J. Clin. Pathol.*, 39, 1066, 1986.
113. **Ferguson, D. J. P., Burns, J., Harrison, D., Jonasson, J. A., and McGee, J. O'D.,** Chromosomal localization of genes by scanning electron microscopy using *in situ* hybridization with biotinylated probes: Y chromosome repetitive sequences, *Histochem, J.*, 18, 266, 1986.
114. **Holland, V. R. Saunders, B. C. Rose, F. L., and Walpole, A. L.,** A safer substitute for benzidine in the detection of blood, *Tetrahedron,* 30, 3299, 1974.
115. **De Serres, F. J. and Ashby, J. Eds.,** *Progress in Mutation Research,* Vol. 1, Elsevier/North Holland Biomedical Press, Amsterdam, 1981.
116. **Dykes, D.,** personal communication, 1987.
117. **Hsu, S-M. and Raine, L.,** Protein A, avidin, and biotin in immunohistochemistry, *J. Histochem. Cytochem.*, 29, 1349, 1981.

118. **Mesulam, M.-M.**, Tetramethyl benzidine for horseradish peroxidase neurohistochemistry: a non-carcinogenic blue reaction-product with superior sensitivity for visualizing neural afferents and efferents, *J. Histochem. Cytochem.*, 26, 106, 1, 1978.
119. **Mesulam, M.-M. and Rosene, D. L.**, Sensitivity in horseradish peroxidase neurohistochemistry: a comparative and quanitative study of nine methods, *J. Histochem. Cytochem.*, 27, 763, 1979.
120. **Bondi, A., Cheiregatti, G., Eusebi, V., Fulchert, E., and Bussolati, G.**, The use of β-galactosidase as a tracer in immunocytochemistry, *Histochemistry*, 76, 153, 1982.
121. **Hemmilä, I.** Fluoroimmunoassays and immunofluorometric assays, *Clin. Chem.*, 31, 359, 1985.
122. **Oi, V. T., Glazer, A. N., and Stryer, L.**, Fluorescent phycobilprotein conjugates for analyses of cells and molecules, *J. Cell Biol.*, 93, 981, 1982.
123. **Kronick, M. N. and Grossman, P. D.**, Immunoassay techniques with fluorescent phycobilprotein conjugates, *Clin. Chem.*, 29, 1582, 1983.
124. **Haugland, R. P.**, *Handbook of Fluorescent Probes and Research Chemicals*, Molecular Probes Inc., Oregon, 1985.
125. **Albertson, D. G.**, Mapping muscle protein genes by *in situ* hybridization using biotinlabeled probes, *EMBO J.*, 4, 2493, 1985.
126. **Pinkel, D., Gray, J. W., Trask, B., van den Engh, G., Fuscoe, J., and van Dekken, H.**, Cytogenetic analysis by *in situ* hybridization with fluorescently labeled nucleic acid probes, *Cold Spring Harbor Symp. Quant. Biol.*, 51, 151, 1986.
127. **Nagata, Y., Yokota, H., Kosuda, O., Yokoo, K., Takemura, K., and Kikuchi,T.**, Quantification of picogram levels of specific DNA immobilized in microtiter wells, *FEBS Lett.*, 183, 379, 1985.
128. **Yokota, H., Yokoo, K., and Nagata, Y.**, A quantitative assay for the detection of hepatitis B virus DNA employing a biotin-labeled DNA probe and the avidin- β-galactosidase complex, *Biochim. Biophys. Acta.* 868, 45, 1986.
129. **Hemmilä, I., Dakubu,S., Mukkala, V.-M., Siitari, H., and Lygren, T.**, Europium as a label in time-resolved immunofluorometric assays, *Anal. Biochem.*, 137, 335, 1984.
130. **Jackson, T. M. and Ekins, R. P.**, Theoretical limitations on immunoassay sensitivity; Current practice and potential advantages of fluorescent Eu^{3+} chelates as non-radioisotopic tracers, *J. Immunol. Methods*, 87, 13, 1986.
131. **Syvänen, A.-C., Tchen, P., Ranki, M., and Söderlund, H.**, Time-resolved fluorometry: a sensitive method to quantify DNA-hybrids, *Nucleic Acids Res.*, 14, 1017, 1986.
132. **Oser, A., Roth, W. K., and Valet,G.**, Sensitive non-radioactive dot-blot hybridization using DNA probes labeled with chelate group substituted psoralen and quanitative detection by europium ion fluorescence, *Nucleic Acids Res.*, in press, 1987.
133. **Dahlén, P.**, Detection of biotinylated DNA probes by using Eu-labeled streptavidin and time-resolved fluorometry, *Anal. Biolchem.*, 164, 78, 1987.
134. **Wetmur, J. G.**, Acceleration of DNA renaturation rates, *Biopolymers*, 14, 2517, 1975.
135. **Amnasino, R. M.**, Acceleration of nucleic acid hybridization rate by polyethylene glycol, *Anal. Biochem.*, 152, 304, 1986.
136. **Roswell, D. F. and White, E. H.**, The chemiluminescence of luminol and related hydrazides, in *Methods in Enzymology*, Vol. 57, DeLuca, M. A., Ed., Academic Press, New York, 1978, 409.
137. **Seitz, W. R.**, Chemiluminescence of enzymically generated peroxide, in *Methods in Enzymology*, Vol. 57, DeLuca, M. A., Ed., Academic Press, New York, 1978, 445.
138. **Wulff, K.**, Luminometry, in *Methods of Enzymatic Analysis*, Vol. 1, 3rd ed., Bergmeyer, J. and Grabl, M., Eds., Verlag Chemie, Weinheim, Germany, 1983, 340.
139. **Campbell, A. K., Hallett, M. B., and Weeks, I.**, Chemiluminescence as an anlaytical tool in cell biology and medicine, in *Methods of Biochemical Analysis*, Vol. 31, Glick, D., Ed., John Wiley & Sons, New York, 1985, 317.
140. **Matthews, J. A., Batki, A., Hynds, C., and Kricka, L. J.**, Enchanced chemiluminescent method for the detection of DNA dot-hybridization assays, *Anal. Biochem.*, 151, 205, 1985.
141. **Leong, M. M. L., Milstein, C., and Pannell, R.**, Luminescent detection method for immunodot, Western, and Southern blots, *J. Histochem. Cytochem.*, 34, 1645, 1986.
142. **Heller, M. J. and Morrison, L. E.**, Chemiluminescent and fluorescent probes for DNA hybridization system, in *Rapid Detection and Identification of Infectious Agents*, Kingsbury, D. T., and Falkow, S., Eds., Academic Press, New York, 1985, 245.
143. **Menissier, J., Hunting, D. J., and De Murcia, G.**, Electron microscopic mapping of single-stranded discontinuities in cauliflower mosaic virus DNA by means of the biotin-avidin technique, *Anal. Biochem.*, 148, 339, 1985.
144. **Varndell, I. M., Polak, J. M., Sikri, K. L., Minth, C. D., Bloom, S. R., and Dixon, J. E.**, Visualization of messenger RNA directing peptide synthesis by *in situ* hybridization using a novel single-stranded cDNA probe: potential for the investigation of gene expression and endocrine cell activity, *Histochemistry*, 81, 597, 1984.

145. **Kress, H., Meyerowitz, E. M., and Davidson, N.,** High resolution mapping of *in situ* hybridized biotinylated DNA to surface-spread *Drosophila* polytene chromosomes, *Chromosoma,* 93, 113, 1985.
146. **Cremers, A. F. M., Jansen in de Wal, N., Wiegant, J., Dirks, R. W., Weisbeek, P., van der Ploeg, M., and Landegent, J. E.,** Non-radioactive *in situ* hybridization: a comparison of several immunocytochemical detection systems using reflection-contrast and electron microscopy, *Histochemistry,* 86, 609, 1987.
147. **Danscher, G. and Nörgaard, J. O. R.,** Light microscopic visualization of colloidal gold on resin-embedded tissue, *J. Histochem. Cytochem.,* 31, 1394, 1983.
148. **Holgate, C. S., Jackson, P., Cowen, P. N., and Bird, C. C.,** Immunogold-silver staining: a new method of immunostaining with enhanced sensitivity, *J. Histochem. Cytochem.,* 31, 938, 1983.
149. **Lowry, O. H., Passonneau, J. V., Schultz, D. W., and Rock, M. K.,** The measurement of pyridine nucleotides by enzymatic cycling, *J. Biol. Chem.,* 236, 2746, 1961.
150. **Self, C. H.,** Enzyme amplification — a general method applied to provide a immunoassisted assay for placental alkaline phosphatase, *J. Immunol. Methods,* 76, 389, 1985.
151. **Stanley, C. J., Johannsson, A., and Self, C. H.,** Enzyme amplification can enhance both the speed and the sensitivity of immunoassays, *J. Immunol. Methods,* 83, 89, 1985.
152. **Johannsson, A., Ellis, D. H., Bates, D. L., Plumb, A. M., and Stanley, C. J.,** Enzyme amplification for immunoassays: detection limit of one hundredth of an attomole, *J. Immunol. Methods,* 87, 7,1986.
153. **Carr, R. I., Mansour, M., Sadi, D., James, H., and Jones, J. V.,** A substrate amplification system for enzyme-linked immunoassays: demonstration of its general applicability to ELISA systems for detecting antibodies and immune complexes, *J. Immunol. Methods,* 98, 201, 1987.
154. **Torrance, L.,** Use of enzyme amplification in an ELISA to increase sensitivity of detection of barley yellow dwarf virus in oats and in individual vector aphids, *J. Virol. Methods.,* 15, 131, 1987.
155. **Gatley, S.,** Enzyme amplification and its relevance as a non-isotopic detection system for the use with nucleic acid probes, in *DNA Probes: A New Technology in its Development and Application,* IBC Technical Services Ltd., London, 1985.
156. **Saiki, R. K., Scharf, S., Faloona, F., Mullis, K. B., Horn, G. T., Erlich, H. A., and Arnheim, N.,** Enzymatic amplification of β-globin genomic sequences and restriction site analysis for diagnosis of sickle cell anemia, *Science,* 230, 1350, 1985.
157. **Mullis, K., Faloona, F., Scharf, S., Saiki, R., Horn, G., and Erlich, H.,** Specific enzymatic amplification of DNA *in vitro:* the polymerase chain reaction, *Cold Spring Harbor Symp. Quant. Biol.,* 51, 263, 1986.
158. **Mullis, K. B. and Faloona, F. A.,** Specific synthesis of DNA *in vitro* via a polymerase-catalysed chain reaction, in *Methods in Enzymology,* Vol. 155, Wu, R., Ed., Academic Press, New York, 1987, 335.
159. **Saiki, R. K., Bugawan, T. L., Horn, G. T., Mullis, K. B., and Erlich, H. A.,** Analysis of enzymatically amplified β-globin and HLA-DQα DNA with allele-specific oligonucleotide probes, *Nature (London),* 324, 163, 1986.
160. **Van Brunt, J. and Klausner, A.,** Pushing probes to market, *Bio/Technology,* 5, 211, 1987.

Chapter 3

NUCLEIC ACID PROBES IN THE DIAGNOSIS OF HUMAN MICROBIAL PATHOGENS

T. Hyypiä, P. Huovinen, M. Holmberg, and U. Pettersson

TABLE OF CONTENTS

I. INTRODUCTION

The development of effective vaccines and antimicrobial drugs against infectious diseases has been among the most successful achievements in modern medicine. The control of these diseases requires efficient diagnostic methods for the evaluation of the prevalence of diseases and for initiation of specific treatment. Virtually all known microbes can be specifically identified today but in many cases further development is needed for more accurate, rapid, easy-to-use, and inexpensive diagnostic assays.

Cell culture facilities are needed for the isolation of viruses in clinical specimens. Although this method is one of the most sensitive ways for the detection of virus, its use is restricted to specialized laboratories. Viruses can be visualized by electron microscopy, which is a rapid but laborious technique in clinical use and requires a trained specialist in addition to expensive equipment. For these reasons, immunoassays are gaining increasing use in modern diagnostic virology. They enable simultaneous testing of large numbers of specimens with standardized reagents. Direct detection of viral antigens in blood, stool, and nasopharyngeal specimens is widely in routine diagnostic use. Immunoassays can also be applied for demonstration of virus specific immunoresponse in blood which is essential especially in cases where detectable virus components have already disappeared at the onset of disease symptoms. In instances where no susceptible cell cultures are available for virus isolation (e.g., hepatitis B and papillomaviruses) and when the production of viral antigens is low (e.g., persistent infections), new assay principles are needed for specific identification of the pathogen.

Conventional bacterial diagnosis is based first on staining and microscopic examination of the sample (secretions, blood, etc.) Direct microscopy is rapid, but often an insensitive and nonspecific method. All bacterial samples are cultured on nonselective or selective media to allow growth of bacterial in the sample. According to the colony appearance, isolates are collected for further studies. Identification of bacterial isolates include fermentation assays, agglutination with antisera, direct immunofluorescence assays, immunocounter electrophoresis, direct chemical reactions with colonies, and growth on selective media or in different environmental conditions. Usually, identification takes at least 1 or 2 days, but the time required for identification depends largely on the growth rate of the bacterial studied and may take even several weeks (e.g., *Mycobacterium tuberculosis).* For some organisms, cultures are not available in bacterial laboratories (e.g., *Mycobacterium leprae* and *Treponema pallidum*). Antibiotics or other bacteriocidal or growth inhibiting agents present in patient samples may also prevent bacterial diagnosis. Serological diagnosis of bacterial pathogens is available only for certain pathogens (e.g., *Yersinia enterocolitica*) because of a lack of a specific antibody response.

Many parasitic diseases caused by protozoans and helminths present special diagnostic problems. The populations in most tropical countries live under a heavy parasite burden. Moreover, several parasitic diseases have a chronic course and the causative agent is not always easy to isolate. Many of these infections are vector borne, and incrimination of vectors is an important part of the diagnostic problem. *In vitro* cultivation is often not possible or is feasible only at central laboratories. The direct microscopic identification of the causative organism is sometimes complicated and tedious, and tropical infections caused by hematogenous, visceral, cutaneous, and intestinal parasitism require different methods for isolation and detection. The detection of specific antibodies in diagnosis and epidemiological screening has, in many cases, been limited due to lack of standardized reagents, especially antigens. The use of recombinant DNA technology to produce purified parasite components will make serodiagnosis more reliable. However, the problems connected with immunodiagnosis are outside the scope of this chapter.

Gene technology offers several methodological advantages when compared to other approaches for microbial diagnosis. Any gene of any known microorganism can be cloned in a vector and produced in large amounts economically and then used in diagnostic assays for the identification of the pathogen. The technical approaches in the tests are presented in other chapters of this book. The application of nucleic acid hybridization methods in detection of human pathogens has received considerable attention during the past few years. The literature on the field has also increased exponentially and, therefore, it is outside the volume of this chapter to cover all published papers. Rather, we have tried to choose elucidative examples of this important application of gene technology. The topic has also been reviewed by other authors including recent articles by Palva[1] and Zwadyk Jr. and Cooksey.[2]

II. DETECTION OF VIRUSES BY HYBRIDIZATION

Viruses were the first microbes which were genetically characterized in great detail. Therefore, it can easily be understood that reagents were soon available for their detection in clinical specimens by hybridization. None of the clinically important virus groups have been neglected in this approach. Our own work has mainly focused on adenoviruses and picornaviruses. They will be presented here as examples to illustrate the general advantages and problems of hybridization assays. The results of detection of herpes, papilloma and hepatitis B virus DNA will then be discussed as examples of potential routine use of the method.

A. Adenoviruses

Adenoviruses are human pathogens responsible for several common diseases including gastroenteritis and respiratory infections. The diagnosis of adenovirus infections can be carried out by isolation of the virus in tissue culture or by electron microscopy. Recently, immunoassays have been successfully applied for detection of viral antigens in stool and nasopharyngeal specimens. Adenoviruses are nonenveloped icosahedral viruses that replicate and assemble their virions in the nucleus of the infected cell. Hexon is the major polypeptide component of the virion. The DNA genome of adenovirus type 2 has a length of 35,937 base pairs. Since the molecular biology of adenovirus is well known, it has been one of the model viruses indirect detection of viral antigens and DNA in clinical specimens.

We tested 24 nasopharyngeal mucus aspirates by enzyme immunoassay (EIA) for adenovirus hexon antigen and by nucleic acid hybridization assay employing both radioactive and enzymatic probe systems.[3] In all three tests, the same 16 specimens were scored positive when the hybridization results were analyzed by autoradiography and by precipitative color reaction (Table 1). The results were similar with whole adenovirus DNA and a cloned fragment as probes. However, when the hybridization reaction with radioactive probe was analyzed using cpm values after scintillation counting, one of the EIA positive specimens became negative and one of the negatives appeared positive. The enzyme reaction measured as absorbance values was positive in ten cases, all of which were also positive by EIA. This can be explained by possible contamination of negative spots by material from highly positive ones when cutting the filter and the lower sensitivity of the enzyme reaction when measured on microtitier wells instead of direct reaction on the filter.

Also, 49 stool specimens from children with acute gastroenteritis were tested both by the hybridization assay and by radioimmunoassay (RIA) for adenovirus hexon antigen.[4] Twenty-four specimens were positive by RIA and 22 by the hybridization test with the whole adenovirus 2 DNA probe. These included one RIA-negative specimen. A "prozone" effect with high quantities of adenovirus DNA appeared, indicating that the specimens should be preferably tested at two dilutions. Phenol extraction was used for the specimen treatment in both these studies. This is a drawback and limits the number of specimens which can be analyzed in the test. Kidd et al.[5] have introduced a method for the detection of enteric

Table 1
EXAMPLES OF THE DETECTION OF ADENOVIRUS DNA IN CLINICAL SPECIMENS BY NUCLEIC ACID HYBRIDIZATION (NAH)

Specimens	Method	Positive by NAH/ reference test[a]	Ref.
Stool	Spot hybridization/ ^{32}P labeled probe	15/18	4
	Spot hybridization/ ^{32}P labeled probe	61/76[b]	5
	Spot hybridization/ Eu^{3+} labeled probe	5/5	11
Nasopharyngeal	Sandwich hybridization/ ^{125}I labeled probe	22/25	10
	Spot hybridization/ biotinylated probe	16/16	3
	Spot hybridization/ Eu^{3+} labeled probe	5/5	11

[a] The specificity was virtually 100% in all the studies.
[b] The sensitivity was 100% when the specimens were phenol extracted prior to testing instead of direct denaturation of the filter.

adenoviruses in stool specimens without preciding phenol treatment. However, the sensitivity in this study was 20% lower when compared to pretreated samples. Also, several other groups have reported successful applications of hybridization methods for the detection of adenoviruses in stool specimens.[6-9]

Virtanen et al.[10] have described a nucleic acid sandwich hybridization test for the detection of adenoviruses in nasopharyngeal mucus aspirates from children with acute respiratory infections. In this assay, binding of the specimen DNA to the solid phase is mediated by a specific DNA fragment. The bound nucleic acid is then detected using a probe which is cloned in a vector that does not cross-react with the solid phase vector (Figure 1A). The reagents included cloned DNA fragments from adenoviruses 2 and 3. The filter DNA reagent and the probe were cloned in plasmid pBR322 and phage M13mp7, respectively. The test was compared with RIA for adenovirus hexon antigen. Twenty-two of the 25 RIA positive specimens also gave a positive signal in the sandwich hybridization test and all the 22 RIA negative specimens were negative by hybridization. In addition, the test could discriminate between adenovirus subgroups B and C. The advantage of this method is the simple treatment of the sample which includes proteinase K treatment in the presence of SDS and boiling prior to hybridization.

Dahlen et al.[11] have recently introduced a hybridization assay for adenoviruses which uses time-resolved fluorescence as a detection method after hybridization. Eu^{3+}-chelates coupled to the probe DNA (Figure 1B) are measured in a fluorometer. The results for detection of adenoviruses in clinical samples by this nonisotopic test are comparable with radioactive probe systems as well as with immunoassays (Table 1).

B. Picornaviruses

The two genera of picornaviruses commonly infecting man are enteroviruses and rhinoviruses. The virions are icosahedral, naked, and contain an RNA genome which can directly act as messenger RNA in eukaryotic cells. The genes for capsid proteins are located at the 5′ terminus of the genome while the 3′ end contains genes coding for functions necessary in virus replication. Enteroviruses consist of polio, coxsackie, echo, and hepatitis A viruses.

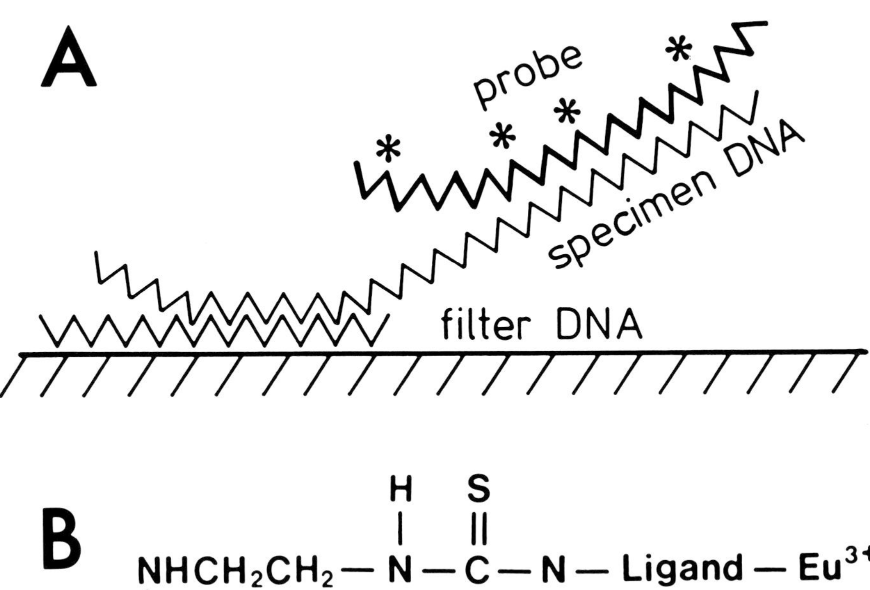

FIGURE 1. A. The principle of the sandwich hybridization assay. B. Structure of a nucleotide labeled with europium (Eu^{3+}) chelate. (From Dahlén, P., Hurskainen, P., Lövgren, T., and Hypiä, T., *J. Clin. Microbiol.*, in press. With permission.)

Viruses from all enterovirus groups except hepatitis A can cause complications of the central nervous system. Polioviruses are most frequently associated with paralysis and echoviruses with encephalitis. Coxsackie B virus infections are a common cause of viral carditis and evidence of their role at the onset of type I diabetes has been reported. Rhinoviruses are causative agents of common cold and they are also involved in infections of lower respiratory tract.

Enteroviruses consist of more than 60 and rhinoviruses of more than 100 serotypes. Therefore, they provide a particularly challenging problem in diagnostic virology. Virus isolation, followed by laborious neutralization or acid lability tests, is used for recognition of these agents in clinical specimens. The classical enterovirus serology is time consuming and often inadequate. Recently, immunoassays for antigen detection and serological assays detecting specific IgM antibodies during the acute phase of human picornavirus infections have been reported.

We have used a coxsackievirus B3 (CB3) derived probe to detect enterovirus RNA in infected cells and in clinical specimens.[12] The probe was a cloned cDNA copy of the 3′ end of CB3 genome consisting of approximately 4300 nucleotides. When cells infected with various strains of enteroviruses were tested, the probe gave a positive signal with coxsackievirus A9, B3, B4, and B5, with echovirus 11 and poliovirus 3 but not with unrelated viruses. The sensitivity of the test was not, however, sufficient for direct detection of

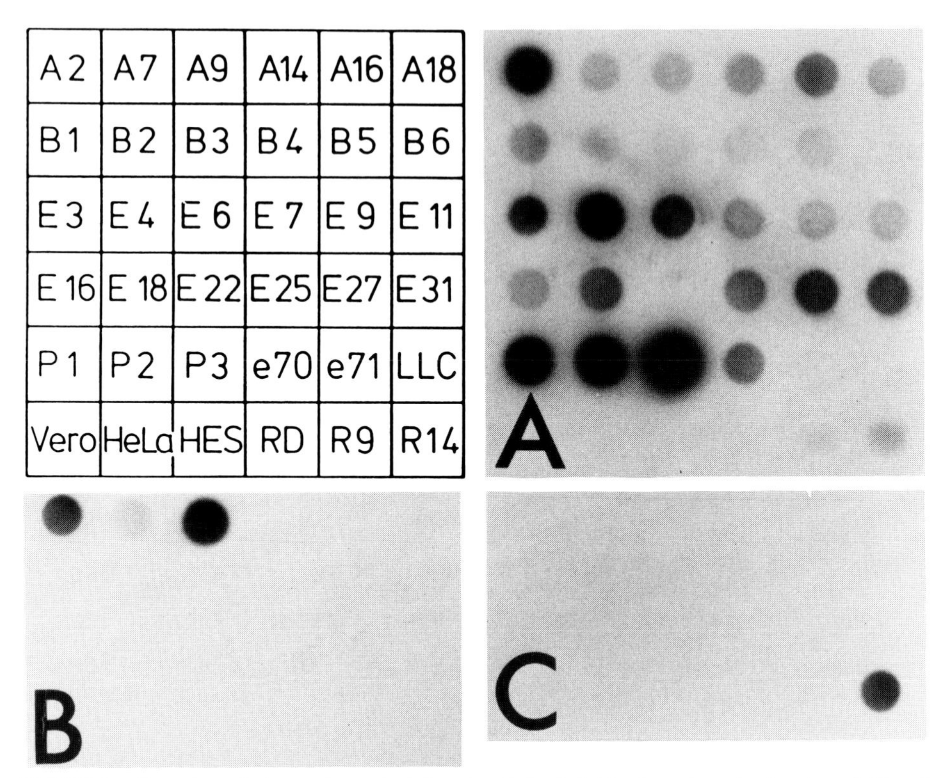

FIGURE 2. Detection of picornaviruses in infected cells by hybridization. A. poliovirus type 3 5′ end probe B. Echovirus 6 3′ end probe C. Rhinovirus 14 3′ end probe. The panel of virus strains on the filter includes coxsackie A (A), B (B), echo (E), polio (P), rhino (R) and newer enteroviruses (e). The rest of the samples consist of uninfected cells which were used to propagate the viruses.

enteroviruses in clinical samples. In a further analysis of 48 enterovirus serotypes by hybridization with coxsackievirus A21 and B3, poliovirus 3 and enterovirus 70 probes, we observed that more than 90% of the strains could be identified with this pattern of reagents and also roughly typed.[13] The broad reactivity of the 5 end probe as well as the specificity of some the 3′ end probes is shown in Figure 2.[195]

Rotbart et al.[14,15] have also described an assay for the detection of enterovirus RNA using cloned cDNA sequences from poliovirus type 1 and CB3 genomes as probes. In hybridization assays these probes could detect enteroviruses both generally and specifically. Tracy[16] has analyzed the cross-reactivity among the CB subgroup by using cloned CB3 cDNA fragments as probes. He was able to localize regions in the genome which were common for the subgroup and specific for the serotype. Assays for the detection of hepatitis A (HAV) RNA by hybridization have also been described.[17-19] Their sensitivity exceeds that of the presently used immunoassays for the detection of HAV components in the specimens, but demonstration of specific IgM antibodies still remains as the method of choice for HAV diagnosis.

Fifty-six serotypes of human rhinoviruses (HRV) have been analyzed by Al-Nakib et al.[20] in a hybridization test which employed HRV14 5′ end sequence as a probe. Fifty-four of the strains gave a positive hybridization signal indicating that this region is highly conserved among rhinoviruses. The strength of the signal varied and it was possible to draw preliminary conclusions on the relationships between the serotypes. However, more work with probes from different regions together with analysis of sequence data is still needed to clarify the genetic relationships between the members of this large virus group.

The assays described above can be used for rapid confirmation of picornavirus growth in cell cultures after virus isolation, and they have also a potential for typing of the viruses.

The most important improvement in the picornavirus diagnosis with hybridization techniques concerns the broad cross-reactivity of genome sequences outside the capsid protein regions, allowing construction of group-common reagents which has not been possible with other methods.

An interesting application of enterovirus hybridization methods concerns detection of virus genomes in heart and skeletal muscle. Bowles et al.[21,22] have demonstrated the presence of sequences reacting with CB2 cDNA probe in 9 of 17 human samples from myocarditis as well as in biopsy specimens from skeletal muscle of myositis patients. Similar results have recently been obtained by the *in situ* hybridization technique from myocardium[23] of CB3 infected mice. The evaluation of the clinical significance of these findings still needs confirmation by more extensive studies.

C. Herpesviruses

The group of herpesviruses consists of any important human pathogens including herpes simplex virus (HSV) 1 and 2, varicella zoster virus (VZV), cytomegalovirus (CMV) and Epstein-Barr virus (EBV). These viruses are enveloped and they contain a double-stranded linear DNA genome. Herpes simplex viruses cause a variety of clinical entities including stomatitis, keratoconjunctivitis, encephalitis, and genital herpes. VZV is responsible for chickenpox and herpes zoster, while CMV causes intrauterine and postnatal infections which can lead to congenital or generalized disease. EBV is the etiological agent of infectious mononucleosis and it is also associated with Burkitt's lymphoma and nasopharyngeal carcinoma in certain geographical regions. It is characteristic of all the herpes group viruses that the primary infection is often followed by a long latency period. Reactivation of the virus may result in severe secondary diseases which are more frequent in immunocompromised patients. HSV, VZV, and CMV can be diagnosed by isolation and HSV and VZV also by antigen detection assays. Serological assays are available for all the members in the group. The clinical problems include diagnosis of herpes simplex encephalitis and development of rapid and reliable methods for cytomegalovirus diagnostics.

Several nucleic acid hybridization assays for herpesviruses have been introduced. One of the earliest reports of the use of the spot hybridization test for the detection of viral DNA sequences was published in 1980 by Brandsma and Miller.[24] They showed that EBV DNA was detectable in cell lines isolated from Burkitt's lymphoma patients. The presence of EBV DNA has also been recognized in specimens from patients with infectious mononucleosis, nasopharyngeal carcinoma, and certain lymphomas.[25] The respiratory tract appears to be an important reservoir of EBV since approximately 50% of bronchial washings harvested from patients without signs of EBV-induced diseases contain detectable amounts of viral DNA.[26] Also throat washings of infectious mononucleosis patients[27] (87%) as well as recipients of renal[27] (50%) and bone marrow[28] (15%) transplants contain EBV DNA.

The nucleic acid hybridization test has also been described for the detection of varicella zoster DNA sequences in clinical specimens.[29] The spot hybridization assay used was comparable to cell culture in sensitivity and specificity. *In situ* hybridization[30] and Southern blotting[31] methods have been successfully used to reveal VZV specific sequences in trigeminal ganglia in patients without recent history or herpes zoster.

In 1983, Chou and Merigan[32] introduced an assay for cytomegalovirus which had a sensitivity of picogram amounts of homologous DNA and detected CMV DNA in isolation positive urine specimens. These observations have been later confirmed by several research groups.[33-37] The sensitivity of these assays has been in the range of 75 to 95% when compared with virus isolation. Also nonisotopic hybridization methods using biotinylated probes have been used for this purpose.[38,39] In addition to urine samples, CMV DNA has been detected in leukocytes[33,40] and in lung autopsy specimens.[41]

Direct demonstration of HSV DNA in clinical specimens using spot hybridization has been reported by Redfield et al.[42] The sensitivity of the test was 78% and the specificity 100% when compared to virus isolation. Applications of *in situ* hybridization using commercially available biotinylated probes for detection of HSV-specific sequences in brain specimens of encephatitis patients[43] as well as in various samples of patients with genital, oral, and other manifestations of HSV infections[44] have recently been described. The spot hybridization test can also be applied in studies of the effects of antiviral compounds on HSV DNA synthesis in infected cells.[45] Typing of herpes simplex viruses can be carried out by several methods, which include restriction enzyme analysis of viral DNA and immunoassays. Many groups have reported the use of nucleic acid hybridization in discrimination between HSV 1 and 2 by whole viral DNA,[46,47] cloned fragments,[48] and synthetic oligonucleotides[49] as probes.

D. Papillomaviruses

Since 1842, several epidemiological studies have supported an infectious etiology in human cervical cancer (for review, see Reference 50). Similar observations have also been reported concerning penile and vulvar cancer. The presence of numerous candidate pathogens has been analyzed in malignant lesions. However, it was not until 1974 that the first reports on the association of papillomaviruses in genital cancer were published.

Papillomaviruses were earlier known to induce mainly benign tumors, including warts of skin and genital mucosa. The warts contain sufficient amounts of material for purification and further analysis of the virus, but it has not been possible to grow this pathogen in cell cultures so far. Papillomaviruses are small naked viruses which contain a circular DNA genome having a length of about 8000 base pairs. Although antiserum reacting with capsid proteins of several animal and human papillomaviruses have been developed, the knowledge of the viral polypeptides is still very restricted and mostly derived from nucleic acid sequence data.

For the reasons mentioned above, the classification of human papillomaviruses (HPV) is based on their nucleic acid homology, i.e., the definition of a new type requires its cross-hybridization with known prototypes to less than 50% by applying hybridization conditions of high stringency. At present, there are nearly 50 known HPV types and the complete nucleotide sequences of at least 8 types have already been published. As a natural consequence of the lack of specific antigen detection systems and the inability to grow the virus in cell cultures, nucleic acid hybridization has become a routine method in HPV diagnosis. At the moment, the detection of HPV DNA in genital specimens is the most important application of this diagnostic technique in modern clinical virology.

Spot, sandwich,[51] filter *in situ*,[52] hybridization, as well as Southern blotting[54] methods, have been used for the identification of HPV types in clinical specimens. For spot hybridization, biopsy or other clinical material is digested with proteinase, phenol extracted, applied onto filter membrane and hybridized. The same specimen on the filter can be successively tested by several HPV probes. The sandwich hybridization method has the advantage that the pretreatment of the sample is rather simple when compared to spot hybridization. In filter *in situ* hybridization, the cells from smears are denatured on the filter and tested by hybridization. This assay also cirumvents the phenol extraction step, but the problem is occasionally the background which may cause difficulties in the interpretation of the results. *In situ* hybridization gives more information, including histology, when compared to the other tests. It is also the most laborious one and presently applied mainly for scientific applications. An example of this technique is presented in Figure 3. Since cross-reactivity at the nucleotide level is observed between some HPV types, Southern blotting technique is needed for exact identification. It can also be used to discriminate whether the viral DNA genome is in episomal or integrated form in the host cell. This method is tedious in routine use because of the gel electrophoresis procedure.

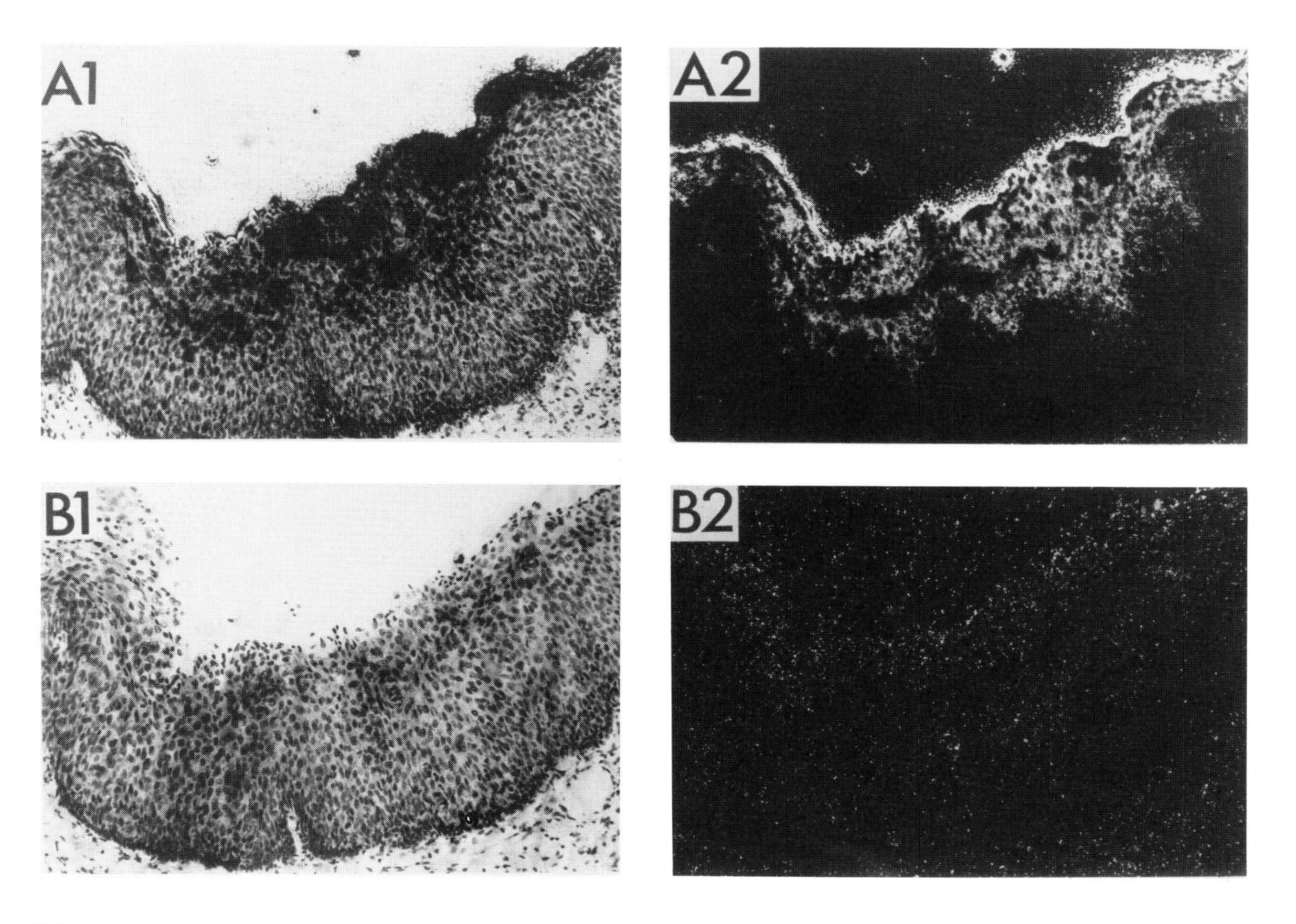

FIGURE 3. Detection of HPV 16 in a cervical biopsy specimen by *in situ* hybridization using ^{35}S-labeled probes. A1 HPV 16 probe; B1 BR 322 probe; A2 and B2 are dark field microscopy pictures of the hybridization results, respectively. (Courtesy of Dr. Pirkko Korkiamäki.)

Largely as a result of the efforts of zur Hausen and his group, previously unknown HPV types were found to be present in various lesions of the genital tract. It is now known that HPV types found in genital lesions include at least 6, 11, 16, 18, and 31[55-58] but there may be several others which have not yet been identified. HPV16 and 18 are present in approximately 80% of invasive squamous cell cancers of the cervix, vulva, and penis while types 6 and 11 are more often associated with genital warts and benign lesions. The differences in the transforming potential of the HPV types are at least partially explained by the fact that in benign lesions the state of viral DNA is episomal, whereas it is usually integrated to the genome of the host cell in invasive cancers.[54] The persisting DNA is actively transcribed suggesting a role of virus proteins in the maintenance of the transformed state.[59]

The original observations concerning the close association of certain HPV types, especially HPV16, with genital malignancies have recently been confirmed in several reports (e.g.,[60-62]). HPV DNA has also been detected in cytologically and histologically normal genital samples[63-65] indicating that further analysis of the mechanisms of malignant transformation are still needed.

Although there is no specific antiviral treatment available for the HPV infections the lesions can be surgically removed. The prognosis is good provided that the lesion is recognized at an early stage. Recurrence occurs relatively often and, therefore, it is important to know the HPV type present for proper treatment and further controls. Probably nonradioactive labeling methods will soon enable this method to become an important complement to the Papanicolau smear technique for screening of genital dysplacias.

E. Hepatitis B Virus

Hepatitis B virus (HBV) causes a disease which is characterized by fever and icterus. The clinical picture varies from asymptomatic infections to fatal fulminant hepatitis. Approximately 10% of the patients remain carriers of the virus excreting infections HBV or viral components in their blood. A strong epidemiological association exists with the carrier state and hepatocellular carcinoma (HCC), which is one of the most common cancers in the world. Therefore, the newly developed HBV vaccines may become the first example of active prevention of human cancer.

HBV consists of a partly double-stranded circular DNA genome which is covered with the core antigen (HBcAg). The capsid is surrounded with an envelope which carries the surface antigen (HBsAg). In addition to complete virus particles the blood of infected patients contains empty envelopes consisting of HBsAg. Due to the absence of cell-culture systems to propagate HBV, the molecular events in the infected cell are not completely understood and the mechanisms which play a role in the malignant transformation of the liver cells are unknown. It is, however, most probable that also in this case the integration of viral DNA to the host cell genome has an important role in the process since integrated sequences are almost always detected in HCC patients positive for HBV markers (for review see Reference 66).

HBsAg appears in the blood of most patients 2 to 8 weeks before the onset of symptoms and remains at detectable levels for about 4 more weeks. Diagnostic tests use detection of HBsAg for screening of blood specimens. The antibody against HBsAG appears during the convalescence and there is often a short period when the antibodies against the HBcAg are the only marker of infection. The e antigen (HBeAg), which is a derivative of HBcAg, is found at the acute phase of the disease and it correlates with the infectivity of the specimens, while the appearance of antibodies against HBeAg are usually a sign of good prognosis. The infectivity, however, can be directly screened by detection of HBV DNA in blood.

In 1982, at least four reports describing hybridization assays for HBV DNA were published.[67-70] It was shown that the test had picogram sensitivity and it detected HBV sequences both in the serum and liver of hepatitis patients. In an analysis of 153 patients at various stages of disease with indetectable HBeAg, Tur-Kaspa et al.[71] detected HBV DNA in 10 patients including 3 with acute hepatitis and 4 asymptomatic carriers. Seven of the 10 HBeAg positive samples tested were also positive by hybridization. Moestrup and co-workers[72] have shown that in acute self-limiting hepatitis, HBV DNA is cleared within a few weeks while in the 43 patients who developed chronic hepatitis, DNA could be detected for more than 6 months in all but one case.

In a material of blood donors in the United Kingdom, Harrison et al.[73] reported that HBV DNA was found in five HBeAg negative serum samples thus indicating that it should be used as a marker for infectivity. Brechot et al.[74] tested 105 serum samples 10 of which were positive for HBV DNA. Six of these had no serological marker for HBV. They also showed that the viral genome was detectable in 52 of 88 HBsAg-negative chronic liver disease patients including 17 of the 20 patients with HCC.

Blum et al.[75] have applied detection of viral antigens by immunohistochemical staining in combination with detection of HBV DNA by *in situ* hybridization. These techniques will be valuable for the detailed characterization of the pathogenesis of HBV infection in tissues. The presence of viral genome in the body is not restricted to liver since HBV genome has been detected in mononuclear blood cells,[76] pancreas, kidney, and skin.[77]

Taken together, these results indicate that detection HBV DNA is not necessary for the diagnosis at the acute phase of infection because immunoassays provide a rapid and simple test for this purpose. However, infectivity can be screened by identifying HBV DNA directly in the specimens and this test is also necessary in some chronic hepatitis and HCC cases which are negative for other viral markers. Therefore, the presence of HBV DNA in the serum should be tested before hepatitis is considered to be caused by non-A non-B hepatitis virus. Since persistence of virus multiplication in the hepatocytes is considered as a factor of poor prognosis, the presence of virus particles can be reliably tested by detecting HBV DNA. It is also a valuable marker in antiviral therapy, because the presently used drugs are only active during virus multiplication.

F. Other Viruses

Spot hybridization assays for the detection of parvovirus DNA in the serum have also been described.[78,79] The sensitivity of the test with a cloned fragment of virus DNA as a probe was 0.5 pg which is equivalent to approximately 10,000 virus particles. This virus does not grow in cell cultures and although the parvovirus specific IgM assay may be more useful in detection of acute infection, the hybridization test is at the moment the only method to screen large numbers of blood specimens (up to 1000 sera per day) for the risk of infectivity. This is necessary especially in cases where the recipient of the blood is suffering from hemolytic anemia.

Flores and co-workers[80] described a hybridization assay for the detection of rotavirus RNS. When a radioactive probe produced by *in vitro* transcription reaction was used, the test had a sensitivity of 8 pg of homologous RNA. In detection of viruses in stool specimens the results were comparable to those obtained by an antigen detection assay. Rotaviruses which have segmented genomes can also be analyzed by Northern blotting.[81] Similarly, Southern blotting analysis of DNA viruses can be used in epidemiological studies.

The infection caused by the human immunodeficiency virus (HIV) is commonly diagnosed by detection of specific antibodies in serum specimens. At the early stages of infection and during the late course of the acquired immunodeficiency syndrome (AIDS), infectious virus isolated in cell cultures or viral antigens detected in blood are the only markers which can be used for diagnosis. Attempts have also been made to use nucleic acid hybridization as a diagnostic tool in HIV sequences, but their use is still restricted to confirmation of virus replication in cell cultures after isolation. The usefulness of nucleic acid probes to directly reveal HIV sequences in clinical samples will most probably require use of efficient *in vitro* amplification systems[82] as is the case with many viruses which are present in proportionally low copy numbers.

III. DETECTION OF BACTERIA BY HYBRIDIZATION

The probes used for studying bacteria by hybridization have included whole bacterial genomes,[83] cryptic plasmids,[84] virulence plasmids,[85] genes encoding enterotoxins[86] or antibiotic resistance enzymes,[87] and chromosomal fragments carrying species specific and usually unknown functions.[88,89] The disadvantage when using cloned chromosomal fragments is that probes which detect unique regions are not as sensitive as probes detecting multicopy plasmids or ribosomal RNA (rRNA). In epidemiological studies the colony hybridization technique using various types of inoculations has been the most sophisticated way to study large numbers of bacterial isolates.[90] A summary of the probes used in detection of different bacteria is presented in Table 2.

Usually, the probes have been labeled using radioisotopes (^{32}P, ^{35}S and ^{125}I), but also nonradioactive probes have been developed.[91-93] There have been difficulties when biotin-labeled probes have been used in colony hybridization, and one explanation for the high background signal found is biotin-like structures in bacterial cells, which bind the avidin-phosphatase complex ad give false-positive signals.[94] These can be avoided using labeling techniques other than biotin-streptavidin-complex. Seriwatana et al. used an alkaline phosphatase-conjugated oligonucleotide probe for the detection of enterotoxigenic *Escherichia coli*.[92] The alkaline phosphatase was covalently linked directly to the C-5 position of a thymidine base through a 12-atom spacer arm.[95] The size of the probe may also be an important factor; using a 17kb biotin-labeled probe, Sethabutr et al.[96] obtained results which were reliable enough for the detection of enterioinvasive *E. coli* and *Shigella* strains by colony hybridization. The amount of bacteria on the filter is also important; Carter et al.[97] used a 998 bp TEM-1 beta-lactamase probe for detection of TEM-1 genes in clinical isolates. An inoculum of 3 to 4 x 10^6 cells gave optimum positive descrimination.

A. Enteric Pathogens

First clinical studies concerning the use of nucleic acid probes in the detection of bacterial pathogens were published in 1980 by Moseley et al.[86] They developed probes for detection of enterotoxigenic *E. coli* that is a major cause of diarrheal disease throughout the world. Detection of these strains was earlier based on the suckling mouse test, rabbit ileal loop test, and guanylate cyclase assay that all are complicated, time consuming, and not suitable for routine purposes. The hybridization technique was found to be a relevant and even more sensitive method and could be used for simultaneous analysis of large numbers of bacterial strains.[98] The investigators used cloned DNA fragments for both heat-labile and heat-stabile enzymes.[86] Later, three different probes and synthetic enzyme-linked oligonucleotide probes were used to improve the results.[99,100] There are now several published papers concerning the use of these probes for the detection of enterotoxigenic *E. coli* strains from water specimens,[98] directly from the stool (the stool blot technique[101]) from different animal sources,[102,103] and in defining epidemiology of this enteric pathogen.[104] The sensitivity of the method is good; a total of 16,000 to 64,000 bacteria or 0.2 to 3% of enterotoxigenic *E coli* in mixed cultures can be detected.[105] Also, an oligonucleotide probe has been developed for the detection of heat-stabile enterotoxin.[106]

Detection and identification of enteric pathogens using DNA probes has been developed also for *Salmonella* spp. [89,107,108] *Vibrio cholerae*,[109] *Yersinia enterocolitica*,[85,110] as well as for *Shigella* spp. and enteroinvasive and hemorrhagic *E. coli*.[111-113] Fitts et al.[89] developed a hybridization test for the presence of *Salmonella* spp. in food products. The probes consisted of chromosomal DNA sequences, which are unique to the genus *Salmonella*. An initial

Table 2
DETECTION OF BACTERIAL DNA SEQUENCES BY HYBRIDIZATION

Organism/function	Sequence used as probes	Ref.
	Enterobacteriaceae	
E. coli		
Enterotoxigenic	Heat-labile and heat-stabile enterotoxin genes	86, 98—106
Enteroinvasive	Plasmid with enteroinvasiviness genes	111, 112
Enterohemorrhagic	Plasmid DNA coding fimbrial antigen	113
Uropathogenic	Gal-Gal-pilus associated chromosomal genes	114
Salmonella typhi	Chromosomal Vi antigen gene	107
Salmonella spp.	Random chromosomal sequences	89, 108
Vibrio cholerae	*E. coli* heat-labile enterotoxin gene	109
Yersinia enterocolitica	Virulence plasmid	85
Shigella spp.	Enteroinvasiveness plasmid genes	111
Enterobacteriaceae	OmpA gene	115
	Anaerobic Bacteria	
Bacteroides spp.	Chromosomal sequences	116
B. fragilis		116
B. thetaiotaomicron		117
B. vulgatus		118
B. uniformis, B. distasonis, B. ovatus		119
	Legionellae and Mycobacteria	
Legionella spp.	DNA probe to the complementary rRNA	120—122
L. pneumophila		123
Mycobacteria	Chromosomal DNA probes	124
M. tuberculosis		
M. avium complex		
M. gordonae		
Mycobacteria	DNA probe to complementary rRNA	125
M. tuberculosis		
M. avium		
M. intracellulare		
	Sexually Transmitted Diseases	
Neisseria gonorrhoeae	Cryptic plasmid	84, 126
	Plasmid coding beta-lactamase	126
Chlamydia trachomatis	Chromosomal and plasmid DNA	127—130
	Other Bacteria	
Mycoplasma	DNA probe to complementary rRNA	131
Pseudomonas aeruginosa	Chromosomal genes	132
P. fluorescens group	cDNA probe for 23S rRNA	133

Table 2 (continued)
DETECTION OF BACTERIAL DNA SEQUENCES BY HYBRIDIZATION

Organism/function	Sequence used as probes	Ref.
Corynebacterium diphteriae	Corynephage	134
Mobiluncus	Chromosomal DNA	135
Listeria monocytogenes	Beta-hemolysin gene	136
Candida spp., *Torulopsis* spp.	Actin protein gene	137
	Antimicrobial Resistance Genes	
Trimethoprim resistance	Trimethoprim resistant dihydrofolate genes, parts of carrying transposon	140—144
Beta-lactamase production	Genes coding beta-lactamase enzymes	146—152
Aminoglycoside resistance genes	Genes coding aminoglycoside modifying enzymes	155
Tetracycline resistance	Genes mediating alteration of cell membrane or transport	161—165
Chloramphenicol resistance	Gene for acetylating enzyme	166

preenrichment step of food samples in nutrient broth was required prior to hybridization. *Salmonella typhi* has been identified using an 8.6 kb viaB probe; this genetic locus is involved in the synthesis of virulence antigen (Vi).[107] As few as 1000 bacterial cells could be detected by this probe and the hybridization reactions were specific. Tompkins et al.[108] used cosmids containing random chromosomal sequences from *S. enteritidis* to differentiate *Salmonella* spp. Kaper et al. hybridized *Vibrio cholerae* strains with an *E. coli* heat-labile enterotoxin gene.[109] All toxigenic *Vibrio cholerae* strains hybridized with the probe used, whereas all nontoxigenic strains failed to demonstrate homology with heat-labile enterotoxin genes.

In the detection of *Yersinia enterocolitica* in food, a pool of fragments of a 44-Mdalton plasmid was used as probe.[85] Most of the virulent cells added to a scallop sample were observed after direct colony hybridization. There was no need for pure cultures or enrichment of food samples. Later, the same research group published a paper where the same probes were used in enumeration of virulent *Y. enterocolitica* in different food products.[110] Depending on the type of food, 66 to 215% of the *Y. enterocolitica* colonies were observed. The indigenious microflora, but not the type of food, had an effect on the enumeration.

A DNA hybridization method for the detection of enteroinvasive *Shigella* and *E. coli* species was developed by Boileau et. al.[111] The same 17 kb probe used in these studies worked also when labeled with biotin.[96] The method was found to be highly specific and sensitive particularly in the characterization of atypical enteroinvasive isolates. For the detection of enteroinvasive strains, two different probes were found to be more effective than the Sereny-test (inoculation of invasive bacteria into the eye of guinea pigs); three Sereny-test negative isolates were found to give a positive hybridization signal.[112]

Urophathogenic *E. coli* cause severe infections because of the presence of virulence factors like hemolysin and adhesins. O'Hanley et al.[114] used hemolysin and Gal-Gal-pilus associated genes in the detection of uropathogenic strains of *E. coli* The probes detected all pyelonephritis strains and all hemolytic strains of *E. coli.*

Enterobacteriaceae can also be detected using a fragment of ompA gene of *E. coli* K 12.[115] This probe did not hybridize with other bacteria, including aerobic and anaerobic species of clinical importance.

B. Anaerobic Bacteria

Conventional procedures to detect anaerobic bacteria in bacteriological laboratories are cumbersome and time-consuming. Detection of anaerobic bacteria from clinical samples needs special care in transportation and certain well-controlled environmental conditions for culture techiques.

DNA hybridization has been used for identification of the most important anaerobic pathogens belonging to the *Bacteroides* species.[116] Kuritza and co-workers cloned HindIII fragments of chromosomal DNA from various *Bacteroides* species. They characterized probes for *Bacteroides fragilis* group and also for *B. fragilis* only.[116] There are also descriptions of probes for *B. thetaiotaomicron.*[117] and *B. vulgatus*[118] that belong to the human colonic flora as well as for *B. uniformis, B. disstasonis,* and *B ovatus.*[119] The methods developed were also used in the enumeration of different *Bacteroides* spp. in fecal samples.[119] All *Bacteroides* spp. except *B ovatus* were found in high numbers in fecal flora.

C. Legionellae and Mycobacteria

Current methods for detection and identification of *Legionella* spp. rely on culture or biochemical procedures as well as on direct immunofluorescent staining. Detection of antigenuria and specific antibody response are also in use.[120] The culture methods require usually several days or even up to 2 weeks to complete. Specimens tested by direct immunofluorescence require well trained personnel; also, the sensitivity of this test varies in different laboratories. Rapid diagnosis is, however, important because the disease may be fatal. DNA probes have been developed for detection of rRNA of *Legionella* spp. either from culture or directly from clinical specimens[121] (Gen-Probe Inc., San Diego, CA). The sensitivity of the test to detect and identify *Legionella* spp. colonies from culture is 98% and specificity 100%.[120] These results have been recently confirmed by Edelstein.[122] A specific probe for *Legionella pneumophila* has also been reported,[123] but use of this probe in screening of suspected *Legionella* colonies is limited because of its genus specificity. The test is rapid, less than 3 h is needed to get positive or negative result. The probes are radiolabeled with ^{125}I, which is a relative disadvantage.

Infections caused by mycobacteria are of growing importance throughout the world because of an increasing number of patients with acquired immunodeficiency syndrome. Most of the mycobacteria grow slowly and require months for isolation. Moreover, identification and susceptibility testing both need an additional time, a minimum of 2 weeks, to complete. Whole chromosomal DNA probes have been used to identify *Mycobacterium tuberculosis* and *Mycobacterium avium* complex.[124] The DNA probe for *M. tuberculosis* was prepared from the closely related *Mycobacterium bovis* (BCG). DNA from three different strains representing the three DNA homology groups in the *M. avium* complex were also used as probes. Identification by the dot-blot method required a maximum of 48 h and identified correctly 93% of cultures grown on solid media. Antibiotic treated cells were also correctly identified. DNA probes complementary to the rRNAs of mycobacterial have also been used as probes.[125] A specific probe for *M. avium* and *M. intracellulare* identified 52 of 56 *M. avium* complex isolates growing on culture media. The probes did not hybridize with any of the *M. tuberculosis* isolates studied. The detection time was only 2 h in this suspension hybridization protocol comparable to that for identification of *Legionella* spp.

D. Sexually Transmitted Diseases

In the detection of *Neisseria gonorrhoeae* in genital specimens a 2.6-Mdalton cryptic plasmid has been successfully used as a probe.[84] However, five of six isolates from patients who were positive by culture but negative by hybridization lacked the gonococcal cryptic plasmid.[84] The sensitivity and specificity of the hybridization method for the detection of *N. gonorrhoeae* is comparable to that with culture isolation.[126] Also, the beta-lactamase production tests have similar sensitivity and specificity when compared to the hybridization method using a 4.4 Mdalton beta-lactamase plasmid.[126]

The genus *Chlamydia* consists of prokaryotic organisms that are obligatory intracellular parasites of eukaryotic cells. *Chlamydia trachomatis* (CT) is currently the most common cause of sexually transmitted diseases in a majority of western countries. The diagnosis of CT is usually carried out by isolation of the microbe in cell cultures or by direct detection of chlamydial antigens in clinical specimens by immunoassays. Strains of *C. psittaci*, which is the second species of the genus, cause respiratory infections (psittacosis).

Hyypiä et al. have used both total DNA isolated from the elementary bodies and a cloned plasmid of the LGV strains for the identification of *Chlamydia trachomatis* in clinical specimens.[127] Also *in situ* hybridization method has been applied for the detection of CT in cervical samples.[128] Since our observations and the results of Palva[129] indicated that the presently used diagnostic systems may fail to identify a significant number of clinical specimens containing chlamydial components, we recently analyzed a material of 135 genital samples by three methods.[130] A total of twenty-one of the specimens were positive by either isolation (16 specimens) or enzyme immunoassay (19 specimens). A positive hybridization signal was observed in 14 cases out of these samples. However, 6 of the specimens that were negative by other tests were positive by hybridization, indicating the need for further studies to analyze these potentially positive specimens and for improvement of diagnostic methods of this clinically important pathogen.

E. Other Bacterial Organisms and Yeasts

Genomic DNA probes for the detection of *Mycoplasma pneumoniae* and *Mycoplasma genitalium* were reported to detect 0.1 ng of specific homologous DNA or 100,000 mycoplasmal colony forming units.[131] A cDNA probe for mycoplasmal rRNA has also been developed by Gen-Probe, Inc., San Diego, CA. The method is 100% sensitive and 100% specific, when mycoplasmas are detected in throat swab specimens from pediatric patients.

An exotoxin gene of *Pseudomonas aeruginosa* has been used in the classification of strains including to this genus.[132] More than 100 different strains of *P. aeruginosa* have been classified using whole bacterial DNA in Southern hybridization. A hybridization probe for the *Pseudomonas fluorescens* group has been also tested; this cDNA probe consists of a fragment of 23S rRNA of *P. aeruginosa* and detects only strains representing the *P. fluorescens* group.[133]

Corynebacterium diphteriae can easily be detected by using corynephage-DNA as a probe.[134] Nontoxigenic strains of *C. diphteriae* did not hybridize with this probe. This test has a potential to replace conventional toxin production tests in epidemiological studies, when need for rapid screening tests to separate toxin-producing strains from nonproducers is apparent.

Mobiluncus is an anaerobic, gram-negative bacillus that has been isolated exclusively from women having bacterial vaginitis.[135] The culture of this bacillus requires an average 37 days to complete, whereas the DNA-probe assay using the whole DNA from *Mobiluncus curtisii* and *Mobiluncus mulieris* needs only 5 days. There was 100% correlation between species identification by the DNA probe and conventional biochemical tests.

Listeria monocytogenes is known to cause meningitis and intrauterine infections. An enzyme called beta-hemolysin plays a role in the virulence of this strain.[136] A 500 bp fragment of beta-hemolysin gene was used to detect hemolytic strains of *L. monocytogenes*. The probe detected only hemolytic *L. monocytogenes* strains among 52 *Listeria* strains from different sources.

Yeasts have usually been identified by staining, colony morphology, and biochemical tests. A DNA probe containing the gene for the protein actin was used successfully in the identification of several medically important yeasts belonging to *Candida* and *Torulopsis* spp.[137] There are, however, no studies published yet where the probes developed have been used for identification of *Candida* spp. in clinical samples.

F. Antimicrobial Resistance Genes

Trimethoprim resistance — Detection of genes mediating antimicrobial resistance are epidemiologically of primary interest. The first probes developed for resistance gene detection were developed to characterize trimethoprim resistance.[87,138] There are two main plasmid-mediated dihydrofolate reductase (DHFR) genes encoding trimethoprim resistance spread in the developed world.[139] Transposon Tn7-mediated DHFR Ia has been detected using several types of probes; the 500 bp HpaI fragment containing the structural gene is the most commonly used one,[140,141] but also other parts of the transposon have been used. [142] DHFR II is mediated by three genes.[139] However, there are not yet reports on specific DHFR II probes; all probes used have contained surrounding vector areas of DHFR II carrying sequences. The intragenic probe might be the best possible choice for detection of genes for DHFR II.[143] In the developing world, different mechanisms are spread; Sundström et al.[144] described the DHFR V to be widely spread in Sri Lanka. In general, more studies are needed to establish the validity of these probes and the frequency of different transferable trimethoprim resistance genes among clinical pathogens.

Beta-lactamase genes — Beta-lactum antibiotics, like penicillins and cephalosporins, are hydrolyzed by more than 25 different plasmid-encoded beta-lactamase enzymes produced by gram-negative bacteria.[145] TEM-1 is the most widely spread one,[145] and produced, e.g., by the well-known artificial pBR322 plasmid. Several different probes to detect the TEM-1 gene have been used.[146-149] We found the intragenic TEM-1 probe to be 100% sensitive and specific in the detection of this gene in clinical bacteria.[150] Jouvenot et al.[151] used a 1 kb TEM-1 probe with about 200 bp extragenic structures and obtained 8 false-negative results and 16 false-positive results by DNA hybridization among 328 ampicillin resistant *E. coli* strains studied. The probes for OXA-1, OXA-2, PSE-1, PSE-2, PSE-4, and SHV-1 plasmid-mediated beta-lactamases have also been published.[147,148,150,152] It has been shown clearly that intragenic probes are essential to specific diagnosis of beta-lactamase genes.[150] However, there are still findings of so called "silent resistance genes" that hybridize with specific probes (e.g., an intragenic OXA-2 probe) but do not produce the corresponding beta-lactamase. More studies are needed to establish the importance of this finding.

Aminoglycoside resistance genes — Adenylation, phosphorylation, and acetylation are the most common bacterial resistance mechanisms to the aminoglycoside class of antibiotics.[153] The enzymes can be determined by an indirect method using a MIC pattern analysis to different aminoglycoside substrates[153] or with an phosphocellulose paper binding assay.[154] The former is a rapid and convenient method, but it is not reliable enough to detect several different mechanisms at the same time in a single isolate. The latter is time consuming and may fail to detect low levels of enzymes produced.[155] Several different probes have been developed to detect these resistance mechanisms.[156-158] Probes for phosphorylation genes APH(3′)-I and APH(3′)-II identified accurately these two genes.[156] Several different genes are found to produce ANT(2″) adenylator enzyme.[157,158] The diversity of genes producing enzymes with the same function seems to be more a problem in the detection of aminoglycoside resistance genes than in the detection of other resistance genes.

Other resistance genes — There are at least eight different tetracyline resistance determinants, which can be differentiated. From Gram-negative bacteria four different groups were defined, one of which was found to be the most frequent among fecal coliforms and also among Haemophilus influenzae strains.[159-161] For Gram-positive bacteria, like streptococci, three different tetracycline resistance determinants have also been defined.[162] One of these is found to be present also in *Mycoplasma* spp.[163] In *Campylobacter* spp. determinants that show no homology to resistance genes reported earlier have also been found.[164,165]

Probes have also been developed for detection of chloramphenicol resistance genes,[166] as well as for genes encoding resistance to the macrolide-lincosamine group of antibiotics.[167]

IV. DETECTION OF PARASITES BY HYBRIDIZATION

The parasitic infections described here represent the most important protozoan diseases of the world and one of the most important group of helminth infections — the filarioses. Malaria, leishmaniasis, trypanosomiasis (African sleeping sickness and Chagas disease) and two types of filarioses (blood and skin) have all been given special priorities by the World Health Organization in research and control programs.

The protozoans are highly complex, unicellular eukaryotes. They also have complex life-cycles, including at least two hosts. Their genomes range in size from 1 to several x 10^7 bp, and are divided into many chromosomes, the exact number of which is usually not known. Most diagnostic probes are directed against highly repeated, satellite-alike chromosal sequences. In the kinetoplastids (leishmania and trypanosomes) there are also high copy numbers of extrachromosomal DNA that can be used as targets for detection. The helminths, being multinuclear, yield more DNA per organism, thus offering prospects to use less abundant sequences as targets.

In some diseases, like leishmaniasis, species and subspecies differentiation is only now becoming feasible when the diagnosis is brought to the genetic level. For other diseases, like malaria for which other diagnostic means are available, nucleic acid hybridization will allow surveys and control programs to be expanded with new possibilities for mass screening.

A. Malaria

Human malaria, caused by the four *plasmodium* species (*P. falciparum, P. malariae, R. vivax,* and *P. ovale*), has until recently been diagnosed exclusively by microscopic detection of parasites in blood smears. Microscopical diagnosis of malaria plays a crucial part in epidemiological investigations and species identification, and assessment of parasite density institutes important criteria for treatment. The disadvantages of microscopy are that the analysis requires a skilled microscopist and is time consuming. Routine examination of a thick blood film takes 5 to 10 min or longer, if low level parasitemias are to be detected reproducibly. This clearly limits the size of a population that can be studied in a malaria survey. The recent development of DNA probes for diagnosis of malaria is therefore an important achievement.

A highly repetitive sequence from the *P. falciparum* genome was first characterized in 1984.[168] It was shown to consist of 21 basepairs long, tandemly repeated motifs, that show some degeneracy (Figure 4). The repeats are organized in large clusters, are represented on most or all chromosomes, and the copy number has been estimated to be 10^4 to 10^5 per organism, thus comprising 1 to 10% of the *P. falciparum* genome.[169,170] It has been detected in all clinical isolates so far tested, numbering several thousands from all continents. Only small differences in amount or distribution of the repeat has been described.[171,172]

No cross-hybridization to other human malarias has been found. A large number of samples has been studied by the dot blot technique, usually with a ^{32}P-labeled probe, and visualization by autoradiography. In a field trial in the Gambia,[173] about 400 samples were analyzed (Table 3). In an initial trial the hybridization assay had a specificity for *P. falciparum* of 100% and a sensitivity of 68%. False-negative results were obtained only on samples with low parasitemia. Assay of red cells collected during an earlier malaria survey, which had been stored for 1 year at —20°C gave a higher level of sensitivity (85%), suggesting a beneficial effect from freezing and thawing. This was confirmed by examining in the same assay red cells processed immediately after collection and after 2 weeks of storage at — 20°C. Freezing and thawing gave a 21% increase in positivity and a sensitivity of 100% was achieved with the frozen samples.

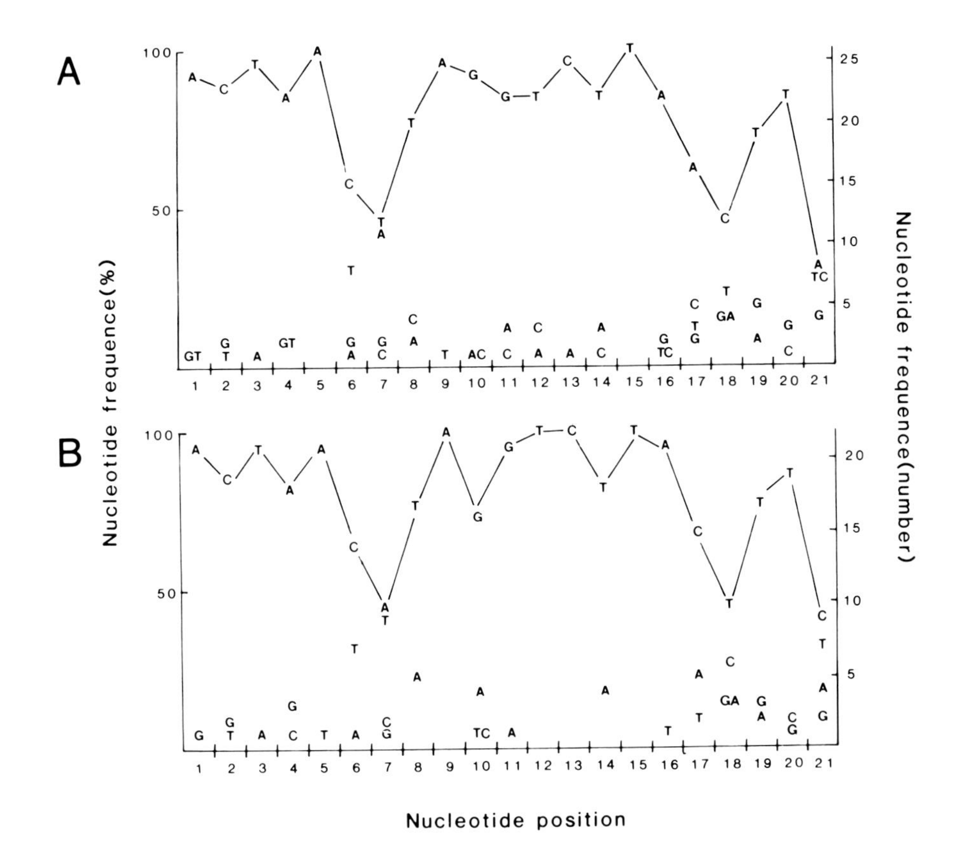

FIGURE 4. A diagrammatic representation of the degree of conservation in different positions of two 21 base pair (A) pRepHind and (B) pRebHpa repetitive clones isolated from *P. falciparum* DNA. (From Åslund, L., Franzen, L., Westin, G., Persson, T., Wigzell, H., and Pettersson, U., *J. Mol. Biol.*, 185, 509, 1985. With permission.)

Table 3
SENSITIVITY OF THE DNA HYBRIDIZATION ASSAY IN RELATION TO LEVELS OF *P. FALCIPARUM* PARASITEMIA

Parasite density		Katchang (fresh samples)			Sarakunda (stored samples)		
		Number of positive		Sensitivity of hybridization assay (%)	Number of positive		Sensitivity of hybridization assay (%)
No. per HPF	per μl[a]	Microscopy	Probe		Microscopy	Probe	
≥10	≥5000	20	20	100	20	20	100
1—9	500—4999	23	23	100	14	14	100
0.1—0.9	51—499	40	22	55	20	19	95
≤0.1	≤50	22	3	14	27	16	59
Total		105	68	65	81	69	85

[a] Approximate values.

From Holmberg, M., Shenton, F., Franzen, L., Janneh, K., Snow, R., Petterson, U., Wigzell, H., and Greenwood, B. M., *Am. J. Trop. Med. Hyg.*, 37, 230, 1987. With permission.

Genomic probes, plasmid probes, *in vitro* synthesized RNA probes, and synthetic oligonucleotides have all been used in these assays.[174-176] In one study by McLaughlin et al.,[175] it was shown that a plasmid probe was more sensitive than a synthetic oligonucleotide, but an amplification of the signal was obtained by sequential hybridization with the plasmid probe followed by hybridization with the oligonucleotide. The reason for this might be either that the oligonucleotide hybridizes with target *P. falciparum* sequences that have lost hybridization plasmid sequences during washing, or because the synthetic DNA probe hybridizes to 21 bp sequences, present on the tails of the long hybridized plasmid probes. An enzyme-conjugated synthetic oligonucleotide with the 21 repeated sequence has also been evaluated. In a small study the sensitivity of this probe was shown to equal that of ^{32}P-labeled probes.[177]

The sensitivities in several studies have ranged between 10^2 to 10^4 parasites detected, which approximately equals that of microscopic examination. The main problem in some studies has been a relatively high number of false-positive results.[196] This seems to depend on the degree of purification of the sample DNA before hybridization, and thus most likely represents a nonspecific reaction. New techniques in sample treatment and assay methods are presently being applied to solve this problem.

Two other plasmodial repeats have been described. One is a 5.8 kb long, transcribed but untranslated repeat, originally isolated from *P. yoelli*. It appears to be present in all plasmodial species so far tested. The copy number can vary between species from 50 to 150 and it thus comprises 1 to 3% of the parasite genomes.[178] The other is a telomeric repeat of 7 bp reiterated at least 70-fold. It was first isolated from *P. berghei*.[179] A 1.4 kb long fragment containing these repeats was shown to hybridize to all resolved chromosomes of *P. yoelli*, *P. chabaudi*, and *P. falciparum*. On the other hand, a cloned subtelomeric repeat from *P. berghei* was specific for this species. Both sequences, therefore have a potential for diagnostic use as broad-reacting plasmodial probes.

The other human malarias, *P. vivax*, *P. malariae* and *P. ovale* have not been studied as extensively. Although likely to contain species-specific repeated elements, none have yet been isolated. There is clearly a need for species differentiating probes in malaria diagnosis.

B. Babesiosis

Babesiosis is a hemoparasitic disease afflicting several animal species; it is caused by different babesia species, and is transmitted by ticks. *B. bovis* infecting cattle is a major problem. Sporadic human cases of *B. microti* and *B. divergens* have been described. Diagnosis using light microscopy of stained blood films, as in malaria, is a reliable but tedious procedure, when many samples are processed. Recently, a *B. bovis*-specific probe has been described, probably containing repeat elements.[180] In dot blot assays this probe is sensitive enough to detect most clinical cases.

C. Leishmaniasis

Leishmaniasis is estimated to infect at least 20 million people, with 400,000 new cases per year. *Leishmania* species infectious for man, cause a wide spectrum of clinical symptoms (Table 4). An early cutaneous lesion is commonly caused by most leishmanial species. However, the disease will later take different courses, depending on the specific species. Early diagnosis can thus be important for correct treatment. For the epidemiologist, incrimination of reservoir host and detection of vectors present similar diagnostic difficulties. Diagnosis is made mainly by identifying the parasite in infected tissue by direct microscopy of stained smears and by culture of passage through laboratory animals. These methods have several limitations; they are unable to distinguish subspecies, and culture or passage through animals is time consuming.

A special feature of kinetoplastids, including trypanosomes and leishmanias, is the ki-

Table 4
LEISHMANIA SPECIES AND FORMS OF HUMAN LEISHMANIASIS

Old world forms

L. major, "wet" cutaneous; widespread in rural areas of Asia and Africa.
L. tropica, "dry" cutaneous, uncommon; urban areas of Europe, Asia and North Africa.
L. aethipica cutaneous, often diffuse; Ethiopia and Kenya
L. donovani donovani, visceral (Kala Azar); Africa and Asia
L. donovani archibaldi, visceral; Kenya
L. donovani sinensis, visceral; China
L. donovani infantum, infantile, visceral; Mediterranean region

New world forms

L. donovani chagasi, visceral; South America
L. braziliensis braziliensis, mucocutaneous (Espundia); South America, especially Brazil
L. braziliensis guyanensis, cutaneous; South America
L. braziliensis panamensis, mucocutaneous; South and Central America
L. mexicana mexicana, cutaneous; South and Central America
L. mexicana amazonensis, cutaneous; South and Central America
L. peruviana, cutaneous; South America, mainly Andrean region

netoplast. The kinetoplast DNA (kDNA) consists of about 50 maxi-circles, which code for mitocondrial genes, and approximately 10,000 to 20,000 highly reiterated mini-circle sequences of 500 to 2500 base pairs. Mini-circle sequences show great divergence between species. Unlike trypanosomes, which have up to 300 sequence classes of mini-circles, leishmanias have a limited number of major sequence classes, which are suitable targets for subspecies diagnosis. This type of differentiation has been obtained by Southern blotting of kDNA, by dot blot of parasite DNA fixed to filters, and by *in situ* hybridization. The DNA diagnosis of leishmaniasis has recently been reviewed by Barker.[181]

By using whole kDNA as a probe, discrimination of the main complexes *L. major, L. tropica, L. donovani, L. mexicana,* and *L. braziliensis* can be made, with some exceptions.[182] Cloned mini-circle sequences can also be used to allow greater discrimination within and between members of the main complexes. Even single schistodemes or geographical isolates have been distinguished with some probes.[183] Probes derived from chromosome sequences have also been used to analyze karyotypic variation between species and isolates.[181]

D. Trypanosomiasis

A satellite fraction of *T. brucei* DNA has been found by centrifugation in sucrose gradients.[184] It is made up of 177 bp long tandem repeats and has basically the same features as satellite DNA from higher eukaryotes; it is highly repetitive, shows a limited sequence divergence within one organism, but with an extreme variation between species (*T. brucei* and *T. cruzi*) and it is located in nuclear DNA. Finally, it has not been shown to be transcribed.

The 177 bp repeat bears no resemblance to kDNA, it is instead located on the minichromosomes.[185] Further studies will show if this repeat can be used to distinguish trypanosomes that are infective to man (the *T. brucei gambiense* and *T. brucei rhodesiense* variants) from those that are not. With a similar repeat, isolated from a strain of the subgenus *Nannomonas,* a trypanosome pathogenic to domestic animals in Africa, two genetically distinct groups of *Trypanosoma (Nannomonas) congolense* have been distinguished.[186]

Although the South American trypanosome *T. cruzi* is placed in the same genus as *T. brucei*, it belongs to a different subgenus *(Schizotrypanum)* and shows a very different behavior in the mammalian host. *T. cruzi* has been shown to contain a satellite DNA that differs totally from the 177 bp repeat of *T. brucei.*[184,187] It has a repeat length of 195 bp and is the most abundant repetitive DNA in this organism, with relatively little sequence heterogeneity. The copy number has been estimated to be 120,000 per genome. No transcripts have been detected although the repeats surprisingly contain long open translational reading frames. The repeats seem to be located on minichromosomes as in *T. brucei*. The organization of trypanosomal repetitive elements on minichromosomes may thus be a general feature of these parasites.

In a dot blot assay, as little as 10 parasite genomes have been detected.[187] Several different *T. cruzi* strains were positive, while there was not cross-hybridization to other related protozoans, with the exception of a weak signal from *Leptomonas collosoma*. Whereas *T. cruzi* kDNA has been shown to display considerable strain-specific variation,[188] the 195 bp satellite-like repeat may be more suitable for diagnostic detection of *T. cruzi*.

E. Filarial Diseases

A major concern in epidemiological studies of filariasis is the correct identification of filarial species and subspecies. The genome of *Brugian filariids* contains families of highly repeated DNA sequences. In *B. malayi,* a 320 bp repeat has been identified.[189] The copy number is estimated to be 3 x 10^4, constituting about 12% of the genome. Sequence analysis has shown a high degree of homology of the repeat between *B. malay* and the cat species *B. pahangi,* but not cross-hybridization to other common filariids. In addition *B. malayi* also contains highly repeated sequences unique to this species.[190] Hybridization experiments show that a probe (pBm15) containing the *B. malayi* specific sequence can detect 150 pg of *B. malayi* DNA, which is equivalent to the DNA content of half a microfilaria. Only a low level of cross-hybridization to *B. pahangi* was seen. The pBm15 probe has also been used to detect infective larvae in mosquitoes.[190]

Oncocerciasis, or river blindness is caused by the parasite nematode *Oncocerca volvulus*. Epidemiological studies have shown differences in pathogenicity between the savanna and the forest forms of oncocerciasis, the savanna form being the more likely to cause blindness. No basic morphological differences between these two forms have been found, but isoenzyme differentiation can be performed.[191] Oncocerciasis is currently diagnosed by the presence of subcutaneous nodules containing adult worms, and/or the detection of microfiariae emerging from skin snips after *in vitro* incubation. Oncocerca larvae are detected in the black vector fly (*Simulium* spp.) by microscopy.

Recently, some probes of diagnostic interest have been described. Two repeated sequences of different specificities have been found, both with a copy number of 100 to 300 or more.[192] One clone (pOV26) is specific for filariae of the genus *Oncocerca;* the other (pOV8) hybridizes to DNA from all parasitic filarial worms tested, including *B. malayi* and *W. bancrofti*. Another recombinant plasmid, containing a two part tandem repeat of 148 and 149 bp sequences respectively has been isolated.[193] It is also *Oncocerca*-specific and seems to differ from the POV8 clone in genomic organization. A sequence, unique to the forest form (and intermediate form) has also been found.[194] All mentioned probes are able to detect as little as the DNA content of one infective larva in a dot blot assay.

V. CONCLUSIONS AND FUTURE PROSPECTS

Nucleic acid hybridization is an extremely powerful technique and the possible applications in basic biomedical research are nearly limitless. Many of the advantages associated with hybridization methods can in various ways be utilized in diagnostic procedures. A major advantage is the enormous specificity of the method, perhaps best illustrated by the fact that a short oligonucleotide, comprising some 20 bases, is capable of finding its position in the mammalian genome, consisting of some 3 x 10^9 nucleotide pairs. The specificity can obviously be taken advantage of in diagnostic procedures as it allows specific identification of microorganisms which are extremely similar. At least in theory a difference in only a few base pairs should be detectable by hybridization. At the same time, it is also possible to construct broad-reacting probes by using long segments of DNA which will ignore regional differences in nucleotide sequence.

Molecular cloning facilitates probe production as large quantities of homogeneous DNA can be manufactured at a moderate cost. Moreover, recently developed methods for production of synthetic oligonucleotides enable the production of highly specific probes. Another advantage, particularly in comparison to the isolation of viruses and some bacteria, is the fact that DNA is a comparatively stable molecule which in most instances can survive in a clinical sample without any particular precautions. The samples can then be tested in an inactivated form which reduces the risk of infections among laboratory personnel.

A useful property of hybridization methods is that they allow for screening of large numbers of samples, obviously a useful property in connection with epidemiological studies. Perhaps the most significant advantage with hybridization technology is associated with the fact that nucleic acid probes offer possibilities to identify properties at the genome level which governs the phenotype of the microorganism, i.e., for instance, pathogenicity, drug resistance, etc.

There are unfortunately several disadvantages associated with probe diagnostics. When performed as a dot blot assay the method is rather cumbersome and time consuming. To asssure high sensitivity, it is usually required to expose an autoradiographic film overnight. Another disadvantage is associated with the use of radioisotopes for detection. ^{32}P is usually preferred to achieve maximum sensitivity. This isotope is unpleasant to work with and has a short half-life, thus requiring continuous production of labeled probes. In recent years, several methods have been introduced which are based on non-radioactive detection of nucleic acids. Such probes are easier to handle, are stable, and can be produced in large quantities to be utilized over a long period of time. However, the sensitivity obtained is usually lower than that obtained with radioactive probes.

Despite their many advantages, hybridization methods are not yet very widely used in routine diagnostic work. The reason is that excellent immunological procedures, based on enzyme immunoassays, radioimmunoassays, immunofluorescence, etc. are already available. The laboratory personnel are usually trained to handle these methods and additional training would be required to switch to hybridization methods. There is, thus, in most cases no need to introduce hybridization methods as an alternative as they usually offer no significant advantages in practical terms. There are, however, a few notable exceptions.

Conventional methods are obviously of no use for detection of microorganisms which produce little or no protein, which is necessary for immune detection. A striking example is provided by the papillomaviruses. Cervical tumors frequently harbor viral DNA in the absence of any significant amount of viral protein. There is then no alternative to the use of hybridization for detection of viral genomes. For other agents, like cytomegaloviruses, cultivation is tedious and weeks may be required to provide a diagnostic answer. Direct detection of viral nucleic acids is then an interesting alternative. In cases, such as chronic hepatitis B virus infections, it may be of clinical relevance to demonstrate that viral genomes are present and not only noninfectious envelopes.

In clinical bacteriology there is a need for more rapid diagnosis because specific treatment is usually available. The advantages of nucleic acid hybridization techniques over conventional bacteriology are in many cases apparent. Rapid, specific diagnosis directly from the sample is the primary aim of developmental work with nucleic acid probes in diagnostic bacteriology. Use of nucleic acid probes offer a great advantage also in epidemiological analysis of infectious agents or in the determination of certain properties of bacterial isolates. Screening tests to determine frequency of resistance genes (e.g., trimethoprim) have been earlier practically impossible to obtain. However, there are not yet good approaches for using probes in the routine susceptibility testing of bacterial isolates available because of the diversity of genes responsible for resistance mechanisms and presence of nonspecific resistance mechanisms.

The need for techniques for mass screening and diagnosis of the main protozoan and helminth diseases makes nucleic acid hybridization assays potentially very useful. The results have been promising, and progress in this field has been rapid. However, there is an urgent need for new probes capable of better species and subspecies differentiation. The implementation of these techniques in monitoring large scale control programs is currently feasible, where samples can be sent to well equipped laboratories to perform the assays. There is still much to be accomplished before these techniques can be applied in routine diagnosis. The assays need to be simplified, before these tests can have a significant impact on the clinical field. Hybridization methods are likely to find wide use in the diagnosis and epidemiological studies of parasitic infections as the currently used methods, i.e., microscopy, require highly skilled personnel and are time consuming.

In order to become a practical and useful diagnostic alternative, development of sensitive nonisotopic probes is urgently needed. Introduction of commercial kits will obviously be beneficial and the possible introduction of methods for amplification of nucleic acids in test tubes[82] may ultimately make hybridization a superior diagnostic method for all areas of infectious diseases.

ACKNOWLEDGMENTS

The authors thank Professor Pekka Halonen and Dr. Maurice Harmon for their comments and Ms. Taina Kivelä for secretarial help during the preparation of the manuscript.

REFERENCES

1. **Palva, A.,** Microbial diagnostics by nucleic acid hybridization. *Ann. Clin. Res.,* 18, 327, 1986.
2. **Zwadyk, P., Jr. and Cooksey, R. C.,** Nucleic acid probes in clinical microbiology, *Crit. Rev. Clin. Lab., Sci.,* 25, 71, 1987.
3. **Hyypiä, T.,** Detection of adenovirus in nasopharyngeal specimens by radioactive and nonradioactive DNA probes, *J. Clin. Microbiol.,* 21, 730, 1985.
4. **Stålhandske, P., Hyypiä, T., Allard, A., Halonen, P., and Pettersson, U.,** Detection of adenoviruses in stool specimens by nucleic acid spot hybridization, *J. Med. Virol.,* 16, 213, 1985.
5. **Kidd., A. H., Harley, E. H., and Erasmus, M. J.,** Specific detection and typing of adenovirus types 40 and 41 in stool specimens by dot-blot hybridization, *J. Clin. Microbiol.,* 22, 934, 1985.
6. **Takiff, H. E., Seidlin, M., Krause, P., Rooney, J., Brandt, C., Rodriquez, W., Yolken, R., and Straus, S. E.,** Detection of enteric adenoviruses by dot-blot hybridization using a molecularly cloned viral DNA probe, *J. Med. Virol.,* 16, 107, 1985.
7. **Schuster, V., Matz, B., Wiegand, H., Traub, B., and Neumann-Haeflin, D.,** Detection of herpes simplex virus and adenovirus DNA by dot-blot hybridization using in vitro synthesized RNA transcripts., *J. Virol. Methods,* 13, 291, 1986.
8. **Niel, C., Gomes, S. A., Leite, J. P. G., and Pereira, H. G.,** Direct detection and differentiation of fastidious and nonfastidious adenoviruses in stools by using a specific nonradioactive probe, *J. Clin. Microbiol.,* 24, 785, 1986.
9. **Hammond, G., Hannan, C., Yeh, T., Fischer, K., Mauthe, G., and Straus, S. E.,** DNA hybridization for diagnosis of enteric adenovirus infection from directly spotted human fecal specimens, *J.Clin. Microbiol.,* 25, 1881, 1987.
10. **Virtanen, M., Palva, A., Laaksonen, M., Halonen, P., Söderlund, H., and Ranki, M.,** Novel test for rapid viral diagnosis: detection of adenovirus in nasopharyngeal mucus aspirates by means of nucleic-acid sandwich hybrization, *Lancet,* i, 381, 1983.
11. **Dahlen, P., Hurskainen, P., Lövgren, T., and Hyypiä, T.,** Time-resolved fluorometry for the identification of viral DNA in clinical specimens, *J. Clin. Microbiol.,* in press.
12. **Hyypiä, T., Stålhandske, P., Vainionpää, R., and Pettersson, U.,** Detection of enteroviruses by spot hybridization, *J. Clin. Microbiol.,* 19, 436, 1984.

13. **Hyypiä, T., Maaronen, M., Auvinen, P., Stålhandske, P., Pettersson, U., Stanway, G., Hughes, P., Ryan, M., Almond, J., Stenvik, M., and Hovi, T.**, Nucleic acid sequence relationships between enterovirus serotypes, *Mol. Cell. Probes*, 1, 169, 1987.
14. **Rotbart, H. A., Levin, M. J., and Villarreal, L. P.**, Use of subgenomic poliovirus DNA hybridization probes to detect the major subgroups of enteroviruses, *J. Clin. Microbiol.*, 20, 1105, 1984.
15. **Rotbart, H. A., Levin, M. J., Villarreal, L. P., Tracy, S. M., Semler, B. L., and Wimmer, E.**, Factors affecting the detection of enteroviruses in cerebrospinal fluid with coxsackievirus B3 and poliovirus 1 cDNA probes, *J. Clin. Microbiol.*, 22, 220, 1985.
16. **Tracy, S.**, A comparison of genomic homologies among the coxsackievirus B group: use of fragments of the cloned coxsackievirus B3 genome as probes, *J. Gen. Virol.*, 65, 2167, 1984.
17. **Jansen, R. W., Newbold, J. E., and Lemon, S. M.**, Combined immunoaffinity cDNA-RNA hybridization assay for detection of hepatitis A virus in clinical specimens, *J. Clin. Microbiol.*, 22, 984, 1985.
18. **Jiang, X., Estes, M. K., Metcalf, T. G., and Melnick, J. L.**, Detection of hepatitis A virus in seeded estuarine samples by hybridization with cDNA probes, *Appl. Environ. Microbiol.*, 52, 711, 1986.
19. **Ticehurst, J. R., Feinstone, S. M., Chestnut, T., Tassopoulos, N. C., Popper, H., and Purcell, R. H.**, Detection of hepatitis A virus by extraction of viral RNA and molecular hybridization, *J. Clin. Microbiol.*, 25, 1822, 1987.
20. **Al-Nakib, W., Stanway, G., Forsyth, M., Hughes, P. J., Almond, J. W., and Tyrrel, D. A. J.**, Detection of human rhinoviruses and their molecular relationship using cDNA probes, *J. Med. Virol.*, 20, 289, 1986.
21. **Bowles, N. E., Richardson, P. J., Olsen, E. G. J., and Archard, L. C.**, Detection of coxsackie-B-virus-specific RNA sequences in myocardial biopsy samples from patients with myocarditis and dilated cardiomyopathy, *Lancet*, i, 1120, 1986.
22. **Bowles, N. E., Dubowitz, V., Sewry, C. A., and Archard, L. C.**, Dermatomyositis, polymyositis, and coxsackie-B-virus infection, *Lancet*, i, 1004, 1987.
23. **Kandolf, R., Ameis, D., Kirschner, P., Canu, A., and Hofschneider, P. H.**, *In situ* detection of enteroviral genomes in myocardial cells by nucleic acid hybridization: an approach to the diagnosis of viral heart disease, *Proc. Natl. Acad. Sci. U.S.A.*, 84, 6272, 1987.
24. **Brandsma, J., and Miller, G.**, Nucleic acid spot hybridization: rapid quantitative screening of lymphoid cell lines for Epstein-Barr viral DNA, *Proc. Natl. Acad. Sci. U.S.A.*, 77, 6851, 1980.
25. **Andiman, W., Gradoville, L., Heston, L., Neydorff, R., Savage, M. E., Kitchingman, G., Shedd, D., and Miller, G.**, Use of cloned probes to detect Epstein-Barr viral DNA in tissues of patients with neoplastic and lymphoproliferative diseases, *J. Infect. Dis.*, 148, 967, 1983.
26. **Lung, M. L., Lam, W. K., So., S. Y., Lam, W. P., Chan, K. H., and Ng, M. H.**, Evidence that respiratory tract is major reservoir for Epstein-Barr virus, *Lancet*, i, 889, 1985.
27. **Diatz-Mitoma, F., Preiksaitis, J. K., Leung, W. C., and Tyrrell, D. L. J.**, DNA-DNA dot hybridization to detect Epstein-Barr virus in throat washings, *J. Infect. Dis.*, 155, 297, 1987.
28. **Ambinder, R. F., Wingard, J. R., Burns, W. H., Hayward, S. D., Saral, R., Perry, H. R., Santos, G. W., and Hayward, G. S.**, Detection of Epstein-Barr virus DNA in mouthwashes by hybridization, *J. Clin. Microbiol.*, 21, 353, 1985.
29. **Seidlin, M., Takiff, H. E., Smith, H. A., Hay, J., and Straus, S. E.**, Detection of varicella-zoster virus by dot-blot hybridization using a molecularly cloned viral DNA probe, *J. Med. Virol.*, 13, 53, 1984.
30. **Hyman, R. W., Ecker, J. R., and Tenser, R. B.**, Varicella-zoster virus RNA in human trigeminal ganglia, *Lancet*, ii, 814, 1983.
31. **Gilden, D. H., Vafai, A., Shtram, Y., Becker, Y., Devlin, M., and Wellish, M.**, Varicella-zoster virus DNA in human sensory ganglia, *Nature (London)*, 306, 478, 1983.
32. **Chou, S. and Merigan, T. C.**, Rapid detection and quantitation of human cytomegalovirus in urine through DNA hybridization, *N. Engl. J. Med.*, 308, 921, 1983.
33. **Spector, S. A., Rua, J. A., Spector, D. H., and McMillan, R.** Detection of human cytomegalovirus in clinical specimens by DNA-DNA hybridization, *J.Infect. Dis.*, 150, 121, 1984.
34. **Virtanen, M., Syvänen, A.-C., Oram, J., Söderlund, H., and Ranki, M.**, Cytomegalovirus in urine: detection of viral DNA by sandwich hybridization, *J. Clin. Microbiol.*, 20, 1083, 1984.
35. **Schuster, V., Matz, B., Wiegand, H., Traub, B., Kampa, D., and Neumann-Haefelin, D.**, Detection of human cytomegalovirus in urine by DNA-DNA and RNA-DNA hybridization, *J. Infect. Dis.*, 154, 309, 1986.
36. **Vonsover, A., Gotlieb-Stematsky, T., Sayar, Y., Bardov, L., Manor, Y., and Siegal, B.**, Detection of CMV in urine: comparison between DNA-DNA hybridization, virus isolation, and immunoelectron microscopy, *J. Virol. Methods*, 16, 29, 1987.
37. **Augustin, S., Popow-Kraupp, T., Heinz, F. X., and Kunz, C.**, Problems in detection of cytomegalovirus in urine samples by dot blot hybridization, *J. Clin. Microbiol.*, 25, 1973, 1987.
38. **Buffone, G. J., Schimbor, C. M., Demmier, G. J., Wilson, D. R., and Darlington, G. J.**, Detection of cytomegalovirus in urine by nonisotopic DNA hybridization, *J. Infect. Dis.*, 154, 163, 1986.

39. **Lurain, N. S., Thompson, K. D., and Farrand, S. K.,** Rapid detection of cytomegalovirus in clinical specimens by using biotinylated DNA probes and analysis of cross-reactivity with herpes simplex virus, *J. Clin. Microbiol.*, 24, 724, 1986.
40. **Martin, D. C., Katzenstein, D. A., Yu, G. S. M., and Jordan, M. C.,** Cytomegalovirus viremia detected by molecular hybridization and electron microscopy, *Ann. Intern. Med.*, 100, 222, 1984.
41. **Churchill, M.A., Zaia, J. A., Forman, S. J., Sheibani, K., Azumi, N., and Blume, K. G.,** Quantitation of human cytomegalovirus DNA in lungs from bone marrow transplant recipients with interstitial pneumonia, *J. Infect. Dis.*, 155, 501, 1987.
42. **Redfield, D. C., Richman, D. D., Albanil, S., Oxman, M. N., and Wahl, G. M.,** Detection of herpes simplex virus in clinical specimens by DNA hybridization, *Diag. Microbiol. Infec. Dis.*, 1, 117, 1983.
43. **Forghani, B., Dupuis, K. W., and Schmidt, N. J.,** Rapid detection of herpes simplex virus DNA in human brain tissue by in situ hybridization, *J. Clin. Microbiol.*, 22, 656, 1985.
44. **Fung, J. C., Shanley, J., and Tilton, R. C.,** Comparison of the detection of herpes simplex virus in direct clinical specimens with herpes simplex virus-specific DNA probes and monoclonal antibodies, *J. Clin. Microbiol.*, 22, 748, 1985.
45. **Gadler, H., Larsson, A., and Sølver, E.,** Nucleic acid hybridization, a method to determine effects of antiviral compounds on herpes simplex virus type 1 DNA synthesis, *Antiviral Res.*, 4, 63, 1984.
46. **Schulte-Holthausen, H. and Schneveis, K. E.,** Differentiation of herpes simplex virus serotypes 1 and 2 by DNA-DNA-hybridization, *Med. Microbiol., Immunol.*, 161, 279, 1975.
47. **Brautigam, A. R., Richman, D. D., and Oxman, M. N.,** Rapid typing of herpes simplex virus isolates by deoxyribonucleic acid: deozyribonucleic acid hybridization, *J. Clin. Microbiol.*, 12, 226, 1980.
48. **Stålhandske, P. and Pettersson, U.,** Identification of DNA viruses by membrane filter hybridization, *J. Clin. Microbiol.*, 15, 744, 1982.
49. **Peterson, E. M., Aarnaes, S. L., Bryan, R. N., Ruth, J. L., and de la Maza, L. M.,** Typing of herpes simplex virus with synthetic DNA probes, *J. Infect. Dis.*, 153, 757, 1986.
50. **zur Hausen, J. and Schneider, A.,** The role of papillomaviruses in human anogenital cancer, in *The Papovaviridae,* Vol. 2, *The Papillomaviruses,* Slazmann, N. P. and Howley, P. M., Eds., Plenum Press, New York, 1987, 245.
51. **Parkkinen, S., Mäntyjärvi, R., Syrjänen, K., and Ranki, M.,** Detection of human papillomavirus DNA by the nucleic acid sandwich hybridization method from cervical scraping, *J. Med. Virol.*, 20, 279, 1986.
52. **Wagner, D., Ikenberg, H., Boehm, N., and Gissman, L.,** Identification of human paillomavirus in cervical swabs by deoxyribonucleic acid in situ hybridization, *Obstet. Gynecol.*, 64, 767, 1984.
53. **Syrjänen, S. and Syrjänen, K.,** An improved in situ DNA hybridization protocol for detection of human papillomavirus DNA sequences in paraffin-embedded biopsies, *J. Virol. Methods,* 14, 293, 1986.
54. **Dürst, M., Kleinheinz, A., Hotz, M., and Gissmann, L.,** The physical state of human papillomavirus type 16 DNA in benign and malignant genital tumors, *J. Gen. Virol.*, 66, 1515, 1985.
55. **Gissman, L., Wolnik, L., Ikenberg, H., Koldovsky, U., Schnurch, H. G., and zur Hausen, H.,** Human papillomavirus type 6 and 11 DNA sequences in genital and laryngeal papillomas and in some cervical cancers, *Proc. Natl. Acad. Sci. U. S. A.*, 80, 560, 1983.
56. **Dürst, M., Gissmann, L., Ikenberg, H., zur Hausen, H.,** A papillomavirus DNA from a cervical carcinoma and its prevalence in cancer biopsy samples from different geographic regions, *Proc. Natl. Acad. Sci. U. S. A.*, 80, 3812, 1983.
57. **Boshart, M., Gissmann L., Ikenberg, H., Kleinheinz, A., Scheurlen,W., and zur Hausen, H.,** A new type of papillomavirus DNA, its presence in genital cancer biopsies and in cell lines derived from cervical cancer, *EMBO J.*, 3, 1151, 1984.
58. **Lorincz, A. T., Lancaster, W. D., and Temple, G. F.,** Cloning and characterization of the DNA of a new human papillomavirus from a woman with dysplasia of the uterine cervix, *J. Virol.*, 58, 225, 1986.
59. **Schwarz, E., Freeze, U. K., Gissmann, L., Mayer, W., Roggenbuck, B., Stremlau, A., and zur Hausen, H.,** Structure and transcription of human papillomavirus sequences in cervical carcinoma cells, *Nature (London),* 314, 111, 1985.
60. **Campion, M. J., McCance, D. J., Cuzick, J., and Singer, A.,** Progressive potential of mild cervical atypia: Prospective cytological, colposcopic, and virological study, *Lancet,* 2, 237, 1986.
61. **Pater, M. M., Dunne, J., Hogan, G., Chatage, P., and Pater, A.,** Human papillomavirus types 16 and 18 sequences in early cervical neoplasia, *Virology,* 155, 13, 1986.
62. **Barrasso, R., De Brux, J., Croissant, O., and Orth, G.,** High prevalence of papillomavirus-associated penile intraepithelial neoplasia in sexual partners of women with cervical intraepithelial neoplasia, *N. Engl. J. Med.*, 317, 916, 1987.
63. **Macnab, J. C. M., Walkinshaw, S. A., Cordiner, J. W., and Clements, J. B.,** Human papillomavirus in clinically and histologically normal tissue of patients with genital cancer, *N. Engl. J. Med.*, 315, 1052, 1986.
64. **Toon, P. G., Arrand, J. R., Wilson, L. P., and Sharp, D. S.,** Human papillomavirus infection of the uterine cervix of women without cytological signs of neoplasia, *Br. Med. J.*, 293, 1261, 1986.

65. **Meanwell, C. A., Cox, M. F., Blackledge, G., and Maitland, N. J.**, HPV16 DNA in normal and malignant cervical neoplasia, *Lancet*, 1, 703, 1987.
66. **Tiollais, P., Pourcel, C., and Dejean, A.**, The hepatitis B virus, *Nature (London)*, 317, 489, 1985.
67. **Berninger, M., Hammer, M., Hoyer, B., and Gerin, J. L.**, An assay for the detection of the DNA genome of hepatitis B virus in serum, *J. Med. Virol.*, 9, 57, 1982.
68. **Kam, W., Rall, L. B., Smuckler, E. A., Schmid, R., and Rutter, W. J.**, Hepatitis B viral DNA in liver and serum of a symptomatic carrier, *Proc. Natl. Acad. Sci. U.S.A.*, 79, 7522, 1982.
69. **Neurath, A. R., Strick, N., Baker, L., and Krugman, S.**, Radioimmunoassays of hidden viral antigens, *Proc. Natl. Acad. Sci. U.S.A.*, 79, 4415, 1982.
70. **Shafritz, D. A., Lieberman, H. M., Isselbacher, K. J., and Wands, J. R.**, Monoclonal radioimmunoassays for hepatitis B surface antigen: demonstration of hepatitis B virus DNA or related sequences in serum and viral epitopes in immune complexes, *Proc. Natl. Acad. Sci. U.S.A.*, 79, 5675, 1982.
71. **Tur-Kaspa, R., Keshet, E., Eliakim, M., and Shouval, D.**, Detection and characterization of hepatitis B virus DNA in serum of HBe antigen-negative HBsAg carriers, *J. Med. Virol.*, 14, 17, 184.
72. **Moestrup, T., Hansson, B. G., Widell, A., Blomberg, J., and Nordenfelt, E.**, Hepatitis B virus-DNA in the serum of patients followed-up longitudinally with acute and chronic hepatitis B., *J.Med. Virol.*, 17, 337, 1985.
73. **Harrison, T. J., Bal, V., Wheeler, E. G., Meacock, T. J., Harrison, J. F., and Zuckerman, A. J.**, Hepatitis B virus DNA and e antigen in serum from blood donors in the United Kingdom positive for hepatitis B surface antigen, *Br. Med. J.*, 290, 663, 1985.
74. **Bréchot, C., Degos, F., Lugassy, C., Thiers, V., Zafrani, S., Franco, D., Bismuth, H., Trépo, C., Benhamous, J.-P., Wands, J., Isselbacher, K., Tiollais, P., and Berthelot, P.**, Hepatitis B virus DNA in patients with chronic liver disease and negative tests for hepatitis B surface antigen, *N. Engl. J. Med.*, 312, 270, 1985.
75. **Blum, H. E., Haase, A. T., and Vyas, G. N.**, Molecular pathogenesis of hepatitis B virus infection: simultaneous detection of viral DNA and antigens in paraffin-embedded liver sections, *Lancet*, ii, 771, 1984.
76. **Pontisso, P., Poon, M. C., Tiollais, P., and Brechot, C.**, Detection of hepatitis B virus DNA in mononuclear blood cells, *Br. Med. J.*, 288, 1563, 1984.
77. **Dejean, A., Lugassy, C., Zafrani, S., Tiollais, P., and Brechot, C.**, Detection of hepatitis B virus DNA in pancreas, kidney and skin of two human carriers of the virus, *J. Gen. Virol.*, 65, 651, 1984.
78. **Anderson, M. J., Jones, S. E., and Minson, A. C.**, Diagnosis of human parvovirus infection by dot-blot hybridization using cloned viral DNA, *J. Med. Virol.*, 15, 163, 1985.
79. **Clewley, J. P.**, Detection of human parvovirus using a molecularly cloned probe, *J. Med. Virol.*, 15, 173, 1985.
80. **Flores, J., Boeggeman, E., Purcell, R. H., Sereno, M., Perez, I., White, L., Wyatt, R. G., Chanock, R. M., and Kapikian, A. Z.**, A dot hybridization assay for detection of rotavirus, *Lancet*, 1, 555, 1983.
81. **Midthun, K., Valdesuso, J. Hoshino, Y., Flores, J., Kapikian, A. Z., and Chanock, R. M.**, Analysis by RNA-RNA hybridization assay of intertypic rotaviruses suggests that gene reassortment occurs in vivo, *J. Clin. Microbiol.*, 25, 295, 1987.
82. **Saiki, R.K., Scharf, S., Faloona, F., Mullis, K. B., Horn, G. T., Erlich, H. A., and Arnheim, N.**, Enzymatic amplification of β-globin genomic sequences and restriction site analysis for diagnosis of sickle cell anemia, *Science*, 230, 1350, 1985.
83. **Hodgson, A. L. M. and Roberts, W. P.**, DNA colony hybridization to identify Rhibzobium strains, *J. Gen. Microbiol.*, 129, 207, 1983.
84. **Totten, P. A., Holmes, K. K., Handsfield, H. H., Knapp, J. S., Perine, P. L., and Falkow, S.**, DNA hybridization technique for the detection of Neisseria gonorrhoeae in men with urethritis, *J. Infect. Dis.*, 148, 462, 1983.
85. **Hill, W. A., Payne, W. L., and Aulisio, C. C. G.**, Detection an enumeration of virulent Yersinia enterocolitica in food by DNA colony hybridization, *Appl. Environ. Microbiol.*, 46, 636, 1983.
86. **Moseley, S. L., Huq, I., Alim, A. R. M. A., So, M., Samadpour-Motalebi, M., and Falkow, S.**, Detection of enterotoxigenic *Escherichia coli* by DNA colony hybridization, *J. Infect. Dis.*, 142, 892, 1980.
87. **Tenover, F. C.**, Minireview. Studies of antimicrobial resistance genes using DNA probes, *Antimicrob. Agents Chemother.*, 29, 721, 1986.
88. **Grimont, P. A., Grimont, F., Desplates, N., and Tchen, P.**, DNA probe specific for *Legionella pneumophila*, *J. Clin. Microbiol.*, 21, 431, 1985.
89. **Fitts, R., Diamond, M., Hamilton, C., and Neri, M.**, DNA-DNA hybridization assay for detection of *Salmonella* spp in foods, *Appl. Environ. Microbiol.*, 46, 1146, 1983.
90. **Grundstein, M. and Hogness, D. S.**, Colony hybridization: a method for the isolation of cloned DNAs that contain a specific gene, *Proc. Natl. Acad. Sci. U.S.A.*, 72, 3961, 1975.
91. **Haas, M. J. and Fleming, D. J.**, Use of biotinylated DNA probes in colony hybridization, *Nucleic Acids Res.*, 14, 3976, 1986.

92. **Seriwatana, J., Echeverria, P., Taylor, D. N., Sakuldaipeara, T., Changchawalt, S., and Chivoratanond, O.,** Identification of enterotoxigenic *Escherichia coli* with synthetic alkaline phosphatase-conjugated oligonucleotide DNA probes, *J. Clin. Microbiol.*, 25, 1438, 1987.
93. **Zwadyk, P., Cooksey, R. C., and Thornsberry, C.,** Commercial detection methods for biotinylated gene probes: comparison with ^{32}P-labeled DNA probes, *Curr. Microbiol.*, 14, 95, 1986.
94. **Hoffman, N. E., Pichersky, E., and Cashmore, A. R.,** A tomato cDNA encoding a biotin-binding protein, *Nucleic Acids Res.*, 15, 3928, 1987.
95. **Jablonski, E., Moomaw, E. W., Tullis, R., and Ruth, J.,** Preparation of oligonucleotide-alkaline phosphatase conjugates and their use as hybridization probes, *Nucleic Acid Res.*, 14, 6115, 1986.
96. **Sethabutr, O., Hanchalay, S., Echeverria, P., Taylor, D. N., and Leksomboon, U.,** A non-radioactive DNA probe to identify Shigella and enteroinvasive *Escherichia coli* in stools of children with diarrhoea, *Lancet*, 2, 1095, 1985.
97. **Carter, G. I., Towner, K. J., and Slack, R. C. B.,** Detection of TEM beta-lactamase genes by non-isotopic spot hybridization, *Eur. J. Clin. Microbiol.*, 6, 406, 1987.
98. **Echeverria, P., Seriwatana, J., Paramaroj, U., Moseley, S. L., McFarland, A., Chityothin, O., and Chaicumpa, W.,** Prevalence of heat-stable II enterotoxigenic *Escherichia coli* in pigs, water, and people at farms in Thailands as determined by DNA hybridization, *J. Clin. Microbiol.*, 19, 489, 1984.
99. **Moseley, S. L., Echeverria, P., Seriwatana, J., Tirapat, C., Chaicumpa, W., Sakuldaipeara, T., and Falkow, S.,** Identification of enterotoxigenic *Escherichia coli* by colony hybridization using three enterotoxin gene probes, *J. Infect. Dis.*, 145, 863, 1982.
100. **Li, P., Medon, P. P., Skingle, D. C., Lancer, J. A., and Symons, R. H.,** Enzyme-linked synthetic oligonucleotide probes: non radioactive detection of enterotoxigenic *Escherichia coli* in faecal specimens, *Nucleic Acids Res.*, 15, 5275, 1987.
101. **Lanata, C. F., Kaper, J. B., Baldini, M. M., Black, R. E., and Levine, M. M.,** Sensitivity and specitivity of DNA probes with the stool blot technique for detection of *Escherichia coli* enterotoxins, *J. Infect. Dis.*, 152, 1087, 1985.
102. **Echeverria, P., Leksomboon, U., Chaicumpa, W., Seriwatana, J., Tirapat, C., and Ropve, B.,** Identification by DNA hybridization of enterotoxigenic *Escherichia coli* in homes of children with diarrhoea, *Lancet*, i, 63, 1984.
103. **Patamaroj, U., Seriwatana, J., and Echeverria, P.,** Identification of enterotoxigenic *Escherichia coli* isolated from swine with diarrhea in Thailand by colony hybridization, using three enterotoxin gene probes, *J. Clin. Microbiol.*, 18, 1429, 1983.
104. **Echeverria, P., Seriwatana, J., Chityothin, O., Chaicumpa, W., and Tirapat, C.,** Detection of enterotoxigenic *Escherichia coli* in water by filter hybridization with three enterotoxin gene probes, *J. Clin. Microbiol.*, 16, 1086, 1982.
105. **Yam, W. C., Lung, M. L., and Ng, M. H.,** Evaluation and optimation ofthe DNA filter assay for direct detection of enterotoxigenic *Escherichia coli* in the presence of stool coliforms, *J. Clin. Microbiol.*, 24, 149, 1986.
106. **Hill, W., Payne, W. L., Zon, G., and Moseley, S. L.,** Synthetic oligodeoxyribonucleotide probes for detecting heat-stable enterotoxin-producing *Esherichia coli* by DNA colony hybridization, *Appl. Environ. Microbiol.*, 50, 1187, 1985.
107. **Rubin, F. A., Kopecko, D. J., Noon, K. F., and Baron, L. S.,** Development of a DNA probe to detect *Salmonella typhi*, *J. Clin. Microbiol.*, 22, 600, 1985.
108. **Tompkins, L. S., Troup, N., Labigne-Roussel, A., and Cohen, M. L.,** Cloned, random chromosomal sequences as probes to identify *Salmonella* species, *J. Infect. Dis.*, 154, 156, 1986.
109. **Kaper, J. B., Moseley, S. L., and Falkow, S.,** Molecular characterization of environmental and nontoxigenic strains of *Vibrio cholerae*, *Infect. Immun.*, 32, 661, 1981.
110. **Jagow, J. and Hill, W.,** Enumeration by DNA colony hybridization of virulent *Yersinia enterocolitica* colonies in artificially contaminated food, *Appl. Environ. Microbiol.*, 51, 441, 1986.
111. **Boileau, C. R., D'Hauteville, H. M., and Sansonetti, P. J.,** DNA hybridization technique to detect *Shigella* species and enteroinvasive *Escherichia coli*, *J. Clin. Microbiol.*, 20, 959, 1984.
112. **Wood, P. K., Morris, J. G., Small, P. L. C., Sethabutr, O., Toledo, M. R. F., Trabulsi, L., and Kaper, J. B.,** Comparison of DNA probes and the sereny test for identification of invasive *Shigella* and *Escherichia coli* strains, *J. Clin. Microbiol.*, 24, 498, 1986.
113. **Levine, M. M., Xu, J-G., Kaper, J. B., Lior, H., Prado, V., Tall, B., Nataro, J., Karch., and Wachsmuth, K.,** A DNA probe to identify enterohemorrhagic *Escherichia coli* of 0157:H7 and other serotypes that cause hemorrhagic colitis and hemolytic uremic syndrome, *J. Infect. Dis.*, 156, 175, 1987.
114. **O'Hanley, P., Low, D., Romero, I., Lark, D., Vosti, K., Falkow, S., and Schoolnik, G.,** Gal-Gal binding and hemolysin phenotypes and genotypes associated with uropathogenic *Escherichia coli*, *N. Engl. J. Med.*, 313, 414, 1985.
115. **Palva, A.,** ompA gene in the detection of *Escherichia coli* and other Enterobacteriaceae by nucleic acid sandwich hybridization, *J. Clin. Microbiol.*, 18, 92, 1983.

116. **Kuritza, A. P., Getty, C. W., Shaughnessy, P., Hesse, R., and Salyers, A. A.,** DNA probes for identification of clinically important Bacteroides species, *J. Clin. Microbiol.*, 23, 343, 1986.
117. **Salyers, A. A., Lynn, S. P., and Gardner, J. F.,** Use of randomly cloned DNA fragments for identification of *Bacteroides thetaiotaomicron, J. Bacteriol.*, 154, 287, 1983.
118. **Kuritza, A. P. and Salyers, A. A.,** Use of a species-specific DNA hybridization probe for enumerating *Bacteroides vulgatus* in human feces, *Appl. Environ. Microbiol.*, 50, 958, 1985.
119. **Kuritza, A. P., Shaughnessy, P., and Salyers, A. A.,** Enumeration of polysaccharide-degrading *Bacteroides* species in human feces by using species-specific DNA probes, *Appl. Environ. Microbiol.*, 51, 385, 1986.
120. **Wilkinson, H. W., Sampson, J. S., and Plikaytis, B. B.,** Evaluation of a commercial gene probe for identification of *Legionella* cultures, *J. Clin. Microbiol.*, 23, 217, 1986.
121. **Kohne, D. W.,** Application of DNA probe tests to the diagnosis of infectious disease, *Am. Clin. Prod. Rev.*, 20, 1986.
122. **Edelstein, P. H.,** Evaluation of the Gen-probe DNA probe for the detection of Legionellae in culture, *J. Clin. Microbiol.*, 23, 481, 1986.
123. **Grimont, P. A. D., Grimont, F., Desplaces, N., and Tchen, P.,** DNA probe specific for *Legionella pneumophila. J. Clin. Microbiol.*, 21, 431, 1985.
124. **Roberts, M. C., McMillan, C., and Coyle, M. B.,** Whole chromosomal DNA probes for rapid identification of *Mycobacterium tuberculosis* and *Mycobacterium avium* complex, *J. Clin. Microbiol.*, 25, 1239, 1987.
125. **Kiehn, T. E. and Edwards, F. F.,** Rapid identification using a specific DNA probe of *Mycobacterium avium* complex from patients with acquired immunodeficiency syndrome, *J. Clin. Microbiol.*, 25, 1551, 1987.
126. **Perine, P. L., Totten, P. A., Holmes, K. K., Sng, E. H., Ratnam, A. V., Widy-Wersky, R., Nsanze, H., Habte-Gabr, E., and Westbrook, W. G.,** Evaluation of a DNA hybridization method for detection of African and Asian strains of *Neisseria gonorrhoeae* in men with urethritis, *J. Infect. Dis.*, 152, 59, 1985.
127. **Hyypiä, T., Jalava, A., Larsen, S. H., Terho, P., and Hukkanen, V.,** Detection of *Chlamydia trachomatis* in clinical specimens by nucleic acid spot hybridization, *J. Gen. Microbiol.*, 131, 975, 1985.
128. **Horn, J. E., Hammer, M. L., Falkow, S., and Quinn, T. C.,** Detection of *Chlamydia trachomatis* in tissue culture and cervical scrapings by *in situ* DNA hybridization *J. Infect. Dis.*, 153, 1155, 1986.
129. **Palva, A.,** Nucleic acid spot hybridization for detection of *Chlamydia trachmatis, FEMS Microbiol. Lett.*, 28, 85, 1985.
130. **Tuokko, H., Ruuska, P., and Hyypiä, T.,** Comparison of nucleic acid hybridization with enzyme immunoassay and isolation in detection of *Chlamydia trachomatic,* submitted for publication.
131. **Hyman, H. C., Yogev, G., and Razin, S.,** DNA probes for detection and identification of *Mycoplasma pneumoniae* and *Mycoplasma genitalium, J. Clin. Microbiol.*, 25, 726, 1987.
132. **Ogle, J. W., Janda, J. M., Woods, D. E., and Vasil, M. L.,** Characterization and use of a DNA probe as an epidemiological marker for *Pseudomonas aeruginosa, J. Infect. Dis.*, 155, 119, 1987.
133. **Festl, H., Ludwig, W., and Schleifer, K. H.,** DNA hybridization probe for the *Pseudomonas fluorescens* group, *Appl. Environ. Microbiol.*, 52, 1190, 1986.
134. **Pappenheimer, A. M.,** Studies on the molecular epidemiology of diphtheria, *Lancet,* ii, 923, 1983.
135. **Roberts, M. C., Hillier, S. L., Schoenknecht, F. D., and Holmes, K. K.,** Comparison of gram stain, DNA probe, and culture for the identification of species of *Mobiluncus* in female genital specimens, *J. Infect. Dis.*, 152, 74, 1985.
136. **Datta, A. R., Wentz, B. A., and Hill, W. E.,** Detection of hemolytic *Listeria monocytogenes* by using DNA colony hybridization *Appl. Environ. Microbiol.*, 53, 2256, 1987.
137. **Mason, M. M., Lasker, B. A., and Riggsby, W. S.,** Molecular probe for identification of medically important *Candida* species and *Torulopsis glabrata, J. Clin. Microbiol.*, 25, 563, 1987.
138. **Burchall, J. J., Elwell, L. P., and Fling, M. E.,** Molecular mechanisms of resistance to trimethoprim, *Rev. Infect. Dis.*, 4, 246, 1982.
139. **Huovinen, P.,** Minireview. Trimethoprim resistance, *Antimicrob. Agents Chemother.*, 31, 1451, 1987
140. **Murray, B. W., Alavarado, T., Kim, K-H., Vorachit, M., Jayanetra, P., Levine, M. M., Prenzel, I., Fling, M., Elwell, L., McCracken, G. H., Madrigal, G., Odio, C., and Trabulsi, L, R.,** Increasing resistance to trimethoprim-sulfamethoxazole among isolates of *Escherichia coli* in developing countries, *J. Infect. Dis.*, 152, 1107, 1985.
141. **Papadopoulou, B., Gerbaud, G., Courvalin, P., Acar, J. F., and Goldstein, F. W.,** Molecular epidemiology of resistance to trimethoprim in enterobacteria isolated in a Parisian hospital. *Ann. Inst. Pasteur Microbiol.*, 137A, 239, 1986.
142. **Pulkkinen, L., Huovinen, P., Vuorio, E., and Toivanen, P.,** Characterization of trimethoprim resistance by use of probes specific for transposon Tn7, *Antimicrob. Agents Chemother.*, 26, 82, 1984.

143. **Lester, S. C., Pla, M. P., Perez, I., Wang, F., Jiang, H., and O'Brien, T. F.**, Antibiotic resistance in *Escherichia coli* of healthy children in the United States, Venezuela and China. Abstr. 27th Interscience Cong. Antimicrobial Agents and Chemotherapy, American Society for Microbiology, New York, 1987.
144. **Sundström, L., Vinayagamoorthy, and Sköld, O.**, Novel plasmid-borne resistance to trimethoprim, *Antimicrob. Agents Chemother.*, 31, 60, 1987.
145. **Medeiros, A. A. and Jacoby, G. A.**, Beta-lactamase-mediated resistance, in *Beta-lactam Antibiotics for Clinical Use*, Queener, S. F., Webber, J. A., Queener, S. W., Eds., Marcel Dekker, New York, 1986, 449.
146. **Cooksey, R. C., Clark, N. C., and Thornsberry, C.**, A gene probe for TEM type beta-lactamases, *Antimicrob. Agents Chemother.*, 28, 154, 1985.
147. **Levesque, R. C., Medeiros, A. A., and Jacoby, G. A.**, Molecular cloning and DNA homology of plasmid-mediated beta-lactamase genes, *Mol. Gen. Gent.*, 206, 252, 1987.
148. **Boissinot, M., Mercier, J., and Levesque, R. C.**, Development of natural and synthetic DNA probes for OXA-2 and TEM-1 beta-lactamases, *Antimicrobial Agents Chemother.*, 31, 728, 1987.
149. **Ouellette, M., Rossi, J. J., Bazin, R., and Roy, P. H.**, Oligonucleotide probes for the detection of TEM-1 and TEM-2 beta-lactamase genes and their transposons, *Can. J. Microbiol.*, 33, 205, 1987.
150. **Huovinen, S., Huovinen, P., and Jacoby, G. A.**, Detection of plasmid-mediated beta-lactamases using DNA hybridization, *Antimicrob. Agents Chemother.*, 32, 1988, in press.
151. **Jouvenot, M., Deshaseaux, M. L., Royez, M., Mougin, C., Cooksey, R. C., Michel-Briand, Y., and Adessi, G. L.**, Molecular hybridization versus isoelectric focusing to determine TEM-type beta-lactamases in gram-negative bacteria, *Antimicrob. Agents Chemother.*, 31, 300, 1987.
152. **Ouellete, M. and Roy, P. H.**, Analysis by using DNA probes of the OXA-beta-lactamase gene and its transposon, *Antimicrob. Agents Chemother.*, 30, 446, 1987.
153. **Miller, G. H., Sabatelli, E. J., Hare, R. S., and Waitz, J. A.**, Survey of aminoglycoside resistance patterns, *Dev. Ind. Microbiol.*, 21, 91, 1980.
154. **Haas, M. J. and Dowing, J. E.**, Aminoglycoside modifying enzymes, *Methods Enzymol.*, 43, 611, 1975.
155. **Gootz, T. D., Tenover, F. C., Young, S. A., Gordon, K. P., and Plorde, J. J.**, Comparison of three DNA hybridization methods for detection of the aminoglycoside 2″-*0*-adenyltransferase gene in clinical bacterial isolates, *Antimicrob. Agents Chemother.*, 28, 69, 1985.
156. **Tenover, F. C., Gootz, T. D., Gordon, K. P., Tompkins, L. S., Young, S., and Plorde, J. J.**, Development of a DNA probe for the structural gene of the 2″-*0*-adenyltransferase aminoglycoside-modifying enzyme, *J. Infect. Dis.*, 150, 678, 1984.
157. **Young, S. A., Tenover, F. C., Gootz, T. D., Gordon, K. P., and Plorde, J. J.**, Development of two DNA probes for differentiating the structural genes of subclasses I and II of the aminoglycoside-modifying enzyme 3′-aminoglycoside phosphotransferase, *Antimicrob. Agents Chemother.*, 27, 739, 1985.
158. **Lee, S. C, Cleary, P. P., and Gerding, D. N.**, More than one DNA sequence encodes the 2″-*0*-adenylytransferase phenotype, *Antimicrob. Agents Chemother.*, 31, 667, 1987.
159. **Mendez, B., Tachibana, C., and Levy, S. B.**, Heterogeneity of tetracycline resistance determinants, *Plasmid*, 3, 99, 1980.
160. **Marshall, B., Tachibana, C., and Levy, S. B.**, Frequency of tetracycline resistant determinant classes among lactose-fermenting coliforms, *Antimicrob. Agents Chemother.*, 24, 835, 1983.
161. **Marshall, B., Roberts, M., Smith, A., and Levy, S. B.**, Homogeneity of transferable tetracycline-resistance determinants in *Haemophilus* species, *J. Infect. Dis.*, 149, 1028, 1984.
162. **Burdett, V., Inamine, J., and Rajagoplan, S.**, Heterogeneity of tetracycline resistance determinants in *Streptococcus*, *J. Bacteriol.*, 149, 995, 1982.
163. **Roberts, M. C., Koutsky, L. A., Holmes, K. K., LeBlanc, D. J., and Kenny, G. E.**, Tetracycline-resistant *Mycoplasma hominis* strains contain streptococcal tetM sequences, *Antimicrobial Agents Chemother.*, 28, 141, 1985.
164. **Tenover, F. C., Williams, S., Gordon, K. P., Nolan, C., and Plorde, J. J.**, Survey of plasmids and resistance factors in *Campylobacter jejuni* and *Campylobacter coli*, *Antimicrob. Agents Chemother.*, 27, 37, 1985.
165. **Taylor, D. E., Garner, R. S., and Allan, B. J.**, Characterization of tetracycline resistance plasmids from *Campylobacter jejuni* and *Campylobacter coli*, *Antimicrobial Agents Chemother.*, 24, 930, 1983.
166. **Roberts, M. C., Actis, L. A., and Crosa, J. H.**, Molecular characterization of chloramphenicol-resistant *Haemophilus parainfluenzae* and *Haemophilus ducreyi*, *Antimicrob. Agents Chemother.*, 28, 176, 1985.
167. **Christie, P. J. and Dunny, G. M.**, Antibiotic selection pressure resulting in multiple antibiotic resistance and localization of resistance determinants to conjugative plasmids in streptococci, *J. Infect. Dis.*, 149, 74, 1984.
168. **Franzen, L, Westin, G., Shabo, R., Åslund, L., Perlman, H., Persson, T., Wigzell, H., and Pettersson, U.**, Analysis of clinical specimens by hybridization with probe containing repetitive DNA from *Plasmodium falciparum*. A novel approach to malaria diagnosis, *Lancet*, i, 525, 1984.

169. **Åslund, L., Franzen, L., Westin, G., Persson, T., Wigzell, H., and Pettersson, U.**, Highly reiterated non-coding sequence in the genome of *Plasmodium falciparum* is composed of 21 base-pair tandem repeats, *J. Mol. Biol.*, 185, 509, 1985.
170. **Oquendo, P, Goman, M., Mackay, M., Langsley, G., Walliker, D., and Scaife, J.**, Characterization of repetitive DNA sequence from the malaria parasite, *Plasmodium falciparum, Mol. Biochem. Parasitol.*, 18, 89, 1986.
171. **Gunthaka, R., Gowda, S., Rao, A., and Green, T.**, Organization of *Plasmodium falciparum* genome 1. Evidence for a highly repeatd DNA sequence, *Nucleic Acids Res.*, 13, 1965, 1985.
172. **Bhasin, V., Clayton, Ch., Trager, W., and Cross, G.**, Variation in the organization of repetitive DNA sequences in the genomes of *Plasmodium falciparum* clones, *Mol. Biochem. Parasitol.*, 15, 149, 1985.
173. **Holmberg, M. Shenton, F., Franzen, L., Janneh, K., Snow, R., Pettersson, U., Wigzell, H., and Greenwood, B. M.**, Use of a DNA hybridization assay for the detection of *Plasmodium falciparum* in field trials, *Am. J. Trop. Med. Hyg.*, 37, 230, 1987.
174. **McLaughlin, G., Collins, W., and Campbell, G. H.**, Comparison of genomic, plasmid, synthetic and combined DNA probes for detecting *Plasmodium falciparum* DNA, *J. Clin. Microbiol.*, 25, 791, 1987.
175. **McLaughlin, G., Breman, J., Collins, F., Schwartz, I., Brandling-Bennett, A., Sulzer, A., Collins, W., Kaseje, D., and Cambell, G.**, Assessment of a synthetic DNA probe for *Plasmodium falciparum* in African blood specimens, *Am. J. Trop. Med. Hy.*, 37, 27, 1987.
176. **Holmberg, M., Björkman, A., Franzen, L., Åslund, L., Lebbad, M., Pettersson, U., and Wigell, H.**, Diagnosis of *Plasmodium falciparum* infection by hybridization assay: specificity, sensitivity, and field applicability, *Bull. W.H.O.* 64, 579, 1986.
177. **McLaughlin, G., Jablonski, E., Ruth, J., and Steketee, R.**, Use of enzyme-linked synthetic DNA in diagnosis of falciparum malaria, *Lancet*, i, 714, 1987.
178. **Vaidya, A. and Arasu, P.**, Tandemly arranged gene clusters of malarial parasites that are highly conserved and transcribed, *Mol. Biochem. Parasitol.*, 22, 249, 1987.
179. **Dore, E., Pace, T., Ponzi, M., Scotti, R., and Frontali, C.**, Homologous telomeric sequences are present in different species of the genus *Plasmodium, Mol. Biochem. Parasitol.*, 21, 121, 1986.
180. **McLaughlin, G., Edling, T., and Ihler, G.**, Detection of *Babesia bovis* using DNA hybridization, *J. Protozool.*, 33, 125, 1986.
181. **Barker, D. C.**, DNA diagnosis of human leishmaniasis, *Parasitol. Today*, 3, 177, 1987.
182. **Arnot, D. and Barker, D.**, Biochemical identification of cutaneous leishmaniasis by analysis of kinetoplast DNA. II. Sequence homologies in *Leishmania* kDNA, *Mol. Biochem. Parasitol*, 3, 47, 1981.
183. **Lopes, U. and Wirth, D.**, Identification of visceral leishmania species with cloned sequences of kinetoplast DNA, *Mol. Biochem. Parasitol.*, 20, 77, 1986.
184. **Sloof, P., Bos, J., Konings, A., Menke, H., Borst, P., Gutteridge, W., and Leon, W.**, Characterization of satellite DNA in *Trypanosoma brucei* and *Trypanosoma cruzi, J. Mol. Biol.*, 167, 1, 1983.
185. **Sloof, P., Menke, H., Caspers, M., and Borst, P.**, Size fractionation of *Trypanosoma brucei* DNA: localization of one 177 bp repeat satellite DNA and a variant surface glycoprotein gene in a mini-chrmosomal DNA fraction, *Nucleic Acids Res.*, 11, 3889, 1983.
186. **Majiwa, P., Masake, R., Nantulya, V., Hamers, R., and Matthyssenns, G.**, *Trypanosoma (Nannomonas) congolense:* identification of two karyotypic groups, *EMBO J.*, 4, 3307, 1985.
187. **Gonzales, A., Prediger, E., Huecas, M., Nogueira, N., and Lizardi, P.**, Minichromosomal repetitive DNA in *Trypanosoma cruzi:* its use in a high-sensitivity parasite detection assay, *Proc. Natl. Acad. Sci. U.S.A.*, 81, 3356, 1984.
188. **Morel, C., Chriari, E., Plessman Camargo, E., Romanha, A., and Simpson, L.**, Strains and clones of *Trypanosoma cruzi* can be characterized by pattern of restriction endonuclease products of kinetoplast DNA minicircles, *Proc. Natl. Acad. Sci, U.S.A.*, 77, 6810, 1980.
189. **McReynolds, L., DeSimone, S., and Williams, S.**, Cloning and comparison of repeated DNA sequences from the human filarial parasite *Brugia malayi* and the animal parasite *Brugia pahangi, Proc. Natl. Acad. Sci, U.S.A.*, 83, 797, 1986.
190. **Sim, B., Shah, J., Wirth, D., and Piessens, W.**, Characterization of the filarial genome, in *Filariasis, Ciba Foundation Symposium 127,* John Wiley & Sons, Chichester, 1987, 107.
191. **Cianchi, R., Karam, M., Henry, M. C., Villami, R., Kumlien, S., and Bulhini, L.**, Preliminary data on the genetic differentiation of *Oncocerca volvulus* in Africa (Nematoda:Filariodea), *Acta Trop.*, 42, 341, 1985.
192. **Perler, F. and Karam, M.** Cloning and characterization of two *Oncocerca volvulus* repeated sequences, *Mol. Biochem. Parasitol.*, 21, 171, 1986.
193. **Shah, J. Karam, M., Piesseus, W., and Wirth, D.**, Characterization of an Oncocerca-specific DNA clone from *Oncocerca volvulus, Am. J. Trop. Med. Hyg.*, 37, 376, 1987.

194. **Erttmann, K., Unnasch, T., Greene, B., Albiez, E., Boateng, J., Denke, A., Ferraroni, J., Karam, M., Schulz-Key, H., and Williams, P.,** A DNA sequence specific for forest form *Oncocerca volvulus, Nature (London),* 327, 415, 1987.

195. **Auvinen, P., Stanway, G., and Hyypiä, T.,** unpublished.

196. **Zolg, W.,** personal communication, 1987.

Chapter 4

NUCLEIC ACID PROBES IN THE DIAGNOSIS OF PLANT VIRUSES AND VIROIDS

J. L. McInnes and R. H. Symons

TABLE OF CONTENTS

I. INTRODUCTION

Plant virologists in the 1980s have available a large number of sophisticated diagnostic procedures for the detection of plant viruses and viroids. Such tests are essential for both the maintenance of pathogen-free seeds and stocks and for the monitoring of virus and viroid infection during crop cultivation. Low levels of these plant pathogens need to be reliably detected by standard procedures while strain identification would be an added advantage. For a review of the serological and biochemical diagnostic techniques currently available, readers are referred to Barnett.[1]

Traditionally, diagnostic techniques for plant pathogens have been biological in nature, involving sap and graft inoculation of indicator plants. These tests, while being sensitive and indicative of the infectious nature of the causal agent, have proved rather inefficient. Symptom expression of a specific agent is dependent on the host genotype and phenotype, so that studies of this nature become labor intensive, often requiring a variety of indicator plants and long incubation periods. Problems arise since some viruses cannot be transmitted by sap inoculation, thus requiring insect vector transmission, and different strains of the same virus can cause markedly different symptoms on the same indicator plant.[2,3]

In the case of viroids, symptom development normally takes longer than is the case with viruses. For example, symptom development of avocado sunblotch viroid (ASBV) can take from 6 months to over 2 years after inoculation of susceptible avocado seedlings and is not 100% reliable.[4-7] Hence, this biological approach for diagnosis is only practical for viroids such as potato spindle tuber viroid (PSTV) that produce symptoms in 2 to 4 weeks. Even in this approach, difficulties can arise. For example, some strains of viroid produce very mild symptoms on tomato plants such that it is difficult to distinguish them from the healthy controls. This was the case with a mild strain of citrus exocortis viroid (CEB).[8] Further, since different viroids can give essentially the same symptoms on tomato seedlings, further tests are needed to identify the actual viroid involved.

Many of the above problems associated with the biological testing of viruses may be overcome by tests based on serology. The most common of these is the enzyme-linked immunosorbent assay (ELISA).[1,9-11] This technique, based on the recognition of coat protein by antibodies, has proved both reliable and versatile. However, ELISA may not always be applicable due to lack of suitable virus-specific antisera, low virus titre, or interference problems during the assay from substances present in the plant extracts. Strain differentiation problems may also occur.[12] In addition, ELISA is not applicable to viroids; these agents, being nonencapsulated pathogenic RNAs, lack a protein coat and so cannot be detected by serological means.

A more recent approach to the diagnosis of plant pathogens utilizes the technique of nucleic acid hybridization.[1,3,13-15] Here, a nucleic acid probe, either labeled with a radioactive isotope (Chapter 1) or labeled nonisotopically (Chapter 2), hybridizes to form a duplex with a target nucleic acid which is exactly complementary to itself. No such double-strand hybrid is formed with other nucleic acids. The specificity, sensitivity, and speed of molecular hybridization allows the method to be a valuable adjunct to the more conventional immunological approaches used in diagnostic plant virology.

II. DOT-BLOT HYBRIDIZATION

A. Introduction

Early studies involving hybridization of plant viral nucleic acids utilized liquid or solution hybridization.[13] In this method, single-stranded ^{32}P-cDNA prepared against a specific viral or viroid RNA was hybridized in solution for 19 to 24 h with a partially purified nucleic acid preparation from infected plants. The hybridization mixture was then treated with nuclease S1 which is specific for single-stranded nucleic acids. Under the conditions used, any unhybridized single-stranded ^{32}P-cDNA was hydrolyzed to mononucleotides or small oligonucleotides while the ^{32}P-cDNA:RNA hybrids were not affected, except for the hydrolysis of any single-stranded nucleic acid tails. In addition, any ^{32}P-cDNA present in mismatched regions of incorrectly hybridized cDNA:RNA hybrids was digested. The remaining stable ^{32}P-cDNA:RNA hybrids were then collected by acid precipitation and counted. A control was always included in which ^{32}P-cDNA was hybridized against its homologous RNA (i.e., the RNA from which the cDNA was prepared) as these control hybridizations only proceeded to 80 to 90% of the theoretical maximum. In addition, an essential backgound control consisted of the ^{32}P-cDNA probe hybridized against a nucleic acid extract prepared from healthy plants.

This technique, originally described for the indexing of ASBV,[4,5] had a number of disadvantages. These included the tedious nature of the technique itself, persistent and variable blank value determinations, and its nonsuitability to large sample analysis. Accordingly, the liquid hybridization technique has now been superseded by the more modern and efficient dot-blot or filter hybridization procedure.

The principle of dot-blot or filter hybridization[14,16,17] is simple: denatured DNA or RNA is immobilized on an inert support (e.g., nitrocellulose or nylon membranes) in such a way that self annealing is prevented, yet bound sequences are available for hybridization with the added labeled-nucleic acid probe. After hybridization, extensive washing of the filter removes unreacted and poorly hybridized probe and detection of bound hybrids is by autoradiography (e.g., ^{32}P-labeled probes) or by enzymatic colorimetric detection (non isotope labeled probes) (Chapter 2, Section V.A).

While dot-blot hybidization is currently the method of choice for plant pathogen diagnosis, it should be noted that it does involve direct application to the filter of *unfractionated* nucleic acid. Dot-blots do not distinguish the size of the molecules hybridizing, so that the hybridization ''signal'' is the sum of all sequences hybridizing to the probe under the conditions used. Two techniques which do allow true qualitative analysis of the nucleic acid species are the Southern[18] and Northern[16,19] blotting protocols. Here, the DNA or RNA species are first fractionated for size analysis by agarose or polyacrylamide gel electrophoresis and then transferred by blotting from the gel to a filter (e.g., nitrocellulose) matrix, which is then used in hybridization.

The basic steps in dot-blot hybridization are now described in more detail.

B. Sample Preparation and Immobilization on Membrane Filters

Of the various types of filters in current use for the immobilization of DNA and RNA (e.g., nitrocellulose, nylon and chemically-activated papers), nitrocellulose (0.45 μm) is the most widely used due to its efficient binding capacity for both types of nucleic acid. Nitrocellulose requires high ionic strength for quantitative binding of both DNA and RNA and the binding efficiency is much reduced at low ionic strength.[18,20] For these reasons and also to maintain a small diameter for the applied spot, nitrocellulose filters are normally treated with 20xSSC (SSC: 0.15 *M* NaCl, 0.015 *M* sodium citrate) prior to spotting of the nucleic acid. After spotting, the nucleic acid sample is immobilized to the nitrocellulose by baking the filter in a vacuum oven for 2 h at 80°C.

Although nitrocellulose (and nylon) filters immobilize the applied nucleic acid, binding by the above conventional procedure is not covalent. This can lead to problems; e.g., the nucleic acid may be gradually leached from the filter, especially if the hybridization involves a long period at a high temperature. In order to overcome this problem, new techniques have been developed for covalent binding of nucleic acid to membranes. One such approach involves cross-linking in ultraviolet light[21] but to date the method is only applicable to nylon filters.

Table 1
LIST OF DOT-BLOT HYBRIDIZATION STUDIES FOR THE DETECTION OF A RANGE OF PLANT VIRUSES, 1983—1987

Plant virus[a]	Probe[b]	Extract	Reported sensitivity[c]	Ref.
BYDV	1	Oats sap	100 pg—1 ng virus	22
	1	Nucleic acid, cereals	N.S.	23
	1	Oats leaf	70 pg virus	24
CaMV	1, 2	Turnip leaf homogenate	5—20 pg virus	25
CMV	3	Cucumber sap or total nucleic acid	N.S.	26
CTV	1	RNA, citrus leaf or bark	N.S.	27
FDV	1	Sugarcane gall, leaf or growing tip	100 pg RNA	28
MSV	1, 2	Insect vector, "squash-blotting"	1 pg DNA	29
PPV	1	RNA, pea or peach leaf sap	100 pg virus	30
PVX	1	Potato sap	50 pg RNA	31—33
TMV	4	Tobacco leaf homogenate	5—20 pg virus	25
	4	Tobacco leaf sap, protoplasts	2.5 pg RNA	34
	5	Tobacco sap or nucleic acid	4-40 ng RNA	35
TRV	4, 6	Tobacco, potato and tulip leaf homogenate Tulip bulb nucleic acid	60 pg RNA	36
TBSV	4	RNA, tomato leaf	N.S.	37

[a] Plant viruses — BYDV: barley yellow dwarf virus; CaMV: cauliflower mosaic virus; CMV: cucumber mosaic virus; CTV: citrus tristeza virus; FDV: Fiji disease virus; MSV: maize streak virus; PPV: plum pox virus; PVX: potato virus X; TMV: tobacco mosaic virus; TRV: tobacco rattle virus; TBSV: tomato bushy stunt virus.

[b] Probes — 1: recombinant plasmid; 2: virion DNA; 3: double-stranded RNA; 4: cDNA; 5: synthetic oligonucleotice; 6: SP6 RNA transcript. All probes were labeled with ^{32}P, except for that of Habili et al.,[23] which was biotin-labeled.

[c] Reported sensitivity — on isolated virus or nucleic acid. N.S. = No statement on sensitivity.

With regard to the nature and purity of the plant nucleic acid extract spotted onto nitrocellulose, samples can vary considerably (Table 1). Strongest hybridization signals are normally obtained using deproteinized samples since protein co-immobilization competes with the nucleic acid for binding sites and also adds to background.[38] For this reason, many dot-blot hybridization protocols in current use employ a phenol purification step. However, as can be seen from Table 1, both plant sap and simple homogenized leaf tissue can be applied to nitrocellulose resulting in viral nucleic acid detection.

Denaturation of samples prior to spotting is an important aspect of the dot-blot method with protocols varying according to the type of nucleic acid involved. For example, plant extracts containing double-stranded viral RNA or DNA need to be denatured before application to nitrocellulose.[25,39] This can be achieved by either heat or brief alkali treatment. The majority of plant viruses, however, possess single-stranded RNAs and no denaturing step appears necessary; in fact, treatment with denaturants such as alkali or glyoxal resulted in a decrease in their level of detection.[25] Surprisingly, viroids with their highly secondary RNA structure, bind to nitrocellulose without the requirement of a denaturation step.[40-42]

A considerable enhancement in hybridization signal strength has been reported by White and Bancroft[43] who briefly treated tissue or cell extracts with a 20% (w/v) formaldehyde/10xSSC solution for 15 min at 65°C prior to spotting on nitrocellulose. When applied to CEV and barley yellow dwarf virus (BYDV) diagnosis,[23,44] similar enhancements in hybridization signal strengths were seen, presumably due to the quantitative denaturation of the plant pathogen RNA and maximum binding to nitrocellulose.

Commercial vacuum manifolds (e.g., Minifold, Schleicher and Schuell; Hybri-Dot, BRL) are now available for the binding of multiple nucleic acid samples to filters. While these

Table 2
LIST OF DOT-BLOT HYBRIDIZATION STUDIES FOR THE DETECTION OF A RANGE OF VIROIDS, 1981—1987

Viroid[a]	Probe[b]	Extract	Reported sensitivity[c]	Ref.
ASBV	1	Avocado leaf	300 pg	46
	1	Avocado leaf	100 pg	47
	2	Partially purified nucleic acid	5 pg	41
	3	Avocado leaf sap	N.S.	35, 48
CCCV	2	Nucleic acid, palm leaf	250 pg	49, 50
CEV	2	*Gynura aurantiaca*, chrysanthemum sap or nucleic acid	50—100 pg	44
	2, 4	Nucleic acid, citrus leaf or bark or nucleic acid, chrysanthemum leaf	N.S.	51, 52
	5	Potato sap, tuber sprouts	80 pg	40, 53
	5	Tomato leaf	100—250 pg	54
PSTV	2	Nucleic acid, potato leaf or tuber	N.S.	55, 56
	2	Nucleic acid, potato or tomato leaf	N.S.	57
	4	Tomato leaf homogenate	1.4 pg	42
	5	Crude nucleic acid, potato or tomato leaf	50 pg	58

[a] Viroids — ASBV: avocado sunblotch viroid; CCCV: coconut cadang cadang viroid; CEV: citrus exocortis viroid; PSTV: potato spindle tuber viroid.
[b] Probes — 1: Viroid RNA; 2: cDNA; 3: synthetic oligonucleotide; 4: SP6 RNA transcript; 5: recombinant plasmid. All probes were labeled with ^{32}P.
[c] Reported sensitivity — on isolated viroid RNA. N.S. = No statement of sensitivity.

devices are possibly time-saving and provide dots of uniform diameter, it should be realized that sensitivity problems can arise. Studies in our laboratory, for example, using a ^{32}P-cDNA probe for detecting PSTV in leaf extracts have shown a three- to fourfold decrease in hybridization signal strength when using a commercial manifold device (0.45 μm nitrocellulose) in place of a direct spotting protocol.[45] Other data showed that this decrease in sensitivity was due to the viroid passing through the filter under suction. Replacing the nitrocellulose in the manifold with 0.025 μm pore size, however, restored the hybridization signal to its original strength. Similar findings have been reported with the detection of BYDV RNA; in comparing direct spotting and vacuum manifold hybridization protocols, it was shown that the former was significantly more sensitive.[23]

C. Preparation of Nucleic Acid Probes

In theory, any nucleic acid may be used as a probe provided that it is sufficiently pure and labeled with a marker which allows its subsequent detection. The presence of contaminating sequences can lead to background hybridization which may make interpretation of results difficult, if not impossible. From Tables 1 and 2, it can be seen that many different probe types have been utilized in plant pathogen diagnosis. Such probes include double- and single-stranded DNAs, double-stranded RNA, single-stranded RNA synthesized *in vitro* and synthetic oligonucleotides. All studies (Tables 1 and 2) have utilized probes labeled with ^{32}P except for those of Habili et al.[23] who used a Photobiotin®-labeled DNA probe.

1. Cloned Double-Stranded DNA Probes

Recombinant double-stranded DNA probes have featured strongly in dot-blot hybridization studies for plant pathogen diagnosis (Table 1). In the case of double-stranded DNA viruses, e.g., cauliflower mosaic virus (CaMV), DNA clones of the viral DNA are readily prepared[25] by the cloning of fragments produced by restriction enzyme digestion. For single-stranded RNA viruses, which make up the bulk of the plant viruses, single-stranded cDNA is first made by copying the RNA using reverse transcriptase in the presence of a random mixture of short DNA fragments[59] or specific oligonucleotide primers[60,61] followed by double-stranded DNA synthesis[62,63] and cloning into a bacterial plasmid vector such as pBR322, pUC, pSP6, etc. These methods give a potentially unlimited supply of easily prepared, will characterized, and uniform double-stranded DNA for the preparation of DNA probes.

Labeling of DNA probes with ^{32}P is usually done by nick translation.[64,65] Nick translation is the net result of three different activities of two different enzymes, DNase I and *E. coli* DNA polymerase I. Briefly, the technique utilizes the ability of the polymerase to combine the sequential addition of nucleotide residues to the 3′-hydroxyl terminus of nicks in double-stranded DNA (generated by a limited digest with DNase I) with the concomitant sequential removal of nucleotides from the adjacent 5′-phosphoryl terminus, in what is really a *path clearing* function.[62,64,65] By the inclusion of one or more α-^{32}P-labeled deoxynucleoside triphosphates in the reaction, the DNA can be labeled to very high specific activity.[62,65] A full account of the nick translation reaction is given in Chapter 1, Section II.A.

Prior to hybridization, the ^{32}P-labeled DNA probe must be heat denatured which gives rise to single-stranded DNA fragments of varying lenth. Because complementary DNA fragments are present in solution during hybridization to the filter-bound target nucleic acid, the hybridization of DNA in solution will be much faster than to the target nucleic acid. However, since these double-stranded fragments formed in solution will have single-stranded tails of varying length because of the varying length of DNA fragments produced by DNase I digestion during nick translation, these single-stranded tails are available for hybridization to the target. There are probably also aggregates or networks formed of these tailed double-stranded fragments and these will serve to enhance the signal strength of probe bound to the target nucleic acid.

We have developed a novel chemical procedure for labeling nucleic acid probes non-isotopically using a photoactivatable analogue of biotin, termed Photobiotin®.[66] Here, the nucleic acid to be labeled is mixed with Photobiotin® and then irradiated briefly with strong white light. This results in the stable linkage of one biotin to every 100 to 150 nucleotides.[66] Such probes have proved as sensitive as radioactive probes in Northern, Southern, and dot-blot analyses.[67,68] A full account of the Photobiotin®-labeling procedure is given in Chapter 2, Section III.A.2.

In the case of DNA probes labeled with Photobiotin®, no DNase I treatment is required and no fragmentation of the DNA results. (This contrasts with the previous situation of DNA probes prepared by nick translation.) Surprisingly, such biotin-labeled probes after heat or alkali denaturation proved efficient as hybridization probes for the detection of BYDV RNA.[23] However, sonication of the probe before use led to a fivefold increase in sensitivity, a result most likely due to the fragmentation of the DNA and the production of tailed double-stranded fragments in solution during hybridization.

2. Single-Stranded DNA Probes

Single stranded ^{32}P-cDNA probes can be prepared by novel strategies using recombinant DNA clones of viruses or viroids in the single-stranded phage M13 vector (Chapter I, Section II.C) or by transcribing purified single-stranded RNA using reverse transcriptase (Chapter I, Section II.H). Using the former approach, Barker et al.[41] prepared a range of ^{32}P-labeled M13 DNA probes for the diagnosis of four viroids in plant tissues; ASBV, CEV, PSTV, and CCCV (coconut cadang cadang viroid). In this work, we reasoned that single-stranded DNA probes should be better than double-stranded ones because there is no complementary DNA strand in solution to effectively compete with the less efficient hybridization to the filter-bound target sequences. In reality, however, single-stranded DNA probes have proved to be no more sensitive than double-stranded probes on the several occasions when comparative studies were done.

3. RNA Probes Prepared from Double-Stranded RNA

In a novel approach to the preparation of ^{32}P-labeled RNA probes, Garger et al.[26] purified double-stranded RNA from plants infected with cucumber mosaic virus (CMV), tobacco mosaic virus (TMV), as well as from plants associated with '447' male sterile *Vicia faba.*[69] The RNA was then denatured and partially hydrolyzed by heating at 100°C for 10 min in 100% formamide. After ethanol precipitation, the RNA fragments were 5'-terminally labeled using γ-^{32}P-ATP and T4 polynucleotide kinase and successfully used in dot-blot hybridizations for the detection of specific viral RNAs (single-stranded replicative forms) and the RNA associated with male sterility.

Double-stranded RNA has a much greater thermal stability than double-stranded DNA under the same ionic conditions; e.g., the Tm of double-stranded RNA in 0.3 *M* NaCl is 95 to 100°C (75 to 85°C in 50% formamide) and of double-stranded DNA is 85 to 95°C (approximately 60°C in 50% formamide).[70] Hence, where the aim is to detect double-stranded RNAs in plant extracts by dot-blot hybridization, it is essential to use vigorous conditions to ensure their denaturation prior to application to the nitrocellulose or nylon filters since double-stranded nucleic acids bind very poorly. Hence, Garger er al.[26] heated plant extracts at 100°C for 3 min followed by rapid cooling, or made the samples to 2.2 *M* formaldehyde, 50% formamide, and incubated for 15 min at 65°C, or added CH_3HgOH (methylmercuric hydroxide) to 10 to 20 m*M* and incubated at room temperature for 5 min. (Methylmercuric hydroxide gives rapid and complete denaturation of double-stranded nucleic acids at concentrations of 5 m*M* and above in the absence of inactivating reagents such as thiols.[71] However, it is highly toxic and great care must be taken in its use.)

Although the preparation of RNA probes from double-stranded RNA has the advantage that no cloning is involved, probes prepared in this way have not often been used for dot-blot hybridization. The approach is certainly restricted to viruses where viral-specific double-stranded RNAs can be readily purified. It is not appropriate for viroids because of the very low concentrations in infected tissues of such RNAs.[72]

4. Single-Stranded RNA Probes

A major advance in RNA probe construction has been the development of specially constructed recombinant DNA plasmids that contain highly specific phage promoter sequences (e.g., SP6, T7, T3) for transcription *in vitro* by DNA-dependent RNA polymerases (Chapter 1, Section III). One such system, that involving SP6 RNA polymerase,[73,74] is described briefly here. Specialized cloning vectors pSP64 and pSP65 have been constructed with the SP6 phage promoter upstream from a multiple cloning site, thereby facilitating insertion of DNA sequences downstream from the SP6 promoter. After preparation of appropriate double-stranded DNA clones of the viral or viroid nucleic acids *with the cloned insert in the correct orientation,* ^{32}P-labeled RNA transcripts can be readily prepared for use as hybridization probes.

More recently, combination vectors containing two phage promoters on the one plasmid have become available commercially. Such vectors, e.g., pSP6/T7, pSP6/T3, and T7/T3 contain the two phage promoters in opposite orientation separated by a multiple cloning site. The use of two promoters allows transcription from either strand to be chosen so that, for example, *sense* or *antisense* ^{32}P-labeled RNA may be produced.

Nonradioactive RNA probes can also be readily prepared, either by incorporation of modified nucleotides (e.g., bio-11-UTP) during the enzymatic synthesis (Chapter 2, Section II.B) or by post-synthesis modification with regents such as Photobiotin® (Chapter 2, Section III.A.2). Added to their attractiveness is the fact that RNA:RNA hybrids are significantly more stable than RNA:DNA hybrids and even more stable than DNA:DNA hybrids.[75-77]

In view of the relatively recent introduction of these RNA transcription systems, there have been only a few reports of the use of such probes for the detection of plant viruses and viroids. For example, ^{32}P-RNA transcripts were used by Linthorst and Bol[36] as a means of detection of tobacco rattle virus (TRV) in potato and tulip. TRV is a bipartite RNA virus. Although the larger RNA (RNA 1) can replicate independently of RNA 2, RNA 2 contains the gene for the coat protein and requires coinfection of RNA 1 for its replication. Hence, hybridization analysis is essential for the detection of RNA 1 in the absence of RNA 2. ^{32}P-RNA transcripts were also used for the sensitive detection of PSTV in partially purified nucleic acid extracts of infected tomato plants.[42]

5. Synthetic Oligodeoxynucleotides

Recent advances in the solid-phase synthesis of oligodeoxynucleotides, computer technology, and instrumentation have provided the means for the routine synthesis of milligram quantities of single-stranded oligodeoxynucleotides of defined sequence.[78,79] Hence, one or more oligodeoxynucleotides probes, 20 to 30 nucleotides long, can be prepared for the detection of specific viral or viroid nucleic acids provided at least part of the sequence is known. Thus, Bar-Joseph et al.[35,48] have diagnosed ASBV, CEV, and TMV by dot-blot hybridization analysis using specific 5′-^{32}P-labeled oligodeoxynucleotides as probes. We have also used this approach for the diagnosis of ASBV, CEV, and chrysanthemum stunt viroid (CSV).[80]

The limitations to the use of synthetic oligodeoxynucleotides as probes should be appreciated in any planning for their use. The main concern is the size of the oligodeoxynucleotides in relation to the total size of the target nucleic acid. For example, a single oligodeoxynucleotide probe of 25 nucleotides represents only 0.4% of a target viral RNA of 6000 nucleotides. The sensitivity of detection of the target will therefore be much lower than if a full-length cDNA probe is used. The sensitivity can be increased by using more than one oligodeoxynucleotide but this has the penalty of a significant increase in the cost of the probes. In the case of viroids, which vary from 246 to 375 nucleotides in length,[81,82] two or three oligodeoxynucleotides 25 nucleotides long represent 13 to 30% of the target and, therefore, offer the potential for a sensitive alternative to the other methods already considered for the preparation of DNA or RNA probes.

Another important problem can arise in the use of synthetic oligodeoxynucleotide probes. All nucleic acids have regions of secondary and tertiary structure of varying strengths and an oligodeoxynucleotide probe must compete against these under the hybridization conditions used. In the case of target RNAs, local RNA:RNA interactions may be more stable than RNA:oligodeoxynucleotide interactions in the same region. If this is the case, then the oligodeoxynucleotide probe will hybridize poorly to the target sequence and be unsuitable for a probe. This problem is more likely to be found in such molecules as viroids where strong secondary and tertiary interactions occur.[81,82] The use of several oligodeoxynucleotides directed at different parts of the target molecule has the potential to alleviate this difficulty.

D. Prehybridization and Hybridization

A variety of protocols are available for the hybridization of a nucleic acid probe in solution to plant nucleic acid immobilized on a nitrocellulose filter. The conditons used vary according to the exact nature of the study and in general are governed by whether the hybridization is of the DNA:DNA, DNA:RNA or RNA:RNA type and whether closely related or distantly related sequences are reacting. Normally, reaction conditions are chosen which allow well-matched (stringent) hybrids to form (high stringency conditions) but occasionally some degree of mismatched (permissive) hybids may be desired (relaxed or low stringency conditions) in order to detect target sequences which are related but not identical. Hybridization stringency can be altered by adjusting salt and/or formamide concentrations and/or by changing the temperature of hybridization.[14,15]

Table 3
HYBRIDIZATION CONDITONS FOR THE DETECTION OF BARLEY YELLOW DWARF VIRUS (BYDV) RNA USING A BIOTIN-LABELED PROBE[23]

Prehybridization solution		Hybridization solution	
Formamide, deionized	50% (v/v)	Formamide, deionized	40% (w/v)
SSC[a]	5 ×	SSC	4 ×
Sodium phosphate, pH 6.5	50 m*M*	Sodium phosphate, pH 6.5	40 m*M*
Salmon sperm DNA, sonicated and denatured	0.025% (w/v)	Salmon sperm DNA, sonicated and denatured	0.02% (w/v)
SDS[b]	0.2% (w/v)	SDS	0.016% (w/v)
EDTA	5 m*M*	EDTA	4 m*M*
Denhardt's solution[c]	1 ×	Denhardt's solution	0.8 ×
		Dextran Sulfate	10% (w/v)

Note: Prehybridization: 4 to 16 h at 42°C. Hybridization: 16 to 23 h at 55°C. Biotin-labeled probe: Routine probe consists of a double-stranded cDNA, 1.7 kb region of the PAV isolate of BYDV in the plasmid pUC8.[23] Probe is labeled by Photobiotin®, sonicated and used at 50 ng/ml. Probe is denatured by heating at 95°C/5 min and snap cooling on ice or by alkali treatment (equal vol. of 100 m*M* NaOH) for 10 min at room temperature. Washes: 2 washes, 15 min each in 2 × SSC, 0.1% SDS at room temperature; 2 washes, 20 min each in 0.1 × SSC, 0.1% SDS at 55%C.

[a] SSC: 0.15 *M* NaCl, 0.015 *M* Na citrate.
[b] SDS: Sodium dodecyl sulfate.
[c] Denhardt's solution: 0.2 mg/ml each of bovine serum albumin, Ficoll 400, and polyvinyl pyrrolidone (M_r 40,000).

The hybridization process itself can be divided into three stages; prehybridization in the absence of probe, hybridization in the presence of probe, and washing to remove unhybridized or poorly hybridized probe. Both the prehybridization and hybridization steps are usually carried out in heat-sealable polythene bags which are immersed in a shaking water bath at the appropriate temperature. The protocols used in our laboratory for BYDV RNA[23] and ASBV[41,83] detection, involving biotin-labeled and ^{32}P-labeled probes, respectively, are described in Tables 3 and 4 and are adapted from Thomas.[16]

In the prehybridization step, nonspecific binding sites on the filter are blocked by incubation with a solution containing bovine serum albumin, Ficoll, polyvinyl pyrrolidone, nonhomologous DNA (salmon sperm DNA, sonicated and denatured to give small single-stranded fragments), formamide, and 0.75 *M* NaCl. Failure to carry out prehybridization results in nonspecific binding of probe and thus high backgrounds.

For hybridization, it is normally necessary to ensure that the added nucleic acid probe, whether it be radioactive or nonradioactive, is in a single-stranded form. For double-stranded DNA probes, this is usually achieved by a heat or alkali denaturation step (see Table 3). In some cases (e.g., upstream M13 DNA probes, Chapter 1, Section II.C.1), no denaturation step with the probe is required. The hybridization solution is essentially the same as the prehybridization solution with respect to most ingredients (Tables 3 and 4). The hybridization solution is 0.8 × concentration of prehybridization solution and has one additional ingredient, dextran sulfate ($10\%^{w}/_{v}$). This anionic polymer enhances the hybridization rate many fold.[14,15] Following hybridization, a series of washing steps are carried out to remove unbound probe and to dissociate unstable hybrids. The temperature and salt concentration of these washes determine which hybrids will be disassociated. Thus, the final stringency of the hybridization is determined by the conditions of *both* the hybridization and washing steps.

III. DIAGNOSIS OF PLANT VIRUSES

Nucleic acid hybridization involving the dot-blot procedure was first applied to plant viral

Table 4
HYBRIDIZATION CONDITIONS FOR THE DETECTION OF VIROIDS USING ^{32}P-LABELED DNA AND RNA PROBES[41,83,84]

Prehybridization solution		Hybridization solution	
Formamide, deionized	50% (v/v)	Formamide, deionized	40% (w/v)
SSC[a]	5 ×	SSC	4 ×
Sodium phosphate, pH 6.5	50 m*M*	Sodium phosphate, pH 6.5	40 m*M*
Salmon sperm DNA, sonicated and denatured	0.025% (w/v)	Salmon sperm DNA, sonicated and denatured	0.02% (w/v)
SDS[b]	0.2% (w/v)	SDS	0.016% (w/v)
EDTA	5 m*M*	EDTA	4 m*M*
Denhardt's solution[c]	1 ×	Denhardt's solution	0.8 ×
		Dextran Sulfate	10% (w/v)

Note: Prehybridization: *DNA probes:* 16 h at 42°C; *RNA probes:* 16 h at 55°C. Hybridization: *DNA probes:* 20—24 h at 55°C; *RNA probes:* 20—24 h at 65°C. ^{32}P-DNA probe: Routine probe consists of a single-stranded ^{32}P cDNA; prepared by use of a downstream primer on an M13 plus clone of viroid.[41,83] Probe has a spec. act. 3 × 10^9 cpm/μg DNA and is used at 0.2—0.5 × 10^6 cpm/ml. Prior to addition to the hybridization solution, the probe is denatured in 50% (v/v) deionized formamide by heating at 80°C for 3 min followed by snap cooling on ice. ^{32}P-RNA probe: Routine probe consists of a ^{32}P-labeled SP6 RNA transcript prepared from a pSP6-4 clone.[84] Probe has a spec. act. 6 × 10^8 cpm/μg RNA and is used at 0.2 — 0.5 × 10^6 cpm/ml. Prior to addition to the hybridization solution, the probe is denatured in 50% (v/v) deionized formamide by heating at 80°C for 1 min followed by snap cooling on ice. Washes: *DNA probes:* 3 washes, 15 min each in 2× SSC, 0.1% SDS at room temperature; 3 washes, 20 min each in 0.1 × SSC, 0.1% SDS at 55°C. *RNA probes:* 3 washes, 15 min each in 2 × SSC, 0.1% SDS at room temperature; 3 washes, 20 min each in 0.1 × SSC, 0.1% SDS at 65°C.

[a] SSC: 0.15 *M* NaCl, 0.015 *M* Na citrate.
[b] SDS: Sodium dodecyl sulfate.
[c] Denhardt's solution: 0.2 mg/ml each of bovine serum albumin, Ficoll 400, and polyvinyl pyrrolidone (M_r 40,000).

diagnosis in 1983.[25,26,31] Table 1 is a list of dot-blot hybridization studies used for the detection of a range of plant viruses over the period 1983 to 1987. As can be seen from Table 1, the most popular method for probe preparation involves the use of recombinant DNA clones. Such clones allow the isolation of large amounts of nucleic acid, so necessary for large scale diagnostic work.

A major drawback to the widespread use of dot-blotting as a routine plant viral diagnostic procedure is the use of ^{32}P for labeling nucleic acid probes. Many laboratories involved in plant viral diagnosis are not equipped to handle this radioisotope, nor can they afford the expense involved in the regular preparation of ^{32}P-labled probes. In an effort to overcome this problem we have demonstrated the use of a nonradioactive probe to detect the RNA of BYDV in plant extracts by dot-blot hybridization.[23] BYDV, the type member of the luteovirus group of plant viruses, causes an important disease of cereals worldwide and occurs in low concentrations in the phloem tissue of infected plants. Isolates of BYDV have been classified into two distinct groups based on immunological and biological properties. One group consists of the serotypes PAV, MAV, and SGV and the other includes the serotypes RPV and RMV.[85]

We have made use of the photoactivatable analogue of biotin, Photobiotin® (Chapter 2, Section III.A.2), to produce a biotinylated recombinant DNA probe in the plasmid, pUC8. This probe, containing a 1.7 kb segment of the PAV isolate of BYDV, has been successfully employed for the routine indexing of PAV isolates (or mixed isolates containing PAV) in extracts of cereal crops.[23]

In examining a range of different isolates (Figure 1), the sensitivity of detection of this

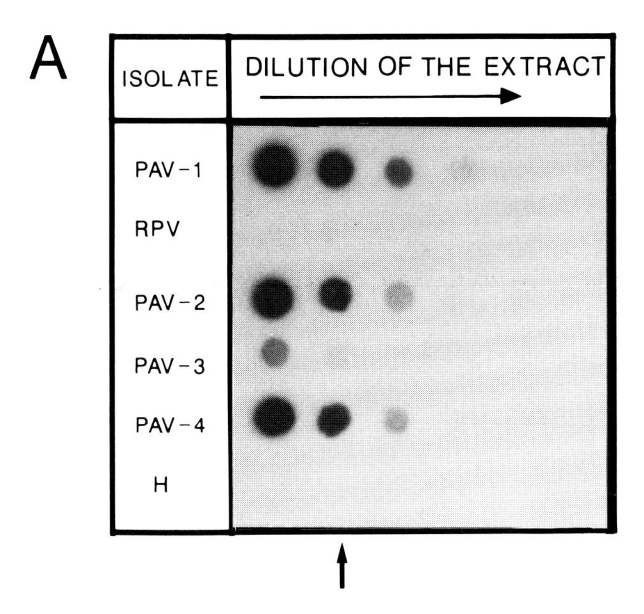

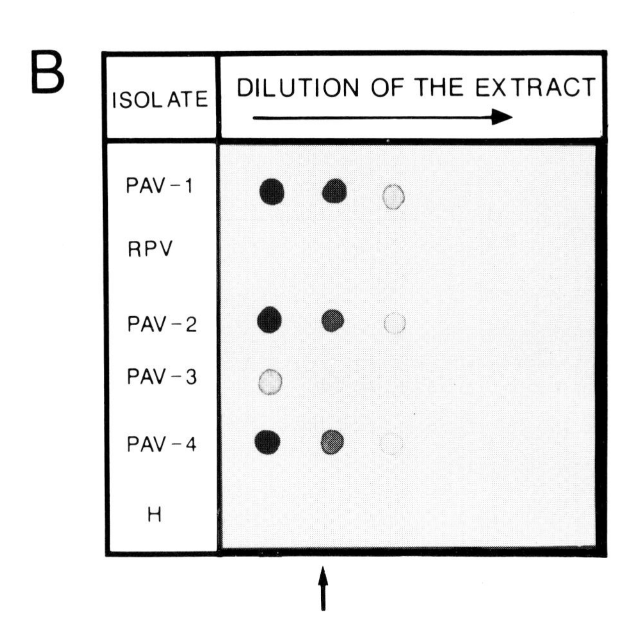

FIGURE 1. Detection of PAV RNA of barley yellow dwarf virus (BYDV) in Australian cereal crops by dot-blot hybridization analysis using a recombinant DNA probe containing a 1.7 kb segment of the PAV isolate of the BYDV genome in the plasmid pUC8.[23,86] Intact leaves from healthy and infected tissue were crushed between the rollers of a sap extractor and nucleic acid extracts were prepared according to Habili et al.[23] Extracts were serially diluted (fivefold dilutions) in 10 × SSC and 4 μl samples spotted directly onto nitrocellulose membranes.[23] The initial dilution spotted for each sample contains nucleic acid extracted from 2 mg of leaf tissue. The filters were prehybridized overnight at 42°C and hybridized for 24 h at 55°C (Table 3). (A) Sonicated, Photobiotin®-labled probe; probe concentration was 50 ng/ml and detection time was 4 h. (B) Nick translated ^{32}P probe (spec. activ. 1.0×10^7 cpm/μg DNA); autoradiography was for 72 h at −70°C using an intensifying screen. All samples, except H, were obtained from field plants showing yellowing symptoms and grown in Adelaide or Canberra. PAV-1, mixed PAV/RPV isolates from oats; PAV-2 and PAV-3, PAV isolates from wheat; PAV-4, PAV isolate from oats; RPV, RPV isolate from oats; H, nucleic acid extract from healthy wheat grown in a glasshouse. The vertical arrow beneath each filter indicates the highest dilution of extract detected in a parallel ELISA study.[87]

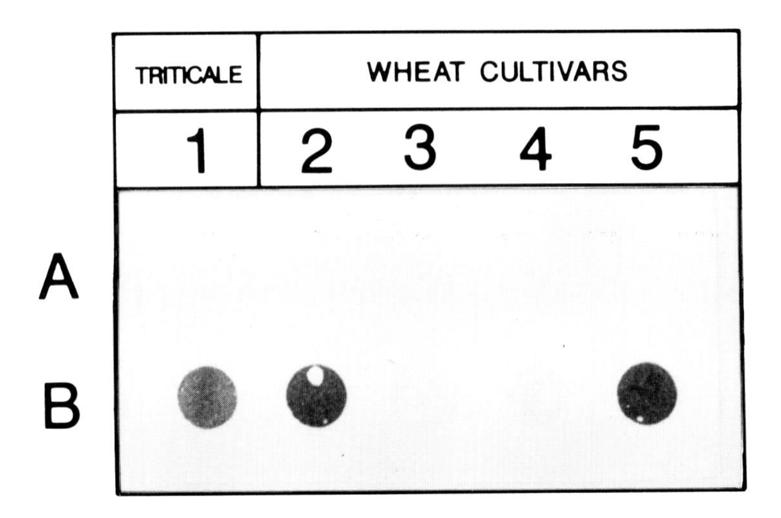

FIGURE 2. Detection of RNA of barley yellow dwarf virus (BYDV) in Australian cereal crops by dot-blot hybridization analysis.[68] Intact leaves from healthy and infected tissue were crushed between the rollers of a sap extractor and nucleic acid extracts were prepared.[23] Extracts were treated with formaldehyde and samples filtered onto nitrocellulose using a vacuum-manifold.[23] Each spot contains extract from 10 mg of leaf tissue. The filter was hybridized (Table 3) with a sonicated, Photobiotin®-labeled pUC8 plasmid containing a 1.7 kb fragment of the PAV isolate of the BYDV genome.[23] Probe concentration was 50 ng/ml, and detection time was 1 h. Sample (1), triticale cultivar; samples (2—5), four different wheat cultivars. Row A, healthy plants. Row B, infected plants. (From McInnes, J. L., Vize, P. D., Habili, N., and Symons, R. H., *Focus (Bethesda Research Laboratories/Life Technologies Inc.)*, 9 (4), 1, 1987. With permission.)

Photobiotin®-labeled probe for BYDV RNA was found to be comparable to that obtained with the same recombinant probe, labeled with ^{32}P by nick translation and used under the same conditions.[86] As can be seen, the probe hybridizes to PAV RNA but not to RPV RNA (other isolates of the virus have not, as yet, been available to test). Plant extracts from healthy tissue gave no detectable hybridization signal with either of the probes and the sensitivity of detection by these dot-blot hybridization procedures was at least 25-fold more sensitive than that obtained in a parallel ELISA study, carried out according to the protocol of Waterhouse et al.[22,87]

The biotinylated PAV probe has also been tested in a typical field study; here, leaf samples from a range of different cereal cultivars, with varying yellowing symptoms, were collected as were corresponding leaf samples from symptomless plants. Extracts were prepared by crushing leaves between the rollers of a sap press (Erich Pollahne, F.R.G.) followed by phenol/chloroform extraction,[23] spotted onto nitrocellulose by vacuum manifold and probed by dot-blot hybridization. The results (Figure 2, line B) indicate that all five different cultivars were BYDV infected. Positive hybridization signals associated with the different cultivars were seen to range from weak to strong, thus indicating different levels of BYDV infection within the plants. All five extracts (Figure 2, line A), prepared from corresponding symptomless cultivars were negative for BYDV.

In a more recent study, we have similarly used Photobiotin® to prepare a biotinylated, nonradioactive probe for the detection of another luteovirus: soybean dwarf virus (SDV), formerly known as subterranean clover red-leaf virus. This recombinant DNA probe contains a 1.4 kb segment of the SDV genome in the plasmid, pUC8,[88] and was sonicated after biotinylation and before use as a probe. Figure 3A shows a typical dot-blot hybridization study for SDV RNA detection; extracts prepared from infected tissue (spots 1 to 5) gave

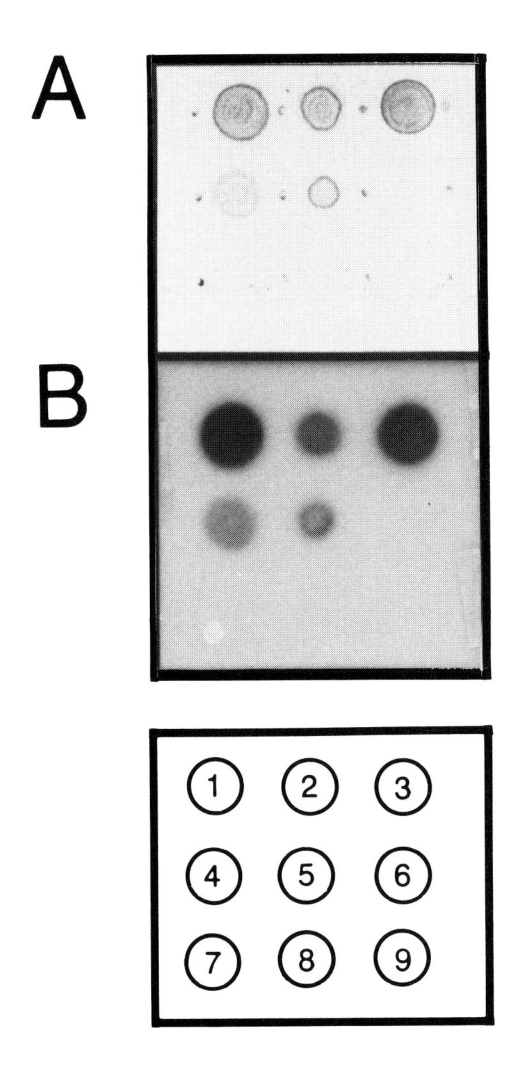

FIGURE 3. Detection of soybean dwarf virus (SDV) RNA in two Australian legumes by dot-blot hybridization analysis using a recombinant DNA probe containing a 1.4 kb segment of the SDV genome in the plasmid, pUC8.[88] Intact tissue from healthy and infected plants was crushed between the rollers of a sap extractor and nucleic acid extracts were prepared according to Habili et al.,[23] except the buffer used was 100 m*M* Tris HCl, pH 8.5, 10 m*M* EDTA, 2.0% SDS, 2.0% β-mercaptoethanol. Extracts were treated with formaldehyde and samples spotted directly onto duplicate nitrocellulose membranes.[23] Each spot contains extract from 10 mg of plant tissue. The filters were prehybridized overnight at 42°C and hybridized for 24 h at 55°C (Table 3). (A) Sonicated, Photobiotin®-labeled probe; probe concentration was 80 ng/ml and detection time was 3 h. (B) Nick-translated ^{32}P probe (spec. activ. 1.1×10^8 cpm/μg DNA); autoradiography was for 19 h at −70°C using an intensifying screen. Samples 1, 4, 5, 8 and 9, snap pea; samples 2, 3, 6 and 7, subterranean clover.

positive hybridization signals while extracts prepared from healthy, symptomless plants (spots 6 to 9) showed no hybridization. An identical result was obtained using the same recombinant DNA probe, labled with ^{32}P by nick translation (Figure 3B).

IV. DIAGNOSIS OF VIROIDS

Since viroids are unencapsidated RNA molecules, immunological methods cannot be used for their detection. The most sensitive, convenient, and specific diagnostic procedure is the standard dot-blot hybridization procedure, already outlined in Section II. Indeed, the first report of the detection of a plant pathogen by the use of dot-blot hybridization was for PSTV.[40] Since then the technique has become widely used in many countries for the routine detection of a range of viroids (Table 2). A major advantage with dot-blot hybridization is the sensitivity of detection; as can be seen from Table 2, viroids can be detected in the low picogram range in many plant extracts.

We have successfully developed a standard dot-blot hybridization procedure with ^{32}P-cDNA probes for the routine detection of several plant viroids.[41] For example, ASBV has been detected down to a level of approximately 20 pg/g fresh weight of leaves, using partially purified nucleic acid extracts. The probe routinely employed was a ^{32}P-labeled single-stranded cDNA and was prepared by the use of a downstream primer on an M13 plus clone of ASBV[41,83] (Chapter 1, Section II.C.2). This approach has proved extremely beneficial over recent years in detecting several viroids in a range of extracts from Australian plants.[41,51,52,57]

As mentioned in Section II.B, a considerable enhancement in hybridization signal strength has been reported by White and Bancroft[43] who briefly treated tissue extracts with a 20% ($^{w}/_{v}$) formaldehyde/10xSSC solution for 15 min at 65°C before application to the nitrocellulose filter. When this procedure was utilized in a PSTV diagnostic dot-blot hybridization assay using a ^{32}P-cDNA probe, low positive hybridization signals were obtained with healthy potato and tomato leaf extracts.[45] Such signals were subsequently shown to be due to low levels of plant ribosomal RNA.

On a more recent note, we have made use of Photobiotin® (Chapter 2, Section III.A.2) to produce a biotinylated recombinant DNA probe for PSTV in the plasmid, pSP64. This nonradioactive probe, containing a full-length cloned monomer insert of PSTV,[89] was successfully used to detect PSTV in partially purified nucleic acid extracts of potato leaf tissue (Figure 4A). The sensitivity of detection using this Photobiotin®-labeled probe was comparable to that obtained with a single-stranded ^{32}P-cDNA probe, prepared by the downstream primer approach of Barker et al.[41] and used under identical conditions (Figure 4B).

V. COMPARATIVE HYBRIDIZATION ANNAYSIS OF VIRAL NUCLEIC ACIDS

Hybridization analysis of plant viral nucleic acids using DNA probes has proved extremely useful for estimating the extent of sequence homology between viral nucleic acids and for probing the finer aspects of the organization of the viral genome. Such data can play an important role in the classification of plant viruses and in considerations of evolutionary relationships. However, it must be appreciated that hybridization analysis can never provide an exact measure of sequence homology between two or more viral RNAs and that this can only be obtained by complete sequence determination. Since the sequence determination of even one viral genome represents a major undertaking, comparative hybridization analysis will continue to play a role in the characterization of plant viruses. It is therefore important to understand the basis of the technique and its scope and limitations. Only a brief discussion is given here and the reader is referred to the review by Gould and Symons[13] and the references quoted here for further details.

Estimates of sequence homology between plant viral nucleic acids have been determined by hybridization in solution[90-95] and by hybridization using nitrocellulose filters.[27,37,96]

A. Hybridization Analysis in Solution

This technique has already been outlined in Section II.A. Briefly, single-stranded ^{32}P- or ^{3}H-cDNA prepared against a specific viral nucleic acid is hybridized in solution with partially purified or purified nucleic acids; remaining single-stranded probe or mismatched regions in hybrids are removed by digestion with the single-stranded specific nuclease S1 and the

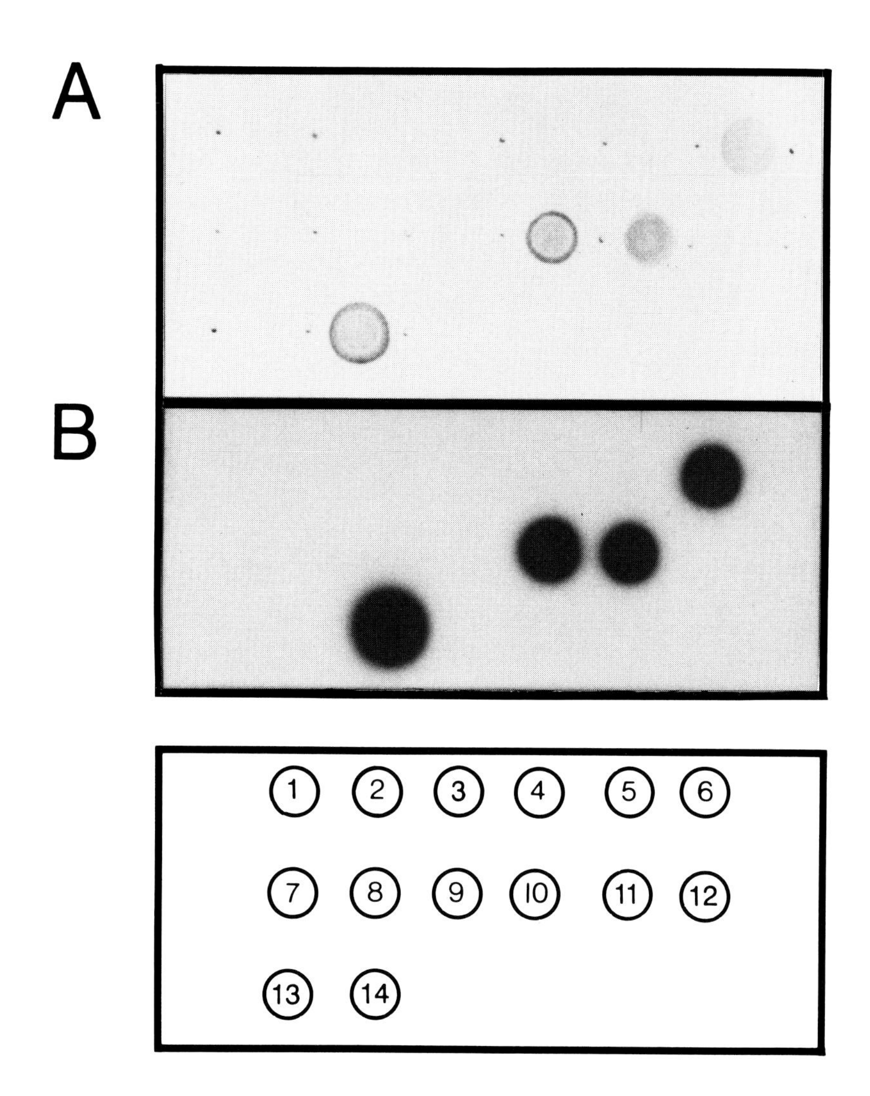

FIGURE 4. Detection of potato spindle tuber viroid (PSTV) in extracts of potato leaves by dot-blot hybridization analysis.[45] Partially purified nucleic acid extracts were prepared from healthy and infected leaves by an extraction procedure using AMES buffer (0.5 *M* sodium acetate, pH 6.0, 10 m*M* $MgCl_2$, 3.0% SDS, 20% ethanol), water-saturated phenol and chloroform. Samples were then spotted directly onto duplicate nitrocellulose membranes.[41] Each spot contains extract from approximately 1.5 mg of leaf tissue. The filters were prehybridized overnight at 42°C and hybridized for 23 h at 55°C (Tables 3 and 4). (A) Sonicated, Photobiotin®-labeled recombinant DNA probe containing a full length cloned monomer insert of PSTV[89] in the plasmid, pSP64. Probe concentration was 50 ng/ml and detection time was 1.5 h. (B) Single-stranded ^{32}P-cDNA probe (spec. activ. 3×10^9 cpm/μg DNA), prepared by the downstream primer approach of Barker et al.;[41] autoradiography was for 3 h at -70°C using an intensifying screen. Samples (1—5, 7—9), different potato cultivars; samples (6, 10, 11, and 14), known PSTV infected potato cultivars kept under controlled conditions; sample (12), extraction buffer; sample (13), healthy tomato tissue.

nuclease-resistant double-stranded nucleic acids are precipitated with acid and counted. Sequence homology can then be estimated using data from appropriate controls.

One of the first applications of the cDNA hybridization technique was the estimation of the sequence homology between the four RNAs (designated RNAs 1 to 4 in order of decreasing Mr) of the Q isolate of CMV.[90] It was shown that there was very little sequence homology between the three genomic RNAs (RNAs 1 to 3) of the Q-CMV isolated, but that the entire sequence of RNA 4 was completely contained within RNA 3 and not present in the other two RNAs.[90]

Comparative hybridization analysis was extended to the four RNAs of alfalfa mosaic virus (AMV).[91] Here, as found with CMV,[90] the complete sequence of AMV RNA 4 was completely contained only in AMV RNA 3; however, AMV RNAs 1 and 2 were shown to be unique, with little or no sequence homology between them or RNAs 3 and 4.[91]

The technique has also found application with viroid-like RNAs. For example, the two small viroid-like RNAs (termed virusoids) of velvet tobacco mottle virus (VTMoV) and solanum nodiflorum mottle virus (SNMV) were characterized by cDNA analysis[97,98] and were shown to be almost but not completely homologous in sequence, a result subsequently confirmed by sequencing.[99]

The examples cited so far have largely concerned sequence homology between the RNAs within a multipartite virus. A natural extension of this approach is to estimate the sequence homology between the genomes of different members of a family of viruses, thus providing information essential for the purpose of classification. Gonda and Symons[92] utilized sequence hybridization analysis to examine the RNAs of three strains (P, Q, and M) of CMV. Although the corresponding RNAs of the P and Q strains were found to be indistinguishable, the corresponding RNAs of the M strain showed only limited sequence homology (15 to 30%, 26 to 56% at high and low stringency, respectively) to the CMV-Q strain RNAs.[92] This result was rather surprising, considering that both viruses are classified as CMV and yet showed only roughly 30% sequence homology between their RNAs.

Solution hybridization analysis with cDNA has also proved helpful in elucidating the Tombamovirus group of viruses. The extent of RNA sequence homology between 13 isolates of tobamoviruses was estimated using ^{3}H-cDNA prepared against 7 of the viral RNAs, and the presence of mismatched regions in cDNA:RNA hybrids investigated by variation of the stringency of the S1 nuclease assay.[93] The data allowed classification of the 13 isolates into 5 groups on the basis of significant sequence homology between members of each group. Members of two of the group were further divided into subgroups on the basis of differences in the hybridization data.[93] Sequence homology studies involving solution hybridization have also been reported for the Potyviruses (potato virus Y group).[94,95]

B. Hybridization Analysis Using Membrane Filters

As an alternative to solution hybridization for the estimation of sequence homology between viral genomes, hybridization to nucleic acids immobilized on membrane filters offers a more rapid and flexible approach.[96,100] Perhaps the main advantages are that a number of samples can be rapidly compared under the same hybridization conditions and melting profiles on the probe-target hybrids can be determined.

As an example of this approach, Gallitelli et al.[37] investigated the relationship among viruses of the Tombusvirus group, of which tomato bushy stunt virus (TBSV) is the type member. Thus, full-length single-stranded RNAs from 11 different viruses were purified by gel electrophoresis and 18 ng of each spotted on nitrocellulose. Hybridization was carried out with ^{32}P-cDNA prepared by random priming (Chapter 1, Section II.B) on ten of the purified RNAs. After stringent hybridization and washing procedures, each spot was cut out and counted and percent sequence homology determined relative to the homologous cDNA:RNA result, taken as 100%. Sequence homologies between different RNAs varied from 0 to 57% as estimated by this method.

As an extension of this method, melting profiles were determined after hybridization by a sequential procedure which involved incubation of each spot three times for 5 min in a defined solution at a given temperature, counting the spot by Cerenkov radiation, and then repeating the procedure with 5°C increments in temperature.[96] The T_m values obtained were consistent with the estimates of sequence homology.[37]

As in the case of estimates of sequence homology determined by the liquid method,[13,92] only rough relative estimates can be obtained by this filter hybridization approach. For example, if only one half of a labeled cDNA molecule is stably hybridized under stringent conditions to the target nucleic acid on the filter and the other half remains single-stranded, then there will be a significant overestimate of sequence homology. It is feasible that the accuracy of such estimates could be improved by the use of nuclease S_1 to remove single-stranded regions of the DNA probe bound to the filter. In addition, the sequence homology estimate will be very dependent, in the absence of a nuclease S_1 step, on the integrity of the target nucleic acid on the filter and especially on the size range of the fragments which make up the DNA probe. For example, a DNA probe containing a lot of short fragments is likely to give a lower estimate of sequence homology than a DNA probe consisting of only one, or very few, long fragments.

Although there are inherent limitations to the use of this filter hybridization approach for accurate estimations of sequence homology, it remains a valuable technique for the rapid, comparative sequence homology analysis of nucleic acids.

VI. FUTURE ASPECTS

A. Introduction

Nucleic acid probes will undoubtedly play a major part in future developments for the diagnosis of plant viruses and probably also of fungal pathogens. As more and more viral genomes are sequenced and characterized and the similarities and differences between strains of each virus are determined, there will be an increasing use of nucleic acid probes in preference at antibody techniques. The great flexibility possible for the preparation of cloned probes, their stability once prepared, and the ease of producing them in large quantities provide significant advantages.

Currently, the use of nucleic acid probes is at an early phase of development. In contrast, antibody-based techniques have been available to the plant pathologist for many years and have enjoyed widespread and successful use. They can be used in simply equipped laboratories with sample preparation usually only requiring expression of sap; the reading of ELISA plates then allows the quantitation and print-out of results. In addition, it is possible to semiautomate the whole process as with "Tape-Press" machine in the Netherlands.[1]

The ELISA technique is critically dependent on both the supply of antibodies and their quality. The production of both polyclonal and monoclonal antibodies has really been a "cottage" industry, in which many researchers have prepared their own from viruses they have purified or have obtained them from colleagues. Problems can obviously arise where the virus used as antigen was poorly characterized or was a mixture of two or more viruses. The increasisng involvement of commercial firms in the marketing of ELISA reagents for plant viruses will help alleviate this problem of quality control.

The more widespread use of nucleic acid probes for routine diagnosis will be dependent, in part, on the development of simpler methods for their use. It is worthwhile considering what the requirements will be for the "second-generation" nucleic acid probes.

B. Future Directions of Probe Technology

1. Preparation of Plant Sample

In order to avoid background problems, especially with nonradioactive probes, it is the usual procedure to prepare a partially purified plant extract for dot-blot hybridization, a process which generally involves a phenol and/or chloroform deproteinization step. However, as with ELISA, unpurified expressed sap can be used with radioactive probes[22,25,26,30,32,40] but there may be a penalty of lower sensitivity of the assay and the potential for background hybridization in healthy samples. Lower sensitivity is not an important consideration where virus or viroid concentration in sap is relatively high, but it is where the concentration is low and it is important to detect all infected plants. Examples of the latter are the detection of ASBV in avocado trees[5,41] and of BYDV in cereals.[22,23] In these cases, a deproteinization step and concentration of the nucleic acids prior to dot-blot assay are needed.

There is certainly a need for simpler extraction procedures, preferably with as few steps as possible and avoidance of the use of phenol for deproteinization. For large numbers of samples, the tissue extraction must be simple since usual laboratory procedures such as homogenization in a blender or grinding in a mortar and pestle are impractical. In our hands, the roller sap extractor (Erich Pollahne, F.R.G.) is very efficient and can be used for soft tissues such as cereals[23] and leaves of wooly trees, such as avocado and oil palm.[101]

2. Nonradioactive Probes

The need for future routine diagnosis to be based on nonradioactive procedures is obvious. It is desirable to develop probe technology where the signal is not an insoluble colored product on a membrane filter but a soluble colored product or some other soluble product, both of which can be measured by suitable means and the result printed out. This approach is essential for the automation of the detection system.

3. Alternatives to Dot-Blots

Dot-blot assays are tedious in requiring the spotting of a small volume of each sample on to a membrane filter, baking to immobilize the nucleic acids, prehybridizing and then hybridizing before washing and color development. The use of plastic bags for the hybridization steps is another part of the procedure which needs eliminating.

a. Two-Phase Sandwich Hybridization

This method was developed by Ranki et al.[102] for the detection and quantitation of nucleic acids in crude clinical samples using adenovirus DNA as a model.[103] As far as we are aware, it has not been used for the detection of plant virus nucleic acids. The general principle of the method is outlined in Figure 5A using nonradioactive single-stranded DNA probes in the phage M13 DNA vector as an example. Two probes are required which hybridize to different, nonoverlapping regions of the target nucleic acid. One probe is bound to a solid support such as nitrocellulose by standard procedures while the other is labeled with biotin using, for example, Photobiotin® (Chapter 2, Section III.A.2). The probe-bound filter is hybridized with the test nucleic acid sample in the presence of the second biotin-labeled probe which can only be bound to the filter via the bridge of target nucleic acid. After the usual washing procedures, the biotin-labeled probe is detected by standard procedures (Chapter 2, Section V.A). It should be appreciated that each sample must by hybridized separately. Thus, Ranki et al.[102] used DNA bound to 10 mm diameter nitrocellulose discs and carried out the hybridization in 0.4 ml in small plastic ampoules at 65°C, usually overnight.

The method has been successfully used in the detection of adenovirus DNA in nasopharyngeal mucus aspirates using ^{125}I-DNA probes,[104] cytomegalovirus in urine,[105] and the *omp*A gene in *E. coli*.[106] Futher examples are given in Chapter 3. In an extension of this approach, Polsky-Cynkin et al.[107] have investigated model sandwich hybridization using DNA probes bound to polypropylene tubes and receptacles and to CNBr-activated Sepharose-4B beads with encouraging results. It appears feasible that this sandwich hybridization technique may have a role in the routine diagnosis of plant viral nucleic acids, especially if it can be readily adapted to nonradioactive detection methods.

b. Single-Phase Sandwich Hybridization

A viable long term alternative is for both the target nucleic acid and the two probes to hybridize *in solution*. Such a system would offer much faster hybridization because the rate

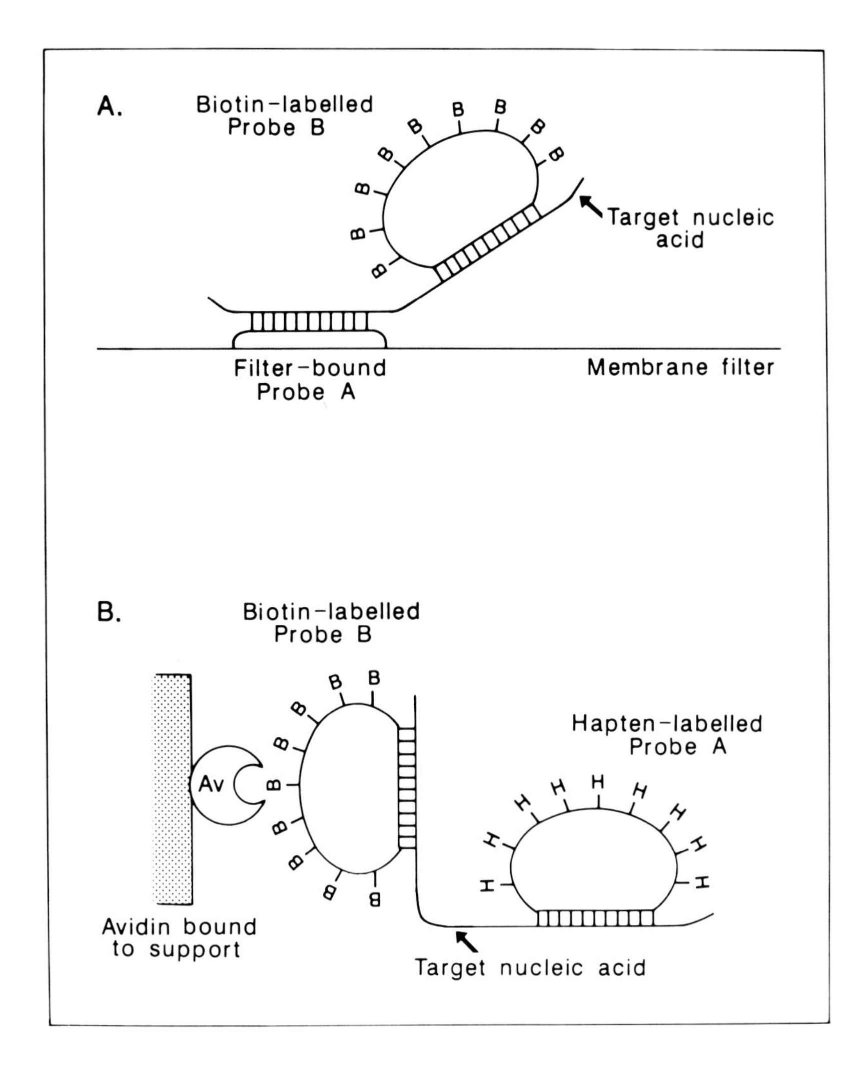

FIGURE 5. Schematic diagrams showing the general principles of two alternative methods of sandwich hybridization: (A) two-phase sandwich hybridization; (B) single-phase sandwich hybridization . The two probes shown here, as examples, are single-stranded DNA probes in the phage M13 DNA vector, one labeled with biotin (B), and the other with a hapten (H). See text for further details.

of hybridization in a single-phase system is a lot greater than in the usual two-phase dot-blot hybridization.[108-111] In addition, there is the potential for greater flexibility;[103,112] e.g., the size of the sample may be increased with consequent increase in final detection signal.

The critical step in this method is to conveniently and efficiently separate the probe: target hybrids from the unhybridized probe prior to signal detection. This can be achieved by adsorption of the hybrid complex to a solid support, followed by thorough washing. In the example in Figure 5B, support-bound avidin is used to adsorb the biotin-labeled probe B in the hybrid complex, an extraction process which we find may only be about 30% efficient.[112] Detection is via the hapten-labeled probe A and, for example, an enzyme coupled antibody.

The system described has considerable potential for the routine diagnosis of plant viruses using nonradioactive probes. Although the hybridization reaction is straight forward, the extraction and detection procedures are where rapid and reliable developments need to be made.

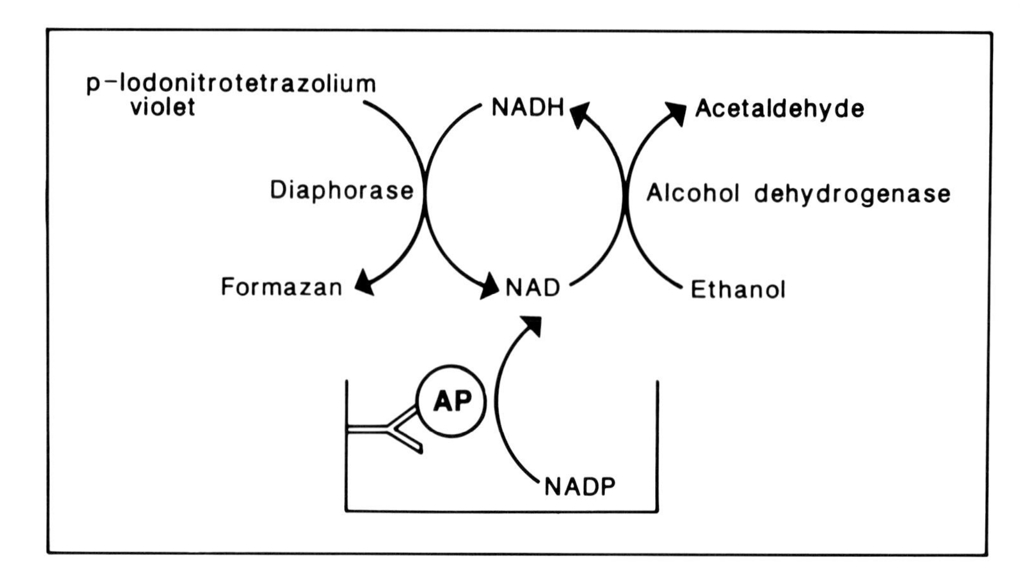

FIGURE 6. Example of the principle of signal amplification by enzymatic cycling.[114] See text for details. AP, alkaline phosphatase which is coupled to target-bound antibody.

4. Increasing the Sensitivity of the Assay

It is somewhat of a surprise that, with all the radioactive and nonradioactive detection methods available (Chapters 1 and 2), the lower level of detection of target nucleic acids is always of the order of 1 pg. Although this sensitivity may be quite adequate for most viruses which are present at relatively high concentrations in plant extracts, even greater sensitivity is needed for viruses and viroids present at low concentrations (see Section VI.B.1) or where there is a limit on the amount of tissue sample available.

a. Amplification of Signal by Enzymatic Cycling

This aspect is considered in more detail in Chapter 2, Section V.E.1. Briefly, the method depends on the use of an indicator enzyme to produce a product which is a catalytic activator for a secondary enzymic system which continually produces and amplifies a detectable signal by the enzymatic cycling of the products.[113-115] In one commercially available system (Figure 6), alkaline phosphatase is the detector enzyme (e.g., coupled to an antibody or avidin) which converts nicotinamide adenine dinucleotide phosphate (NADP) to NAD. NAD is the catalytic activator for the enzymatic cycling. It is converted to NADH in the presence of ethanol by alcohol dehydrogenase and is in turn reconverted to NAD by diaphorase in the presence of a colorless substrate which is converted to a colored product. (Figure 6).

This type of amplification system has yet to be used in the diagnosis of plant viruses using nucleic acid probes. However, it has been successfully used for the detection of BYDV in plants and in vector aphids using a double antibody sandwich ELISA system.[116] In this amplified ELISA, it was essential to have the indicator alkaline phosphatase coupled to a monoclonal antibody to BYDV since polyclonal antibodies gave unacceptable high blank values. The amplified ELISA gave a significant increase in sensitivity over conventional ELISA; for example, Torrance[116] found that only the amplified ELISA could reliably detect the BYDV-B isolate in oat sap.

It is feasible that the adaptation of the enzymatic cycling technique to the detection of nucleic acid probes will make a significant contribution to the development of the next generation of nucleic acid probes for plant viruses.

b. Enzyme Amplification of Target Nucleic Acid

Another approach is to amplify the target nucleic acid *prior* to detection by hybridization.[117-119] In this method, known as the polymerase chain reaction (PCR), a small part of the target nucleic acid (about 100 to 150 nucleotides) is amplified up many thousands of times such that the amplified product can be visualized by staining of polyacrylamide gels. It can successfully amplify up a single short sequence in the complex human genome.[117] Further details are given in Chapter 2, Section V.E.2.

This technique has yet to be applied to plant virus detection. However, it is likely to have limited application in view of the extra work required and the lack of a need for such large amplification of target sequences.

5. Increased Versatility in Detection Systems

At present, nucleic acid probes are not competitive enough to displace the ELISA systems for the detection of viruses which are present at moderate or high levels in infected plants. However, they do have the potential to offer greater flexibility and ease of preparation compared to the preparation and characterization of specific antibodies. As more and more plant virus genomes become sequenced, it should prove far easier and more economical to prepare DNA clones to various parts of the genome than to raise reliable antibodies, and especially monoclonal antibodies. It is feasible that new approaches may allow the probe technology to provide better alternatives than antibody based systems. Two areas of future interest are as follows.

a. Dip Stick Hybridization

The aim here would be to use a two-phase sandwich hybridization approach (Section VI.B.3) whereby one probe would be coupled to a plastic stick for hybridization in a plant extract. The method for the preparation of this extract would require the elimination or inactivation of nucleases. Given a reasonable concentration of target nucleic acid, it is feasible that hybridization times could be as short as 15 to 30 min and the whole assay, including leaf extraction, complete in less than 2 h.

b. Multiple Identification of Targets

In many infections in field or garden samples, more than one virus is involved. At present, the usual approach is to test for these one at a time. However, it is desirable to allow multiple identification of viruses in each sample and techniques should be developed which can do this. Each crop in a given area or country usually has a limited range of viral pathogens so that multiple detection need only be aimed at those pathogens usually present. In addition, numerous plant virus diseases have been shown to be seed-transmitted.[120] For example, in the case of potato seed testing, the important viruses to be diagnosed are the potato viruses Y, S, M, A, X and potato leafroll virus (PLRV).[121-123] A most attractive approach would seem, therefore, to produce a commercial test-kit which would enable multiple identification of these contaminating potato viruses from one test sample. Taking this approach one step further, one can also envisage a test-kit being available in the future which would allow the multiple identification of *individual strains* of a particular virus (e.g., potato virus X).[124]

ACKNOWLEDGMENTS

The writing of this chapter was supported by the Centre of Gene Technology which was set up by a Commonwealth Government Grant in the Department of Biochemistry of Adelaide University. We thank Nuredin Habili for discussions, Dr. W. Gerlach, Dr. A. Miller, and Dr. P. Waterhouse for BYDV and SDV cDNA clones and BYDV infected leaf samples, Ruth Evans for diagram preparations, Jennifer Cassady and Tammy Edmonds for photography, and Ros Murrell for typing the manuscript.

REFERENCES

1. **Barnett, O. W.,** Application of new test procedures to surveys: merging the new with the old, in *Developments in Applied Biology 1. Developments and Applications in Virus Testing,* Jones, R. A. C. and Torrance, L., Eds., Association of Applied Biologists, Wellesbourne, U.K., 1986, 247.
2. **Francki, R. I. B., Gould, A. R., and Hatta, T.,** Variation in the pathogenicity of three viruses of tomato, *Ann. Appl. Biol.,* 96, 219, 1980.
3. **Symons, R. H.,** Diagnostic approaches for the rapid and specific detection of plant viruses and viroids, in *Plant-Microbe Interactions, Vol. 1, Molecular and Genetic Perspective,* Kosuge, T. and Nester, E. W., Eds., Macmillan, New York, 1984, chap. 5.
4. **Allen R. N. and Dale, J. L.,** Application of rapid biochemical methods for detecting avocado sunblotch disease, *Ann. Appl. Biol.,* 98, 451, 1981.
5. **Palukaitis, P., Rakowski, A. G., Alexander, D. McE., and Symons, R. H.,** Rapid indexing of the sunblotch disease of avocados using a complementary DNA probe to a avocado sunblotch viroid, *Ann. Appl. Biol.,* 98, 439, 1981.
6. **Da Graca, J. V. and Van Vuuren, S. P.,** Use of high temperature to increase the rate of avocado sunblotch symptom development in indicator seedlings, *Plant Dis.,* 65, 46, 1981.
7. **Allen, R. N., Palukaitis, P., and Symons, R. H.,** Purified avocado sunblotch viroid causes disease in avocado seedlings, *Aust. Plant Pathol.,* 10, 31, 1981.
8. **Visvader, J. E. and Symons, R. H.,** Eleven new sequence variants of citrus exocortis viroid and the correlation of sequence with pathogenicity, *Nucleic Acids Res.,* 13, 2907, 1985.
9. **Clark, M. F. and Bar-Joseph, M.,** Enzyme immunosorbent assays in plant virology, in *Methods in Virology,* Vol. 7, Maramorosch, K. and Korowski, H., Eds., Academic Press, New York, 1984, 51.
10. **Cooper, J. I. and Edwards, M. L.,** Variations and limitations of enzyme-amplified immunoassays, in *Developments in Applied Biology. 1. Developments and Applications in Virus Testing,* Jones, R. A. C. and Torrance, L., Eds., Association of Applied Biologists, Wellesbourne, U.K., 1986, 139.
11. **Mowat, W. P. and Dawson, S.,** Detection and identification of plant viruses by ELISA using crude sap extracts and unfractionated antisera, *J. Virol. Methods,* 15, 233, 1987.
12. **Bar-Joseph, M., Garnsey, S. M., Gonsalves, D., Moscovitz, M., Purcifull, D. E., Clark, M. F., and Loebenstein, G.,** The use of enzyme-linked immunosorbent assay for detection of citrus tristeza virus, *Phytopathology,* 69, 190, 1979.
13. **Gould, A. R. and Symons, R. H.,** A molecular biological approach to relationships among viruses, *Annu. Rev. Phytopathol.,* 21, 179, 1983.
14. **Meinkoth, J. and Wahl, G.,** Hybridization of nucleic acids immobilized on solid supports, *Anal. Biochem.,* 138, 267, 1984.
15. **Hames, B. D. and Higgins, S. J., Eds.,** *Nucleic Acid Hybridization. A Practical Approach,* IRL Press, Oxford, U.K., 1985.
16. **Thomas, P. S.,** Hybridization of denatured RNA transferred or dotted to nitrocellulose paper, in *Methods in Enzymology,* Vol. 100, Wu, R., Grossman, L., and Moldave, K., Eds., Academic Press, New York, 1983, 225.
17. **Anderson, M. L. M. and Young, B. D.,** Quantitative filter hybridization, in *Nucleic Acid Hybridization. A Practical Approach,* Hames, B. D. and Higgins, S. J., Eds., IRL Press, Oxford, U.K., 1985, chap. 4.
18. **Southern, E. M.,** Detection of specific sequences among DNA fragments separated by gel electrophoresis, *J. Mol. Biol.,* 98, 503, 1975.
19. **Alwine, J. C., Kemp, D. J., Parker, B. A., Reiser, J., Renart, J., Stark, G. R., and Wahl, G. M.,** Detection of specific RNAs or specific fragments of DNA by fractionation in gels and transfer to diazobenzyloxymethyl paper, in *Methods in Enzymology,* Vol. 68, Wu, R., Ed., Academic Press, New York, 1979, 220.
20. **Nagamine, Y., Sentenac, A., and Fromageot, P.,** Selective blotting of restriction DNA fragments on nitrocellulose membranes at low salt concentrations, *Nucleic Acids Res.,* 8, 2453, 1980.
21. **Church, G. M. and Gilbert, W.,** Genomic sequencing, *Proc. Natl. Acad. Sci. U.S.A.,* 81, 1991, 1984.
22. **Waterhouse, P. M., Gerlach, W. L., and Miller, W. A.,** Serotype-specific and general luteovirus probes from cloned cDNA sequences of barley yellow dwarf virus, *J. Gen. Virol.,* 67, 1273, 1986.
23. **Habili, N., McInnes, J. L., and Symons, R. H.,** Nonradioactive, photobiotin-labelled DNA probes for the routine diagnosis of barley yellow dwarf virus, *J. Virol. Methods,* 16, 225, 1987.
24. **Barbara, D. J., Kawata, E. E., Ueng, P. P., Lister, R. M., and Larkins, B. A.,** Production of cDNA clones from the MAV isolate of barley yellow dwarf virus, *J. Gen. Virol.,* 68, 2419, 1987.
25. **Maule, A. J., Hull, R., and Donson, J.,** The application of dot hybridization to the detection of DNA and RNA viruses in plant tissues, *J. Virol. Methods,* 6, 215, 1983.
26. **Garger, S. J., Turpen, T., Carrington, J. C., Morris, T. J., Jordan, R. L., Dodds, J. A., and Grill, L. K.,** Rapid detection of plant RNA viruses by dot blot hybridization, *Plant Mol. Biol. Rep.,* 1, 21, 1983.

27. **Rosner A. and Bar-Joseph, M.,** Diversity of citrus tristeza virus strains indicted by hybridization with cloned cDNA sequences, *Virology,* 139, 189, 1984.
28. **Skotnicki, A. H., Dale, J. L., and Skotnicki, M. L.,** Detection of Fiji disease virus in infected sugarcane by nucleic acid hybridization, *J. Virol. Methods,* 13, 71, 1986.
29. **Boulton, M. I. and Markham, P. G.,** The use of squash-blotting to detect plant pathogens in insect vectors, in *Developments in Applied Biology. 1. Developments and Applications in Virus Testing,* Jones, R. A. C. and Torrance, L., Eds., Association of Applied Biologists, Wellesbourne, U.K., 1986, 55.
30. **Varveri, C., Ravelonandro, M., and Dunez, J.,** Construction and use of a cloned cDNA probe for the detection of plum pox virus in plants, *Phytopathology,* 77, 1221, 1987.
31. **Flavell, R. B., Kemble, R. J., Gunn, R. E., Abbott, A., and Baulcombe, D.,** Applications of molecular biology in plant breeding: the detection of genetic variation and viral pathogens, *CIBA Foundation Symp.,* 97, 198, 1983.
32. **Baulcombe, D., Flavell, R. B., Boulton, R. E., and Jellis, G. J.,** The sensitivity and specificity of a rapid nucleic acid hybridization method for the detection of potato virus X in crude sap samples, *Plant Pathol.,* 33, 361, 1984.
33. **Boulton, R. E., Jellis, G. J., Baulcombe, D. C., and Squire, A. M.,** The application of complementary DNA probes to routine virus detection, with particular reference to potato viruses, in *Developments in Applied Biology 1. Developments and Applications in Virus Testing,* Jones, R. A. C. and Torrance, L., Eds., Association of Applied Biologists, Wellesbourne, U.K., 1986, 41.
34. **Sela, I., Reichman, M., and Weissbach, A.,** Comparison of dot molecular hybridization and enzyme-linked immunosorbent assay for detecting tobacco mosaic virus in plant tissues and protoplasts, *Phyopathology,* 74, 385, 1984.
35. **Bar-Joseph, M., Segev, D., Blickle, W., Yesodi, V., Franck, A., and Rosner, A.,** Application of synthetic DNA probes for the detection of viroids and viruses, in *Developments in Applied Biology 1. Develpments and Applications in Virus Testing,* Jones, R. A. C. and Torrance, L., Eds., Association of Applied Biologists, Wellesbourne, U.K., 1986, 13.
36. **Linthorst, H. J. M. and Bol, J. F.,** cDNA hybridization as a means of detection of tobacco rattle virus in potato and tulip, in *Developments in Applied Biology 1. Developments and Applications in Virus Testing,* Jones, R. A. C. and Torrance, L., Eds., Association of Applied Biologists, Wellesbourne, U.K., 1986, 25.
37. **Gallitelli, D., Hull, R., and Koenig, R.,** Relationships among viruses in the Tombusvirus group: nucleic acid hybridization studies, *J. Gen. Virol.,* 66, 1523, 1985.
38. **Bresser, J., Hubbell, H. R., and Gillespie, D.,** Biological activity of mRNA immobilized on nitrocellulose in NaI, *Proc. Natl. Acad. Sci. U.S.A.,* 80, 6523, 1983.
39. **Bar-Joseph, M., Rosner, A., Moscovitz, M., and Hull, R.,** A simple procedure for the extraction of double-stranded RNA from virus-infected plants, *J. Virol. Methods,* 6, 1, 1983.
40. **Owens, R. A. and Diener, T. O.,** Sensitive and rapid diagnosis of potato spindle tuber viroid disease by nucleic acid hybridization, *Science,* 213, 670, 1981.
41. **Barker, J. M., McInnes, J. L., Murphy, P. J., and Symons, R. H.,** Dot-blot procedure with [^{32}P]DNA probes for the sensitive detection of avocado sunblotch and other viroids in plants, *J. Virol. Methods,* 10, 87, 1985.
42. **Lakshman, D. K., Hiruki, C., Wu, X. N., and Leung, W. C.,** Use of [^{32}P]RNA probes for the dot-hybridization detection of potato spindle tuber viroid, *J. Virol. Methods,* 14, 309, 1986.
43. **White, B. A. and Bancroft, F. C.,** Cytoplasmic dot hybridization: simple analysis of relative mRNA levels in multiple small cell or tissue samples, *J. Biol. Chem.,* 257, 8569, 1982.
44. **Flores, R.,** Detection of citrus exocortis viroid in crude extracts by dot-blot hybridization: conditions for reducing spurious hybridzation results and for enhancing the sensitivity of the technique, *J. Virol. Methods,* 13, 161, 1986.
45. **Habili, N., McInnes, J. L., and Symons, R. H.,** unpublished data, 1987.
46. **Rosner, A., Spiegel, S., Alper, M., and Bar-Joseph, M.,** Detection of avocado sunblotch viroid (ASBV) by dot-spot self-hybridization with a [^{32}P]-labelled ASBV-RNA, *Plant Mol. Biol.,* 1, 15, 1983.
47. **Spiegel, S., Alper, M., and Allen, R. N.,** Evaluation of biochemical methods for the diagnosis of the avocado sunblotch viroid in Israel, *Phytoparasitica,* 12, 37, 1984.
48. **Bar-Joseph, M., Segev, D., Twizer, S., and Rosner, A.,** Detection of avocado sunblotch viroid by hybridization with synthetic oligonucleotide probes, *J. Virol. Methods,* 10, 69, 1985.
49. **Mohamed, N. A. and Imperial, J. S.,** Detection and concentration of coconut cadang-cadang viroid in coconut leaf extracts, *Phytopathology,* 74, 165, 1984.
50. **Imperial J. S., Bautista, R. M., and Randles, J. W.,** Transmission of the coconut cadang-cadang viroid to six species of palm by inoculation with nucleic acid extracts, *Plant Pathol.,* 34, 391, 1985.
51. **Schwinghamer, M. W. and Broadbent, P.,** Association of viroids with a graft-transmissible dwarfing symptom in Australian orange trees, *Phytopathology,* 77, 205, 1987.

52. **Schwinghamer, M. W. and Broadbent, P.,** Detection of viroids in dwarfed orange trees by transmission to chrysanthemum, *Phytopathology,* 77, 210, 1987.
53. **Diener, T. O. and Owens, R. A.,** Detection of viroids in plants, in *Rapid Detection and Identification of Infectious Agents,* Kingsbury, D. T. and Falkow, S., Eds., Academic Press, New York, 1985, 139.
54. **Macquaire, G., Monsion, M., Mouches, C., Candresse, T., and Dunez, J.,** Spot hybridization: application to viroid identification, *Ann. Inst. Pasteur/Virol.,* 135E, 219, 1984.
55. **Harris, P. S., James, C. M., Liddell, A. D., and Okely, E.,** Viroid detection potato quarantine: the cDNA probe and other methods, in *Proc. British Crop Protection Conference. Pests and Diseases,* Vol. 1, Rudd-Jones, D., Ed., BCPC, Surrey, U.K., 1984, 187.
56. **Harris, P. S. and James, C. M.,** Exclusion of viroids from potato resources and the modified use of a cDNA probe, Bull. OEPP/EPPO Bull., 17, 51, 1987.
57. **Schwinghamer, M. W. and Scott, G. R.,** Survey of New South Wales potato crops for potato spindle tuber viroid with use of a ^{32}P-DNA probe, *Plant Dis.,* 70, 774, 1986.
58. **Bernardy, M. G., Jacoli, G. G., and Ragetli, H. W. J.,** Rapid detection of potato spindle tuber viroid (PSTV) by dot blot hybridization, *J. Phytopathol.,* 118, 171, 1987.
59. **Taylor, J. M., Illmensee, R., and Summers, J.,** Efficient transcription of RNA into DNA by avian sarcoma virus polymerase, *Biochim. Biophys. Acta,* 442, 324, 1976.
60. **Feinberg, A. P. and Vogelstein, B.,** A technique for radiolabeling DNA restriction endonuclease fragments to high specific activity, *Anal. Biochem.,* 132, 6, 1983.
61. **Feinberg, A. P. and Vogelstein, B.,** A technique for radiolabeling DNA restriction endonuclease fragments to high specific activity: addendum, *Anal. Biochem.,* 137, 266, 1984.
62. **Maniatis, T., Fritsch, E. F., and Sambrook, J.,** *Molecular Cloning. A Laboratory Manual,* Cold Spring Harbor Laboratory, Cold Spring Harbor, N.Y., 1982.
63. **Davis, L. G., Dibner, M. D., and Battey, J. F.,** *Basic Methods in Molecular Biology,* Elsevier, New York, 1986.
64. **Kelly, R. B., Cozzarelli, N. R., Deutscher, M. P., Lehman, I. R., and Kornberg, A.,** Enzymatic synthesis of deoxyribonucleic acid. XXXII. Replication of duplex deoxyribonucleic acid by polymerase at a single strand break, *J. Biol. Chem.,* 245, 39, 1970.
65. **Rigby, P. W. J., Dieckmann, M., Rhodes, C., and Berg, P.,** Labeling deoxyribonucleic acid to high specific activity *in vitro* by nick translation with DNA polymerase I, *J. Mol. Biol.,* 113, 237, 1977.
66. **Forster, A. C., McInnes, J. L., Skingle, D. C., and Symons, R. H.,** Non-radioactive hybridization probes prepared by the chemical labelling of DNA and RNA with a novel regent, photobiotin, *Nucleic Acids Res.,* 13, 745, 1985.
67. **McInnes, J. L., Dalton, S., Vize, P. D., and Robins, A. J.,** Non-radioactive photobiotin-labeled probes detect single copy genes and low abundance mRNA, *Bio/Technology,* 5, 269, 1987.
68. **McInnes, J. L., Vize, P. D., Habili, N., and Symons, R. H.,** Chemical biotinylation of nucleic acids with photobiotin and their use as hybridization probes, *Focus (Bethesda Research Laboratories/Life Technologies, Inc.),* 9 (4), 1, 1987.
69. **Grill, L. K. and Garger, S. J.,** Identification and charcterization of double-stranded RNA associated with cytoplasmic male sterility in *Vicia faba, Proc. Natl. Acad. Sci. U.S.A.,* 78, 7043, 1981.
70. **Cox, K. H., DeLeon, D. V., Angerer, L. M., and Angerer, R. C.,** Detection of mRNAs in sea urchin embryos by *in situ* hybridization using asymmetric RNA probes, *Dev. Biol.,* 101, 485, 1984.
71. **Bailey, J. M. and Davidson, N.,** Methylmercury as a reversible denaturing agent for agarose gel electrophoresis, *Anal. Biochem.,* 70, 75, 1976.
72. **Sanger, H. L.,** Minimal infectious agents: the viroids, in *The Microbe 1984, Part I, Viruses,* Mahy, B. W. J. and Pattison, J. R., Eds., Cambridge University Press, Cambridge, England, 1984, 281.
73. **Melton, D. A., Krieg, P. A., Rebagliati, M. R., Maniatis, T., Zinn, K., and Green, M. R.,** Efficient *in vitro* synthesis of biologically active RNA and RNA hybridization probes from plasmids containing a bacteriophage SP6 promoter, *Nucleic Acids Res.,* 12, 7035, 1984.
74. **Krieg, P. A. and Melton, D. A.,** *In vitro* RNA synthesis with SP6 RNA polymerase, in *Methods in Enzymology,* Vol. 155, Wu, R., Ed., Academic Press, New York, 1987, 397.
75. **Birnstiel, M. L., Sells, B. H., and Purdom, I. F.,** Kinetic complexity of RNA molecules, *J. Mol. Biol.,* 63, 21, 1972.
76. **Casey, J. and Davidson, N.,** Rates of formation and thermal stabilities of RNA:DNA and DNA:DNA duplexes at high concentrations of formamide, *Nucleic Acids Res.,* 4, 1539, 1977.
77. **Gray, D. M., Liu, J-J., Ratliff, R. L., and Allen, F. S.,** Sequence dependence of the circular dichroism of synthetic double-stranded RNAs, *Biopolymers,* 20, 1337, 1981.
78. **Gait, M. J., Ed.,** *Oligonucleotide Synthesis. A Practical Approach,* IRL Press, Oxford, U.K., 1984.
79. **Caruthers M. H.,** Gene synthesis machine: DNA chemistry and its uses, *Science,* 230, 281, 1985.
80. **Visvader, J. E., McInnes, J. L., and Symons, R. H.,** unpublished data, 1987.
81. **Keese, P. and Symons, R. H.,** The structure of vircids and virusoids, in *Viroids and Viroid-Like Pathogens,* Semancik, J. S., Ed., CRC Press, Boca Raton, FL, 1987, chap. 1.

82. **Keese, P., Visvader, J. E., and Symons, R. H.,** Sequence variability in plant viroid RNAs, in *RNA Genetics: Variability of RNA Genomes,* Vol. III, Domingo, E., Holland, J. J., and Ahlquist, P., Eds., CRC Press, Boca Raton, FL, 1988, chap. 4.
83. **Bruening, G., Gould, A. R., Murphy, P. J., and Symons, R. H.,** Oligomers of avocado sunblotch viroid are found in infected avocado leaves, *FEBS Lett.,* 148, 71, 1982.
84. **Hutchins, C. J., Keese, P., Visvader, J. E., Rathjen, P. D., McInnes, J. L., and Symons, R. H.,** Comparison of multimeric plus and minus forms of viroids and virusoids, *Plant Mol. Biol.,* 4, 293, 1985.
85. **Rochow, W. F. and Duffus, J. E.,** Luteoviruses and yellows *diseases,* in *Handbook of Plant Virus Infections and Comparative Diagnosis,* Kurstak, E., Ed., Elsevier/North Holland, Amsterdam, 1981, chap. 7.
86. **Habili, N.,** unpublished data, 1987.
87. **Waterhouse, P. M.,** unpublished data, 1987.
88. **Habili, N., Hercus, T., McInnes, J. L., and Symons, R. H.,** unpublished data, 1987.
89. **Van Wezedbeek, P., Vos, P., Van Boom, J., and Van Kammen, A.,** Molecular cloning and characterization of a complete DNA copy of potato spindle tuber viroid RNA, *Nucleic Acids Res.,* 10, 7947, 1982.
90. **Gould, A. R. and Symons, R. H.,** Determination of the sequence homology between the four RNA species of cucumber mosaic virus by hybridization analysis with complementary DNA, *Nucleic Acids Res.,* 4, 3787, 1977.
91. **Gould, A. R. and Symons, R. H.,** Alfalfa mosaic virus RNA. Determination of sequence homology between the four RNA species and a comparison with the four RNA species of cucumber mosaic virus, *Eur. J. Biochem.,* 91, 269, 1978.
92. **Gonda, T. J. and Symons, R. H.,** The use of hybridization analysis with complementary DNA to determine the RNA sequence homology between strains of plant viruses: its application to several strains of cucumoviruses, *Virology,* 88, 361, 1978.
93. **Palukaitis, P. and Symons, R. H.,** Nucleotide sequence homology of thirteen tobamovirus RNAs as determined by hybridization analysis with complementary DNA, *Virology,* 107, 354, 1980.
94. **Abu-Samah, N. and Randles, J. W.,** A comparison of the nucleotide sequence homologies of three isolates of bean yellow mosaic virus and their relationship to other potyviruses, *Virology,* 110, 436, 1981.
95. **Reddick, B. B. and Barnett, O. W.,** A comparison of three potyviruses by direct hybridization analysis, *Phytopathology,* 73, 1506, 1983.
96. **Kafatos, F. C., Jones, C. W., and Efstratiadis, A.,** Determination of nucleic acid sequence homologies and relative concentrations by a dot hybridization procedure, *Nucleic Acids Res.,* 7, 1541, 1979.
97. **Gould, A. R.,** Studies on encapsidated viroid-like RNA. II. Purification and characterization of a viroid-like RNA associated with velvet tobacco mottle virus (VTMoV), *Virology,* 108, 123, 1981.
98. **Gould, A. R. and Hatta, T.,** Studies on encapsidated viroid-like RNA. III. Comparative studies on RNAs isolated from velvet tobacco mottle virus and solanum nodiflorum mottle virus, *Virology,* 109, 137, 1981.
99. **Haseloff, J. and Symons, R. H.,** Comparative sequence and structure of viroid-like RNAs of two plant viruses, *Nucleic Acids Res.,* 10, 3681, 1982.
100. **Howley, P. M., Israel, M. A., Law, M-F., and Martin, M. A.,** A rapid method for detecting and mapping homology between heterologous DNAs: evaluation of polyoma virus genomes, *J. Biol. Chem.,* 254, 4876, 1979.
101. **Habili, N., Karageorgos, L., McInnes, J. L., and Symons, R. H.,** unpublished data, 1987.
102. **Ranki, M., Palva, A., Virtanen, M., Laaksonen, M., and Söderlund, H.,** Sandwich hybridization as a convenient method for the detection of nucleic acids in crude samples, *Gene,* 21, 77, 1983.
103. **Syvänen, A-C.,** Nucleic acid hybridization: from research tool to routine diagnostic method, *Med. Biol.,* 64, 313, 1986.
104. **Virtanen, M., Palva, A., Laaksonen, M., Halonen, P., Söderlund, H., and Ranki, M.,** Novel test for rapid viral diagnosis: detection of adenovirus in nasopharyngeal mucus aspirates by means of nucleic-acid sandwich hybridization, *Lancet,* 1, 381, 1983.
105. **Virtanen, M., Syvänen, A-C., Oram, J., Söderlund, H., and Ranki, M.,** Cytomegalovirus in urine: Detection of viral DNA by sandwich hybridization, *J. Clin. Microbiol.,* 20, 1083, 1984.
106. **Palva, A.,** *omp*A gene in the detection of *Escherichia coli* and other *Enterobacteriaceae* by nucleic acid sandwich hybridization, *J. Clin. Microbiol.,* 18, 92, 1983.
107. **Polsky-Cynkin, R., Parsons, G. H., Allerdt, L., Landes, G., Davis, G., and Rashtchian, A.,** Use of DNA immobilized on plastic and agarose supports to detect DNA by sandwich hybridization, *Clin. Chem.,* 31, 1438, 1985.
108. **Nobrega, F. G., Dieckmann, C. L., and Tzagoloff, A.,** A rapid method for detecting specific RNA transcripts by hybridization to DNA probes in solution, *Anal. Biochem.,* 131, 141, 1983.
109. **Kennell, D.,** Principles and practices of nucleic acid hybridization, in *Progress in Nucleic Acid Research and Molecular Biology,* Vol. 11, Davidson, J. N. and Cohn, W. E., Eds., Academic Press, New York, 1971, 259.

110. **Spiegelman, G. B., Haber, J. E., and Halvorson, H. O.,** Kinetics of ribonucleic acid - deoxyribonucleic acid membrane filter hybridization, *Biochemistry,* 12, 1234, 1973.
111. **Flavell, R. A., Birfelder, E. J., Sanders, J. P. M., and Borst, P.,** DNA-DNA hybridizations on nirtocellulose filters. 1. General considerations and non-ideal kinetics, *Eur. J. Biochem.,* 47, 535, 1974.
112. **McInnes, J. L. and Symons, R. H.,** unpublished data, 1986.
113. **Lowry, O. H., Passonnueau, J. V., Schulz, D. W., and Rock, M. K.,** The measurement of pyridine nucleotides by enzymatic cycling, *J. Biol. Chem.,* 236, 2746, 1961.
114. **Self, C. H.,** Enzyme amplification — a general method applied to provide an immunoassisted assay for placental alkaline phosphatase, *J. Immunol. Methods,* 76, 389, 1985.
115. **Bates, D. L.,** Enzyme amplification in diagnostics, *Trends Biotechnol.,* 5, 204, 1987.
116. **Torrance, L.,** Use of enzyme amplification in an ELISA to increase sensitivity of detection of barley yellow dwarf virus in oats and in individual vector aphids, *J. Virol. Methods,* 15, 131 1987.
117. **Saiki, R. K., Scharf, S., Faloona, F., Mullis, K. B., Horn, G. T., Erlich, H. A., and Arnheim, N.,** Enzymatic amplification of β-globin genomic sequences and restriction site analysis for diagnosis of sickle cell anemia, *Science* , 230, 1350, 1985.
118. **Mullis, K., Faloona, F., Scharf, S., Saiki, R., Horn, G., and Erlich, H.,** Specific enzymatic amplification of DNA *in vitro:* the polymerase chain reaction, *Cold Spring Harbor Symp. Quant. Biol.,* 51, 263, 1986.
119. **Mullis, K. B. and Faloona, F. A.,** Specific synthesis of DNA *in vitro* via a polymerase-catalysed chain reaction, in *Methods in Enzymology,* Vol. 155, Wu, R., Ed., Academic Press, New York, 1987, in press.
120. **Bos, L.,** Seed-borne viruses, in *Plant Health and Quarantine in International Transfer of Genetic Resources,* Hewett, W. B. and Chiarappa, L., Eds., CRC Press, Cleveland, 1977, 39.
121. **Jones, R. A. C.,** Tests for transmission of four potato viruses through potato true-seed, *Ann. Appl. Biol.,* 100, 315, 1982.
122. **Jeffries, C. J.,** The Scottish seed potato classification and the production of nucleus stocks using micropropagation, Mono. No. 33, BCPC Symp. Healthy Planting Material, 1986, 239.
123. *Boehringer Mannheim GmbH Biochemica, Phytodiagnostica, Catalog,* 1987/1988.
124. **Torrance, L., Larkins, A. P., and Butcher, G. W.,** Characterization of monoclonal antibodies against potato virus X and comparison of serotypes with resistance groups, *J. Gen. Virol.,* 67, 57, 1986.

Chapter 5

DETECTION OF SPECIFIC DNA AND RNA SEQUENCES IN TISSUES AND CELLS BY *IN SITU* HYBRIDIZATION

E. J. Gowans, A. R. Jilbert, and C. J. Burrell

TABLE OF CONTENTS

I. INTRODUCTION

In situ cytohybridization was developed in the late 1960s and early 1970s to detect specific nucleic acid sequences in cytological preparations[1,2] and has subsequently been adapted to permit detection of either DNA or RNA sequences in cell smears and tissue sections. The method has also been referred to as hybridocytochemistry[3] or hybridization histochemistry,[4] but most workers simply refer to it as *in situ* hybridization (ISH), and we shall use this term in this chapter. The technique has attracted a great deal of attention reflecting the potential for unique information that the method can provide, but sadly, has also suffered from a "Cinderella" image caused by a lack of consensus concerning methodology. As a result, there is a plethora of methods which claim success, not all of which are reproducible in different hands; however, we have not attempted a comprehensive review, but we have aimed to provide an overview of different strategies for ISH and have included proven methods that are known to be reproducible in different laboratories.

ISH is ideally suited not only to the detection of cellular nucleic acid sequences which are nonuniformly distributed in tissues or cells, but also to the detection of viral nucleic acid sequences which are often focal in distribution. Although much of our ISH experience has been gained in the study of virus infections in tissues, many of the approaches and concepts are generally applicable. The detection of nucleic acids by ISH is conceptually analogous to the detection of antigens by immunochemical means in tissues and cells, and, consequently, must satisfy the same primary objective, viz., to reflect accurately the inter- and intracellular distribution of target molecules in the sample. Similarly, strict formal controls for reaction specificity are required (see below). In general, ISH may be used to detect DNA corresponding to normal or abnormal genes to identify the chromosomal location of particular DNA sequences, or to assess the level of expression of these genes by mRNA detection. In particular, mRNA is a common target for *in situ* hybridization reactions in studies of gene expression and cell differentiation.

In the special case of virus-infected cells, the worker may choose to detect either viral genomic nucleic acid or RNA transcripts. Furthermore, because both DNA- and RNA-viruses may be either double- or single-stranded (ds or ss), tissue fixation protocols can be designed to preserve the natural configuration (i.e., ds or ss) of the nucleic acid within the tissue. This necessitates that an optional denaturation step be introduced in order to discriminate between ds and ss forms of nucleic acid target. In addition, use of strand-specific probes[7,8] allows identification of the relative amounts of positive- and negative-sense target molecules (DNA or RNA) that may be present. This kind of analysis may be used to identify viral nucleic acid replicative intermediates.[7,9]

Finally, ISH is especially valuable where the histological identity and relationships of positive cells within a tissue are sought, particularly where target molecules are nonrandomly distributed throughout the tissue. In contrast Southern-, Northern- or dot-blot hybridization assays will determine the average content of target molecules per cell in the extracted tissue. Thus, ISH may detect specific target sequences that are undetected by conventional hybridization assays where the specific sequences are focally distributed in a small number of cells that contain significant levels of target molecules.[10]

II. EQUIPMENT REQUIRED

ISH includes aspects of molecular biology, histochemistry, histology, and the laboratory must have access to basic equipment used in these areas. Good quality microscope slides produce lower backgrounds than inferior slides, but in any case before use, slides should be dipped in a solution which (1) ensures that the tissue sections adhere to the slide, (2) reduces the nonspecific binding of probes, and (3) provides a firm substrate for the nuclear emulsion if autoradiographic methods to detect bound probes are used. The slides are first washed in chromic acid solution, washed exhaustively in distilled water, and coated with poly-L-lysine,[6] gelatin:chrome alum,[11] or 3′ aminopropyltriethoxysilane[12] as described. We routinely use gelatin:chrome alum-coated slides with good results.

The hybridization step itself is performed under sterile siliconized coverslips; ca. 13-mm coverslips are cleaned in chromic acid and siliconized using commercially available solutions, as recommended by the supplier. We have used a number of brands, but prefer Prosil (Fisher) as it provides a cleaner finish than most. Siliconization prevents the probe from attaching to the coverslips and ensures easy removal of coverslips after hybridization, with minimal damage to the tissue section.

III. SAMPLE PREPARATION

Traditionally, frozen sections have been used for ISH studies, although fixed, paraffin wax-embedded (FPE) tissues can, in some instances, be quite satisfactory. Pieces of tissue, fixed in formalin or ethanol:acetic acid (EAA) prior to paraffin wax-embedding, provide better morphology than is usually possible with frozen tissues. FPE blocks are conveniently stored at room temperature. However, FPE tissues restrict probe access during the hybridization step and a protease digestion step (see below) is necessary for optimal signal:noise ratios. Furthermore, EAA may partly denature ds nucleic acids during fixation while formalin-fixed tissues show reduced reaction sensitivity. For these reasons, frozen sections still remain the choice where simplicity and maximum sensitivity are required, while FPE sections are preferable where histological detail is an important part of the information being sought.

A. Frozen Tissues

Pieces of tissue up to 0.5 cm^3 are immersed in OCT (Miles Scientific) and snap frozen on a piece of cork or similar support by holding the sample above the surface of liquid nitrogen until it is two thirds frozen when it is plunged into the nitrogen. The sample may then be stored frozen at −70°C until required, but best results are obtained when the tissue is fresh. Freezing the samples on a piece of cork, etc. ensures that the tissue may be mounted and removed from the cryostat without actually handling and damaging the tissue itself. 5 to 6 μm Sections are mounted on gelatin-coated microscope slides and air dried for 30 min prior to fixation. We have found that this air-drying step helps in section adherance, contributes to low backgrounds, and increases the sensitivity, probably as a result of improved probe access. This step is followed by fixation designed not only to fix nucleic acids, but also to bind the section firmly to the slide; glutaraldehyde is used for DNA (Table 1) and paraformaldehyde for RNA (Table 2). Paraformaldehyde has recently been shown to be the optimum fixation method for RNA detection.[13,14] These methods are simple, as we feel, particularly for RNA, that complex fixation protocols only lead to loss of target; both methods

Table 1
GENERAL FIXATION PROTOCOL FOR THE DETECTION OF DNA BY *IN SITU* HYBRIDIZATION

I.A Cell Smears or Monolayers

Air dry, 30 min, RT
EAA (3:1), 10 min, RT
Rinse in 70% ethanol
Hydrate to PBS
Proceed to II

I.B. Frozen Sections

Snap freeze (see text), store at 70°C; cut sections; air dry 30 to 45 min RT
Proceed to II

I.C FPE Tissue Blocks

Fix biopsy size samples in 10% buffered formalin for 12 to 24 h or in freshly prepared EAA (3:1) for 15 to 20 min
Dehydrate tissues and process into paraffin wax using a rapid procedure
Cut sections, dry overnight at 37°C
Dewax and rehydrate
Proceed to II

II. All Samples

Fix in 0.1% glutaraldehyde in PBS 30 min, 4°C
Wash in PBS 2 × 5 min, RT
Digest with 0.5 to 1.0 μg/ml Proteinase K[a] in 20 m*M* Tris-HCl pH 7.4, 2 m*M* $CaCl_2$, 10 m*M* EDTA, 15 min, 37°C
Digest with 100 to 200 μg/ml RNase A in 2 × SSC[b], 30 min, 37°C
Wash in 2 × SSC[c], 5 min, RT Proteinase K[c], as above, 5 min, 37°C
Denature target[d], 0.1 × SSC, 5 min, 100°C
Plunge into ice cold 0.1 × SSC
Refix in 0.1% glutaraldehyde in PBS, 15 min, 4°C
Rinse in 0.1 × SSC, briefly
Wash in 0.25% acetic anhydride in 0.1 *M* triethanolamine pH 8.0 10 min, RT
Wash in 2 × SSC, 2 × 5 min, RT
Dehydrate and dry

[a] Increased to 1.0 μg/ml for EAA-fixed tissues and 10 to 50 μg/ml for formalin-fixed tissues.
[b] Optional.
[c] Required if RNase A step included.
[d] Omitted if ssDNA is the target of the reaction.

include optional denaturation steps which are necessary if dsDNA or dsRNA is the target of the reaction.

B. Fixed, Paraffin Wax-Embedded Tissues

Fixed, paraffin wax-embedded tissues provide greater convenience, better morphology and, in the case of formalin-fixed tissues, the opportunity to perform retrospective *in situ* by hybridization studies. Methods for EAA- and formalin-fixed tissues are included in Tables 1 and 2. In our hands, EAA-fixed, paraffin wax-embedded tissues show similar sensitivity to frozen sections for the detection of DNA; both these methods produce higher hybridization signals for DNA than is possible in formalin-fixed, paraffin wax-embedded tissue. Neither EAA- nor formalin-fixed tissues are optimal for RNA detection although adequate signals may be obtained with careful technique (Table 2).

Table 2
GENERAL FIXATION PROTOCOL FOR THE DETECTION OF RNA BY *IN SITU* HYBRIDIZATION IN FROZEN OR FPE SECTIONS

I. Cells and Frozen Sections

Air dry, 30 to 45 min, RT
Fix in 4% paraformaldehyde, 5 m*M* $MgCl_2$ in PBS, 5 min, RT
Store slides in 70% ethanol at 4°C for up to 1 week, if necessary
Wash in 2 × SSC
Denature (if target is ds viral RNA) in 90% formamide in TE, 15 min at 70°C; plunge into 70% ethanol
Wash in 2 × SSC
Wash in 0.25% acetic anhydride in 0.1 *M* triethanolamine, pH 8.0, 10 min, RT
Wash in 2 × SSC, 2 × 5 min, RT; dehydrate and dry

II. FPE Tissue

EAA- and formalin-fixed paraffin-wax-embedded tissue may be used to detect RNA, although neither is as sensitive as the above method. Nevertheless, the following protocol has been used successfully to detect hepatitis delta virus (HDV) RNA in formalin-fixed tissue.

Air dry, overnight, 37°C
Dewax sections in xylene, 2 × 5 min, RT
Fix in 4% paraformaldehyde, 5 m*M* $MgCl_2$ in PBS — 5 min, RT
Wash in PBS
Treat with 0.2*M* HCl, 20 min, RT
Wash in DDW
Digest in 50 μg/ml Proteinase K in 20 m*M* Tris pH 7.4, 2 m*M* $CaCl_2$, 10 m*M* EDTA, 15 min, 37°C
Wash in DDW
Wash in 0.25% acetic anhydride in 0.1 *M* triethanolamine pH 8.0, 10 min, RT
Wash in DDW
Dehydrate and dry

Care must be taken to avoid overfixation with EAA that leads to extraction of DNA, while overfixation with formalin results in a further decrease in the hybridization signal. Digestion of the tissue sections with pronase or Proteinase K improves probe access leading to increased hybridization signals, but care must be taken to avoid overdigestion which results in loss of target sequences (Table 3) and poor morphology. It is recommended that each new batch of Proteinase K be titrated to determine optimum conditions and that stock Proteinase K solutions are then stored in aliquots at −70°C. The use of plastic resin-embedding media provides thin sections and good morphology, but probe access is extremely limited, often resulting in decreased hybridization signals.[16]

Irrespective of which fixation protocol is used, frozen and paraffin wax-embedded tissue sections are acetylated prior to hybridization,[17] a step which we have found reduces the background with all commonly used radiolabeled probes.

Although we have no experience with the method, perfusion of laboratory animal tissues with paraformaldehyde, followed by cryoprotection with sucrose prior to freezing, has been reported to provide improved morphology at the cost of a reduction in detectable mRNA.[18]

C. Cells and Cell Cultures

Either trypsinized cell suspensions or suspensions cultures are washed in PBS (0.15 *M* NaCl, 7m*M* Na_2HPO_4, 3m*M* NaH_2PO_4, pH 7.2) and 0.5 to 1 × 10^6 cells deposited on gelatin-coated slides by cytocentrifugation or smearing. The cells are air dried for 30 min, then fixed for DNA or RNA as described in Tables 1 and 2, respectively. We have found it necessary to use an EAA prefixation step, which improves probe access, followed by a

Table 3
THE EFFECT OF DIFFERING PRONASE CONCENTRATIONS ON (A) PROBE AND NUCLEASE ACCESS AND (B) CELLULAR DNA STABILITY, IN TISSUE SECTIONS

A. The Efficiency of *In Situ* Hybridization and Nuclease Digestions is Dependent on an Optimal Pronase Concentration

Pronase concentrations (μg/ml)	Virus DNA in sections treated with	
	RNase	DNase
10 (RT)[a]	+	+
10	+	±
50	+ +	−
100	+	−

[a] Hepatitis B virus-infected liver tissue was fixed in EAA and paraffin wax-embedded. The tissue sections were digested with pronase as described in the fixation protocol for DNA (37°, 15 min), except in row 1 which shows the results of digestion at room temperature. Sections were then digested with RNase or DNase as described in the text and hybridized with a ^{3}H-labeled nick-translated hepatitis B virus DNA probe. In the above table, a pronase concentration of 50 μg/ml is optimal for both probe access and DNase digestion.

B. Loss of Cellular DNA with Increasing Pronase Concentration

Pronase concentration (μg/ml)	Total grain count	Average grains per cell	Percent retained
Control	971	32.3	100
0	862	28.7	88.7
10	908	30.2	93.5
50	1011	33.7	104.1
100	692	23.0	71.2
200	610	20.3	62.8

Human hepatoma cells were labeled with ^{3}H-thymidine, removed from the culture and the cell pellet resuspended in a fibrin clot.[15] This was fixed in ethanol:acetic acid and processed into paraffin wax. 5 μm sections were treated exactly as tissues for *in situ* hybridization (Table 1) using varying levels of Pronase, except that radiolabeled probe was omitted from the hybridization mixture. The degree of loss of ^{3}H-labeled DNA was then determined by autoradiography.

The number of grains in 30 cells were counted to obtain the total grain count. Control sections were simply dewaxed before autoradiography. Cellular DNA was lost from the sections after digestion with >50 μg/ml pronase, under the conditions of digestion used.

glutaralydehyde stabilization step for DNA detection. If the spatial relationship of positive cells in culture is to be examined, e.g., to examine cell-cell spread of virus or the intracellular distribution of mRNA, then it is necessary to examine cells grown directly on slides or coverslips[19,20] as this information is lost after trypsinization.

IV. CHOICE OF INDICATOR MOLECULE

We believe that isotopic detection methods offer several advantages over the use of nonisotopic methods at the present time. Although rapid advances are being made in this

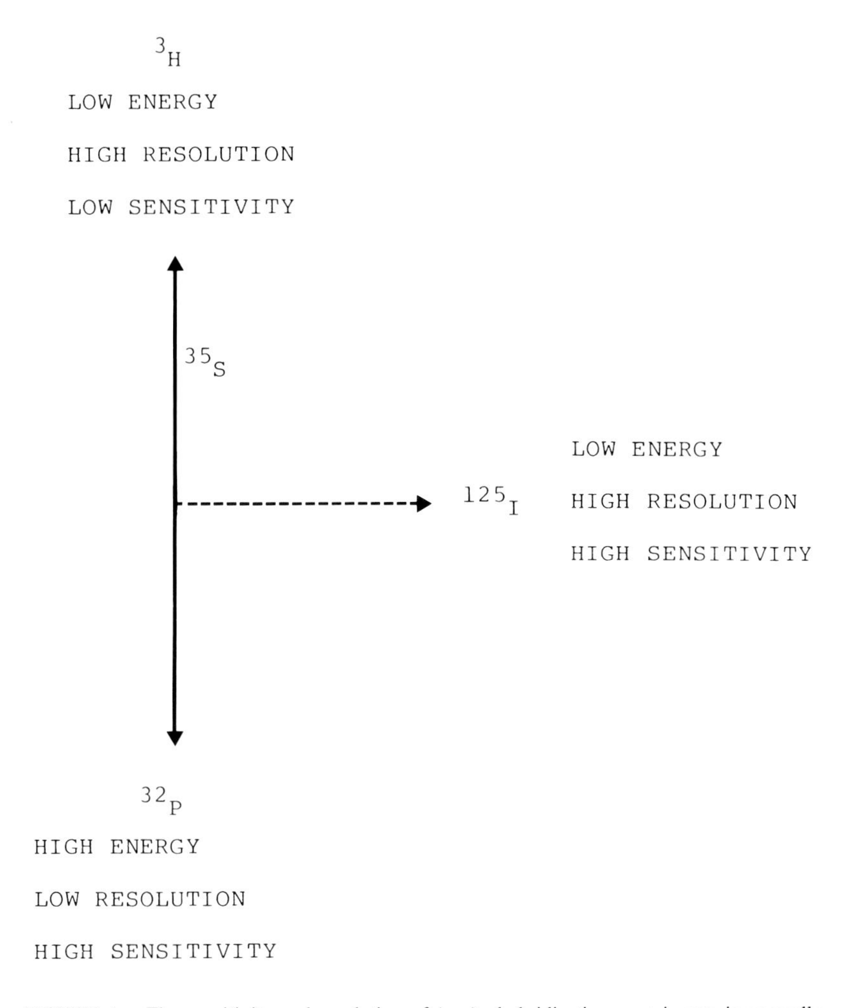

FIGURE 1. The sensitivity and resolution of *in situ* hybridization experiments is generally related to the energy of the isotope used to label the probe.

latter area (see Chapter 2), we have limited experience of their use in ISH reactions and do not propose to discuss these probes further. The main advantages of isotopic methods are (1) higher sensitivity, (2) ease of monitoring probe preparation and quantitation of probe specific activity, (3) ease of quantitation of the signal, and (4) the greater versatility of isotopic labeling methods (Chapter 1 and Reference 21).

Contrary to popular belief, although ^{3}H offers the best autoradiographic resolution, isotopes with higher energies are able to achieve the single cell resolution required by most researchers. There is an adverse relationship between radioisotope energy and the corresponding autoradiographic resolution after ISH reactions, resulting from the longer path length of high-energy particles (Figure 1). However, ^{125}I, a gamma emitter, is an exception to this rule, since the appearance of grains in autoradiographic nuclear emulsions with this isotope depends on low energy secondary β-emissions resulting from ^{125}I decay.[11]

Thus, for optimum resolution, the choice is ^{3}H and for optimum sensitivity, ^{32}P. However, in practice, ^{32}P provides higher resolution than was previously thought possible, particularly with thin films of emulsion; ^{125}I provides both high resolution and high sensitivity, and is the isotope of our choice for most *in situ* hybridization reactions.[7,8,23,24] ^{35}S provides a useful compromise.

V. CHOICE OF PROBE

Most researchers have access to recombinant probes of known sequence, and these have enormous advantages over (even highly purified) naturally derived nucleic acid. Thus, for amplification purposes, the nucleic acid of interest is cloned into a plasmid or viral vector. It is advisable to excise the specific sequences from the vector to avoid contamination of the probes with vector sequences, in order to reduce the backgrounds and possible cross reactions. Although probes may be labeled by several methods (Chapter 1, Reference 2), we have restricted our probe preparation methods to nick translation,[25] primer extension by Klenow after random priming[26] or after specific priming of M13 phage DNA,[27] and, recently, transcription of RNA probes (riboprobes) from recombinant plasmids combining SP 6- and T7-RNA polymerase promoter sites.[7,8,28] Regardless of the method chosen, care should be taken to achieve optimal probe specific activity (e.g., 1 to 2 $\times$ 10^9 dpm/μg for ^{32}P, 5 $\times$ 10^8 dpm/μg for ^{125}I, 1 to 2 $\times$ 10^8 dpm/μg for ^3H). If this is not done, a major loss in sensitivity may result without the worker being aware of this; conversely, if high specific activity probes lead to high backgrounds, a major improvement in signal:noise ratio may be acheived if background grains can be reduced.

^{125}I-labeled probes prepared by nick translation or by random primer extension produce similar results in ISH reactions. M13 probes (and riboprobes) have the advantage of being asymmetric (i.e., only one strand of the duplex is present) and therefore strand specific. For most purposes, a probe length of 100 to 400 bases is optimal. This provides a compromise between the reduced tissue access obtained with longer probes on one hand and the reduced signal per hybridized probe molecule with shorter probes on the other hand. DNA probes prepared by Klenow reactions may need to be digested with DNase to optimal length, while the DNase concentration in nick translation reactions can normally be adjusted so that the reaction product is already this length.

The advantages of increased sensitivity and strand specificity previously reported for riboprobes[29] have prompted us to convert almost entirely to this system with good results. ^{125}I-CTP may be incorporated using either SP6- or T7-RNA polymerase reactions generating probes of $\leq$9 kb.[30] The probes are reduced to 100 to 400 bases by mild alkali hydrolysis as described.[31] Briefly, the transcripts are digested with DNase 1 to remove the template, ethanol precipitated, redissolved in 90 μl of DDW, then 10 μl of 1 *M* sodium bicarbonate is added and the mixture incubated at 60°C for approximately 5 min. This time may need to be adjusted. One tenth volume of 3 *M* sodium acetate is then added and the probes ethanol precipitated. Probes prepared in this manner produced specific labeling with high resolution after ISH. Riboprobes may also be labeled with ^{32}P, ^{35}S, and ^3H and have the added advantage that a post-hybridization RNase digestion step will remove nonspecifically bound, single-stranded probes and thus reduce the background (see below).

The use of gene- or stand-specific probes is necessary in some instances to identify viral nucleic acid replicative intermediates, in order to distinguish infected cells which are supporting active virus replication from cells which are either latently infected or have accumulated virus passively.

We have used this approach to determine that the hepatocyte cytoplasm is the intracellular site of hepatitis B virus (HBV) DNA replication.[9] As the HBV genome contains a ss region, we identified permissively infected hepatocytes by the presence of ss DNA from a region of the genome that is normally ds in the virion (Table 4), because the cells showed a strong hybridization signal without a prior denaturation step. A similar approach using gene-specific probes from human papiloma virus type 16 (HPV 16) was used to detect the varying extent of HPV genome expression in different epithelial cells.[32]

Strand specific probes for hepatitis delta virus (HDV) RNA have also been used to detect HDV-genomic and -antigenomic RNA in the nuclei of infected hepatocytes,[7] strongly sug-

Table 4
THE DETECTION OF HBV DNA REPLICATIVE INTERMEDIATES IN THE CYTOPLASM OF INFECTED HEPATOCYTES

Pretreatment	Probe 1	Probe 2
Undenatured sections	111 ± 18	128 ± 18
Denatured sections	267 ± 28	134 ± 11

Frozen sections of human liver were fixed in glutaraldehyde as described in Table 1 and selected sections denatured in 0.1 × SSC at 100°C for 5 min. The sections were then hybridized with ^{3}H-labeled probes from the HBcAg region that is normally ds (probe 1) or from the HBsAg region that is normally ss (probe 2) in the virion. After hybridization and washing, the autoradiographic grain density in infected cells was quantitated using reflected light photometry.[11]

Although probes 1 and 2 (which were similar in length and specific activity) produced similar signals on undenatured sections, in denatured tissue sectons only P1 showed a 2 × increase in the signal. This suggests that sequences coresponding to the HBcAg gene were present as both ss and ds, forms, while sequences corresponding to the HBsAg were only ss i.e., part of the target DNA was similar to mature virion DNA (HBcAg gene ds; HBsAg gene ss) and part was wholly ss. Thus, it is likely that these cells contained HBV DNA replicative intermediates.

gesting that HDV RNA replication is a nuclear event. Relative quantitation showed that the ratio of HDV RNA genomic:antigenomic species was approximately 20 to 30:1. Examples of the detection of HBV DNA and HDV RNA by ISH are shown in Figure 2.

VI. HYBRIDIZATION

A. Theoretical Considerations

ISH follows the same general principles as a solution and filter hybridization;[33] a standard reaction is therefore performed optimally at a temperature of around Tm − 25°C (Tm is the temperature at which 50% of hybrids dissociate). The reaction temperature is reduced to a level compatible with the preservation of histological detail by the addition of 50% formamide to the hybridization mixture. Thus, for typical DNA-DNA hybridization reactions, the temperature is 37°C, for RNA-DNA it is 44°C, and for RNA-RNA it is 50°C.

Unless background levels are unaccountably and unacceptably high, there is no need for a prehybridization step, as the hybridization mixture (Table 5) is designed to minimize nonspecific probe binding. The aim of the reaction, to produce optimal signal:noise ratios, is accomplished by the addition of a minimum probe concentration to achieve saturation of target sequences, as excess probe will invariably lead to higher backgrounds. We have found that using 0.25 ng of nick translated DNA probes, or 5 × 10^{4} cpm of riboprobes, labeled with ^{125}I per slide appears to saturate typical targets between 24 to 40 h and 3 to 16 h hybridization time, respectively. It is likely that probes radiolabeled with other isotopes will behave in a similar manner. However, these probes quantities may need to be increased for sections with a very high target content.

In practice, nick translated DNA probes may be unlikely to achieve complete target saturation even at high concentrations,[29] because the rate of probe self-annealing in solution is faster that hybridization rate between target sequences in the issue section and the probe. (The kinetics of the reaction may be further complicated by possible self-annealing of ds target sequences, although two reports suggest that is does not occur in practice.[34,35]) On

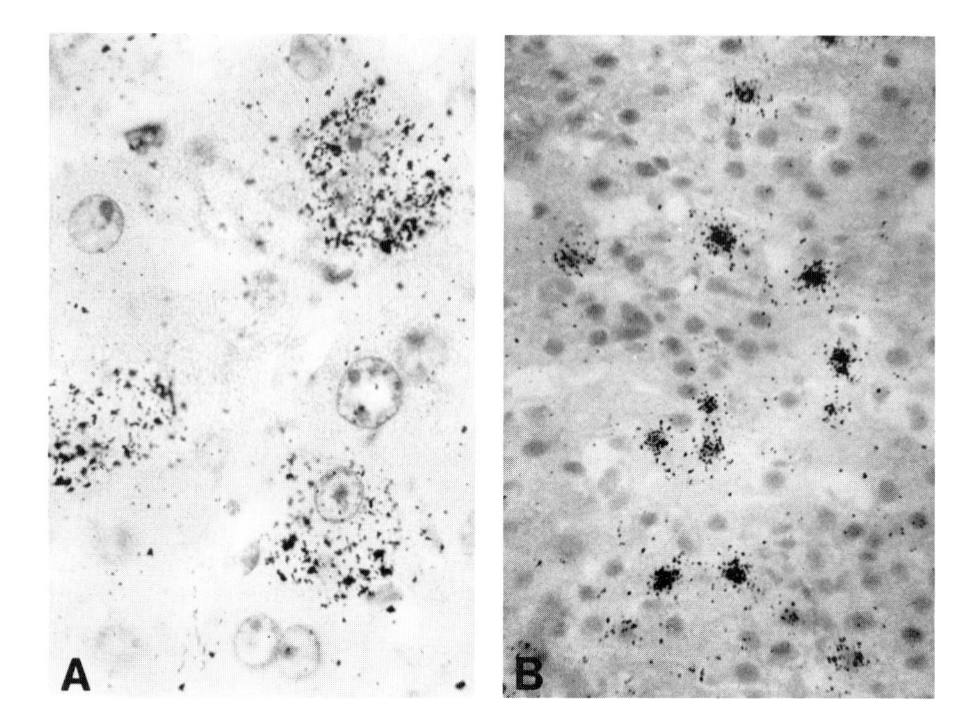

FIGURE 2. Examples of virus nucleic acids detected in frozen tissue and in paraffin wax-embedded tissues previously fixed in ethanol:acetic acid (EAA) or formalin. A. Hepatitis B virus (HBV)-infected human liver biopsy fixed in EAA and subsequently embedded in paraffin wax. Sections (5 μm) were cut treated for DNA target as described in Table 1 and hybridized with hepatitis B virus DNA probe that was radiolabeled with ^{3}H by nick translation. The autoradiographic exposure time was 1 week. In this sample, approximately 2% of hepatocytes contained HBV DNA. B to D: Frozen sections of hepatitis delta virus-infected human liver (B, C) or uninfected control chimpanzee liver (D). Sections were fixed for RNA as described in Table 2 and hybridized with ^{125}I-labeled riboprobes comprising the total hepatitis delta virus genome. Sections B and D were hybridized with strand-specific probes to detect genomic virus RNA and C was hybridized with a strand-specific probe to detect antigenomic virus RNA after a tissue denaturation step. The autoradiographic exposure was 24 h. Approximately 50% of hepatocytes contained HDV RNA. E. Formalin-fixed paraffin wax-embedded human liver infected with delta hepatitis virus. Sections were fixed as described in Table 2 and hybridized with a ^{125}I-labeled riboprobe to detect genomic hepatitis delta virus RNA. The autoradiographic exposure was a 24 h. HDV RNA was detected in 50% of hepatocytes present in a few cirrhotic nodules in an otherwise uninfected cell population.

the other hand, asymmetric RNA or DNA probes which do not self-anneal should be able to saturate the target more completely, probably accounting in part for their greater sensitivity. However, the self-annealing property of DNA probes has been used to advantage by the addition of 10% dextran sulfate to the hybridization mix.[36] This promotes the formation of bound probe hyperpolymers ("networks") with the possibility of increased sensitivity. In our hands, dextran sulfate leads to unacceptably high tissue backgrounds with control probes (pBR322, λ DNA) and we cannot recommend its use. However, we have found that polyethylene glycol,[37] another aqueous exclusion medium, is more useful in network formation, although we routinely use neither. Network formation leads to clusters of grains which are more difficult to read and quantitate, and unless an increased signal:noise ratio is also achieved, use of this approach is of marginal benefit.

We find that addition to the hybridization mix of 20 m*M* DTT made freshly from powder on the day of hybridization, is essential to reduce the background with ^{35}S probes and, in

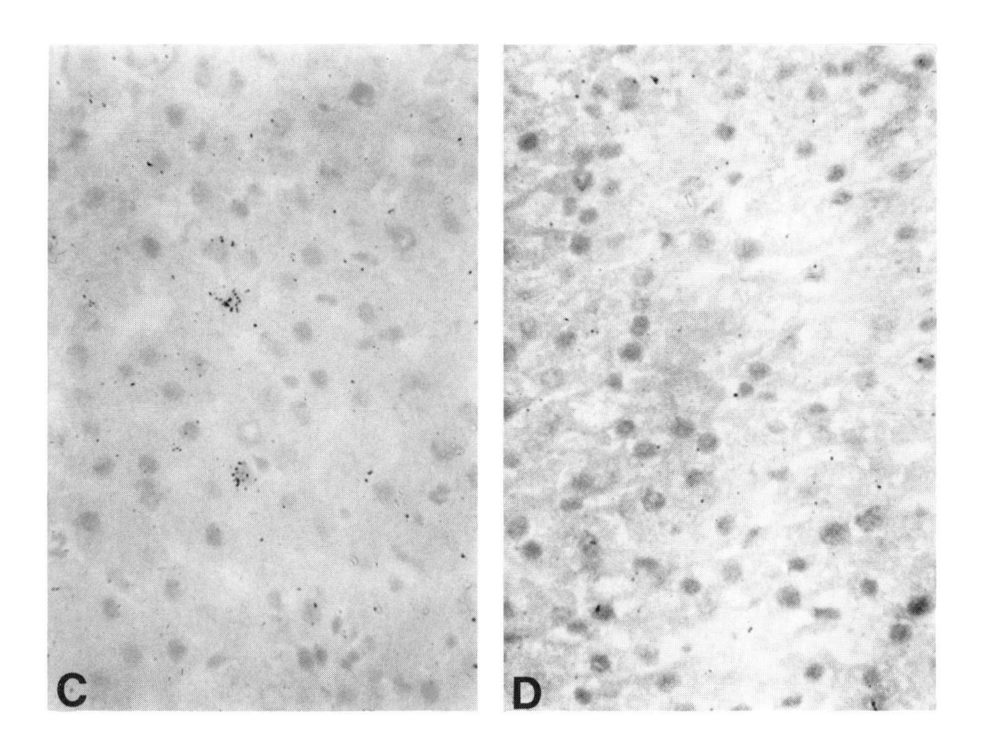

FIGURE 2C. FIGURE 2D.

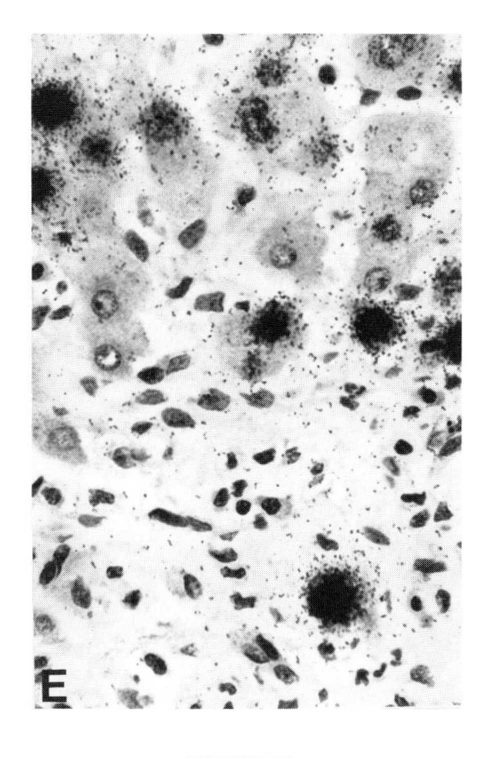

FIGURE 2E.

Table 5
COMPOSITION AND PREPARATION OF HYBRIDIZATION MIXTURE SUITABLE FOR THE DETECTION OF DNA OR RNA BY *IN SITU* HYBRIDIZATION

The final hybridization mixture contains 50% deionized formamide,[38] 2 × SSC, 100 m*M* Tris-HCl pH 7.6, 10 m*M* Na_2HPO_4, 10 m*M* NaH_2PO_4, 0.02% Ficoll, 0.02% polyvinyl pyrollidone (PVP), 500 μg/ml sheared, denatured salmon sperm DNA, 500 μg/ml tRNA, 1.25 mg/ml nuclease-free BSA, 20 m*M* DTT, and 1 U/μl RNasin. RNasin may be omitted from DNA-DNA reactions. This is assembled in an Eppendorf tube as follows:

Concentrated stock probe in DDW	2 to 5 μl
Deionized formamide	50 μl
Sheared carrier DNA (10 mg/ml)	5 μl
tRNA (10 mg/ml)	5 μl

Vortex and heat to 100°C for 5 min for DNA probes or 80°C for 2 min for RNA probes. Plunge immediately into ice-cold water.

Add:	Hybridization mix (5×)	20 μl
	DTT (800 mM)	2.5 μl
	RNasin (40 U/μl)	2.5 μl
	Nuclease free BSA (50 mg/ml)	2.5 μl
	DDW to	100 μl

This mixture should contain approximately 5 × 10^4 cpm/2.5 μl which is applied to each sample. Concentrated stock probe is added initially accordingly. The 5× hybridization mixture contains:

10×SSC	This mixture is filter sterilized through 0.22 μm nitrocellulose, aliquoted, and stored at −20°C.
500 m*M* Tris-HCl pH 7.6	
50 m*M* NaH_2PO_4	
50 m*M* Na_2HPO_4	
0.1% Ficoll	
0.1% PVP	

practice, DTT is added irrespective of the isotope in use. The same hybridization mixture is used for both DNA and RNA probes and is described in Table 5.

B. Practical Application

The surface of the microscope slide supporting the sample is gently blown by a stream of air to remove any dust; this is achieved by attaching a cotton-plugged glass pasteur pipette to a 10-ml pipette bulb and ensures that the coverslip makes good contact with the tissue section. Part (2.5 μl) of the final hybridization mix (Table 5) is then pipetted over the surface of the section and a 13-mm sterile, siliconized coverslip overlaid carefully to exclude the formation of air bubbles. The slide is then incubated flat in a bath of paraffin oil as described[39] for the required time and temperature.

C. Washing

The slides are removed from the paraffin oil, placed in slide racks, drained of oil for 5 to 10 min, the excess oil removed by washing in chloroform for 2 × 10 min, and the slides air dried for 10 to 15 min. At this stage, the coverslips are still firmly in position over the samples. The slides are transferred to a large volume of 2 × SSC (1 × SSC = 0.15 *M* NaCl, 0.15 *M* Na_3 citrate, pH 7.0) (4 to 6 1/30 slides) and washed with stirring for 1 h. If ^{125}I-labeled probes are used, the inclusion of 100 μ*M* potassium iodide further reduces the background.[24,40] Similarly, 1% sodium thiosulfate and 14 m*M* 2-mercaptoethanol are added to the wash solution if ^{35}S probes are used.[18]

During the first few minutes of aqueous washing, the siliconized coverslips become detached from the slides without causing damage to the tissue section. The sections are then transferred to fresh 2 × SSC and washed for a further 2 × 1 h and then in 0.1 × SSC for 2 × 1 h. A high stringency wash may be necessary to reduce background, especially if riboprobes are used; for RNA-RNA hybrids this step is commonly performed at 70°C in 0.1 × SSC for 10 min and may be altered if necessary or correspondingly decreased for RNA-DNA and DNA-DNA hybrids. The required level of stringency may be determined by reference to a post-hybridization melt curve previously constructed to determine the Tm of the hybrids formed *in situ* (see below). If background grains continue to be unacceptably high, then nonspecifically bound ss riboprobes may be digested specifically with RNase while ds RNA-RNA hybrids ramain intact. However, although previous reports used 20 μg/ml RNase[6] and 100 μg/ml RNase,[41] in our hands, these levels also digested newly formed hybrids, and 1μg/ml RNase in 5 × SSC for 15 min at room temperature was found to be efficient. Therefore it is advisable to titrate different RNase preparations before use. After washing, the samples are dehydrated through ascending grades of ethanol before autoradioraphy.

VII. AUTORADIOGRAPHY AND STAINING

A. Autoradiography

Although X-ray film[42] and dried nuclear emulsion-coated coverslips[43] have been used successfully to provide an estimate of the autoradiographic exposure time required, these methods are not commonly used and most workers, including us, prefer to hybridize 2 to 3 additional sections which are used as test exposures. Provided that samples are hybridized with test and control probes, then these slides can be quite invaluable. The objective, to achieve the best possible signal:noise ratio in a system that does not permit multiple reexposures, is best accomplished by duplicate slides which are developed at regular intervals.

Access to a darkroom is necessary for autoradiography, although this need not be sophisticated. The minimum requirement is a light-tight room equipped with a flat bench and a light-trap entry system so that slides may be left in complete darkness for 1 to 2 h after dipping.

Kodak and Ilford are the main manufacturers of a range of nuclear emulsions, and the theoretical choice is covered comprehensively by Rogers.[11] A number of workers continue to advocate the addition of scintillant to nuclear emulsions as originally described,[44] in an effort to increase the autoradiographic efficiency. We are unable to substantiate these reports, and it is likely that the explanation for the increased efficiency is due to complete drying by the scintillant solvent of an otherwise incompletely dried emulsion layer that may suffer from characteristic latent-image fading.[45] This explanation has been verified recently.[5]

Nuclear emulsions must be handled carefully to reduce background grains caused by extraneous light or mechanical stress. We use Ilford K2 or L4 emulsion, containing 1% glycerol to minimize mechanical stress, with the addition of 300 m*M* ammonium acetate as recommended for L4 only[5] (this is incompatible with K2). Mechanical stress may result not only in higher grain backgrounds, but also, in extreme instances, in complete or partial loss of the emulsion from the slide. Mechanical stress is most easily avoided by ensuring that emulsion-coated slides are not subjected to abrupt changes, e.g., pH, rapid air drying, water to alcohol, etc., during processing.

A suitable amount of Ilford emulsion for the immediate experiment is mixed with an equal volume of distilled water containing 2% glycerol (and 600 m*M* NH_4OAc if desired) in a darkroom equipped with safelights recommended by Ilford, and melted at 44°C for 30 min. The diluted emulsion is mixed carefully to avoid air-bubble formation, and individual slides are dipped vertically, drained for a few seconds (excess emulsion is wiped from the back), and laid flat on a level, cold plate[11] for 20 to 30 min. The slides are then removed, laid flat on the bench in total darkness for 1 to 2 h until completely dry, packed into a black, light-tight box, and stored at 4°C. The inclusion of dessicant, e.g., silica gel, in this box is not recommended for Ilford emulsions, but is for Kodak emulsions. After a suitable exposure time, the slides are warmed to room temperature, developed for 4 to 6 min in Kodak D19, then fixed for 5 min in Ilford Hypam diluted 1/4 with distilled water. The slides may then be removed from the darkroom if necessary, the fixer slowly replaced with running tap water and washed in running tap water for 15 min to ensure complete removal of the fixer.

B. Staining

This is most conveniently performed immediately after autoradiographic processing described above, although it may be possible to stain the sections prior to autoradiography, thus avoiding the problems created by an overlying photographic emulsion.[11] The tissues may be stained with Giemsa or toluidine blue, but hematoxylin and eosin (H & E) is the most versatile method for general histopathological interpretation. However, care must be taken in the choice of hematoxylin as most hematoxylin solutions are approximately pH 2.0 and will quickly reduce the size of developed silver grains leading to complete loss of grains within a few minutes. Grain etching is easily avoided by the use of rapid progressive hematoxylin solution (Table 6). Care must also be taken after hematoxylin staining to dehydrate the emulsion gradually, as the abrupt transition from aqueous hematoxylin to alcoholic eosin often completely removes the emulsion. H & E staining as described in Table 6 should not only result in well-stained tissue, but also in a well-defined emulsion layer with sharp border which corresponds to the depth of the emulsion in the dipping chamber. A change in this pattern suggests that the emulsion layer has been handled incorrectly with possible damage to the grains.

VIII. REACTION CONTROLS

There is a vital need for proper controls because of the high potential for artifacts and nonspecific signals in any ISH reaction. The most commonly used controls are (1) nuclease digestions, (2) control probes, (3) control tissues, (4) post-hybridization melt curve, and (5) saturation of the probe with specific nucleic acid sequences.

A. Nuclease Digestions

The section may be digested with RNase or DNase as part of the tissue fixation protocol (see Tables 1 and 2) to remove nucleic acid sequences of the species that do not comprise the target; or a duplicate section may be digested with RNase or DNase to remove the species comprising the target. The sections are digested for 30 min at 37°C in 50 to 100 μg/ml DNase 1 in 50 m*M* Tris-HCl pH 7.5, 7 m*M* $MgCl_2$, or in 100 μg/ml RNase A (previously boiled for 5 min to destroy any residual DNase activity) in 2 × SSC. The efficiency of nuclease digestions is high in frozen sections but may be very low in FPE tissues, particularly if no protease digestion step is included (see Table 3). Protease digestion therefore improves both probe and nuclease access. Digestion of sections with RNase prior to hybridization with riboprobes is not recommended as a control. We have found that, despite a vigorous post-RNase washing step, sufficient RNase remains tissue-associated to subsequently digest the probe, since even DNA targets are undetected by specific riboprobes after prior tissue digestion by RNase. This problem may be overcome by using micrococcal nuclease for predigestion of the target; this enzyme is calcium dependent, has maximum activity of pH 9.0, and is inhibited by NaCl concentrations >0.1 *M*. Subsequent treatment of the tissue inhibits residual nuclease activity.[46]

B. Control Probes

An equal number of tissue sections should be hybridized with test and control probes to

Table 6
METHOD TO STAIN SAMPLES BY HEMATOXYLIN AND EOSIN AFTER *IN SITU* HYBRIDIZATION

Method

Rapid hematoxylin, 1 min;
rinse in 0.1 × SSC with agitation, 30 s;
wash in 2 × SSC, 15 min;
dehydrate through ascending ethanol solutions to 100%;
alcoholic eosin, 1 min;
rinse in alcohol;
wash in ethanol, 2 × 1 min;
soak in xylene, 2 × 10 min;
mount in Permount or DPX.

Formulae

Hematoxylin

Distilled water	1 l
Aluminium ammonium sulfate	50 g
Hematoxylin	4 g
Sodium iodate	0.3 g
Citric acid	1.5 g
Chloral hydrate	75 g

Fully dissolve each component *in the above order*. The solution is stable and may be kept for up to 1 year.

Eosin

95% Alcohol	2750 ml
1 % Aqueous eosin	350 ml
1% Aqueous phloxine	35 ml
Glacial acetic acid	14 ml

Mix well, store at 4°C.

ensure that each autoradiographic exposure time is provided with negative control slides. Ideal controls for DNA probes are either pBR322- or λ-DNA, i.e., common vector sequences, or an unrelated cloned DNA sequence which is readily available. For strand-specific RNA probes, transcripts which are complementary to the test probe, i.e., of the same polarity as the target, are the best controls. These latter control probes are prepared easily if the cDNA is cloned into a gemini vector containing both SP6- and T7-RNA promoter sites. We find it particularly useful to mount two tissue sections, separated by a space of 30 to 40 mm on each slide, and to hybridize these with test and control probes, respectively.

C. Control Tissues

Any tissue that is known not to express the RNA of interest or is not virus-infected.

D. Post-Hybridization Melt Curve

The Tm of hybrids formed *in situ* may be determined by constructing a melt curve. Duplicates of hybridized tissue sections are washed for 5 to 10 min at increasing temperatures, such that the higher temperatures cause complete dissociation of hybrids formed *in situ*. The autoradiographic signal in these sections is then plotted against the temperature, and a typical sigmoidal curve results.[8,47,48] The Tm of hybrids will be lower if probe and target are not completely homologous and will also depend on the G + C content of the nucleic acid, the salt concentration of the washing medium, and to a lesser extent on probe length. Tissue sections treated in this way to melt hybrids formed *in situ* demonstrate the level of nonspecific probe binding that has occurred, since probes bound by mechanisms other than base pairing are not removed by heating in a predictable manner with a steep thermal transition.

E. Probe Competition

The specificity of the reaction may be demonstrated by abolition of the reaction if the hybridization mixture includes a vast excess of unlabeled nucleic acid corresponding to the target sequences sought.[48] This approach is rather circular if both the excess unlabeled nucleic acid and the probes are prepared from the same source, but may be useful in some circumstances.

IX. CORRELATION WITH PROTEIN MARKERS

There is often a clear need to determine if antigenic peptides, hormones, or viral gene products detected intracellularly by immunohistochemical techniques represent *de novo* synthesis or merely deposition and passive accumulation. Conversely, it may be useful to confirm that specific mRNA, detected by ISH, is translated into protein product. The correlation of nucleic acids detected by ISH with proteins detected by immunohistochemistry, provides the means to answer these questions. These experiments may be performed on sequential sections or, more satisfactorily, by the simultaneous detection of both markers in a single tissue section.

Simultaneous detection may be performed on either frozen or FPE tissues. It is important to establish the relative performance of both the ISH and immunohistochemical assays in a simultaneous detection protocol, by comparison with each of these assays on duplicate slides under optimal conditions as a single reaction; some decrease in performance has been seen by us in combined assasys. These are two possible approaches, viz., to perform ISH followed by immunohistochemistry or to perform immunohistochemistry followed by ISH. Our initial attempts to detect viral antigens after ISH were unsuccessful, but we were able to detect viral DNA after immunohistochemistry[24] and have subsequently used this procedure successfully on a number of occasions[7,8,24] to detect viral DNA or RNA.

Paraformaldehyde fixed, frozen sections or FPE sections previously fixed in EAA or formalin are treated with hydrogen peroxide to destroy endogenous peroxidase and incubated in horseradish peroxidase-conjugated antibody or sequentially in multi-step PAP reagents.[7,24,49] To ensure the subsequent detection of RNA by ISH, these reagents should be filter sterilized, diluted to working strength with sterile PBS, and should contain 1 U/μl RNasin and 1 m*M* DTT.[7] The sections are washed in PBS, incubated in diaminobenzidine (DAB) solution[24] for 3 to 10 min, washed, acetylated, and treated with 0.2% Triton X-100.[50] This latter step reduces nonspecific binding of probes and increases the hybridization signal, which are increased and decreased, respectively, by the DAB reaction product. The sections are washed, dehydrated, and hybridized as described above; the only concession from the normal protocol is that sections must be stained with hematoxylin only rather than hematoxylin and eosin (H & E), described above.

X. INTERPRETATION OF RESULTS

Of all the facets of ISH, examination and subsequent interpretations of the final product may present the greatest difficulties, unless the target concentration is high when the identification of positive cells is relatively easy. It may be necessary to perform grain counts over 50 to 100 negative cells in order to set a normal range which may then be used to identify positive cells with low concentrations of target. In any case, unless the distribution and frequency of positive cells are similar in duplicate slides, further work will be required to achieve technical reproducibility.

High autoradiographic grain backgrounds may be uniformly distributed or restricted to the tissue sample: the latter situation is related to probe/tissue interactions and the former, emulsion related. This may be confirmed by examination of an emulsion-coated control blank slide.

Although apparently obvious, progressively longer exposures of positive samples should produce correspondingly higher grain densities, provided that local saturation of available silver grains has not occurred. If longer exposures do not lead to increased grain densities, latent image fading[11,44] caused by inadequate drying of the emulsion may be responsible. Autoradiographic reproducibility may be controlled in different experiments by the use of ^{3}H- or ^{125}I-labeled microscales (Amersham International plc).

XI. QUANTITATION

At the present time, the ability to quantitate nucleic acid sequences in single cells is one of the main advantages that isotopic methods have over nonisotopic detection methods. Three major methods of quantitation are used, viz., theoretical, comparative, and relative.

A. Theoretical Quantitation

This is fraught with difficulties, as there are a number of variable parameters that may affect the calculations; the most important of these are (1) autoradiographic efficiency (number of grains × 100/number of radioactive disintegrations), which has been reported to be 7% for ^{3}H[48] and 25% for ^{125}I[11] and (2) hybridization efficiency (the ratio of probe sequences bound × 100/target sequences present), reported to vary from 10%[35] to 100%.[39] Nevertheless, a reasonably accurate estimate may be achieved. For example, using a ^{3}H-probe with a specific activity of 2×10^{8} dpm/μg, 12 grains per cell per week would correspond to 2×10^{-2} dpm of hybridized probe or 4×10^{-10} μg of target DNA per cell, assuming an autoradiographic efficiency of 7% and a hybridization efficiency of 100%. However, this calculated target content may be low by a factor of 10, if 10% rather than 100% is the assumed hybridization efficiency.

In attempting to calculate target sequence copy number per cell, two additional factors should not be overlooked: (1) the volume of a sectioned cell will typically be considerably less than that of the intact cell, e.g., in 3 μm sections, only 22% of the volume of an intact hepatocyte (mean diameter 20.4 μm) will actually be present within the section[48] and (2) with a low energy β-emitting isotope (tritium) with maximum path length 3 μm, we have demonstrated that no progressive increase in autoradiographic signal occurred as section thickness was increased beyond 4 μm, presumably since disintegrations from lower levels of the section were unable to reach the emulsion.[48]

B. Comparative Quantitation

This may be performed by comparing the autoradiographic grain count with a calibration curve[5] derived from ISH reactions on samples of different target content examined under identical conditions, where the target content has been independently determined (e.g., by filter hybridization assays performed on extracted nucleic acids). This approach is probably the most reliable way to achieve absolute quantitation. It does, however, require: (1) the availability of several tissue samples containing differing levels of target, fairly uniformly distributed throughout the tissue and (2) careful attention to the large number of variables involved.

C. Relative Quantitation

This allows reliable comparison between the amount of target in different cells on the same slide or on different slides within the one experiment. However, it does not allow determination of the absolute genome copy number, and such comparisons between different experiments require meticulous control of variables and clear proof of technical reproducibility before they can be considered valid. Inclusion of standard positive control slides in each experiment is of value.

The information which forms the basis for the above methods may be gathered by: (1) grain counts (2) measurement of incident light that is reflected from silver grains; this has previously been shown to bear a direct relationship to grain counts,[11] and (3) the use of computer-controlled image analysis systems.[51] Despite the fact that quantitation in ISH is often tedious and time consuming, the information derived is often otherwise unattainable and may be extremely valuable.

XII. CHOICE OF STRATEGY AND SUMMARY

The worker must first evaluate whether ISH, or a conventional hybridization assay on extracted nucleic acids, is the required approach. The latter methods yield more quantifiable data and do not require histological techniques, and are probably preferable where target sequences are uniformly distributed throughout the tissues.

Next, the degree of histological preservation required, and requirements for storage of tissues, will determine whether frozen or fixed paraffin wax-embedded tissues are to be preferred. The type of target sought will govern choice of sample preparation techniques prior to hybridization, while the level of resolution required will determine which isotope label is chosen. Choice between the various methods for preparing and labeling probes will largely depend on the familiarity and individual preference for various techniques in the individual laboratory. If simultaneous antigen detection is required, techniques for this must be included. In any case, setting up an ISH technique will require some attention to probe characterization (e.g., length and specific activity) and access to known positive and negative control material is essential in order to validate the methodology chosen.

As can be concluded from the above, many variations on a standard ISH technique are possible and can yield successful results. Choice of technique is to some extent dependent on the desired type of information. The laboratory must then be prepared to carry out a limited range of investigations to validate the particular set of conditions chosen and to fine tune the main variable involved.

ACKNOWLEDGMENTS

We wish to thank Professor B. P. Marmion for his advice and encouragement. Much of the work described in this chapter was carried out with support from the National Health and Medical Research Council of Australia. E. J. G. wishes to thank Dr. J. Gerin, Division of Molecular Virology and Immunology, Georgetown University, for the opportunity to spend time in his laboratory, where the HDV RNA *in situ* hybridization assay was developed. Batches of ^{125}I-CTP used to label riboprobes were generous gifts from Amersham International plc and NEN. We are grateful to Mrs. Dawn Campbell for typing the manuscript.

REFERENCES

1. **Gall, J. F. and Pardue, M. L.,** Formation and detection of RNA-DNA hybrid molecules in cytological preparations, *Proc. Natl. Acad. Sci. U.S.A.,* 63, 378, 1969.
2. **John, H. A., Birnstiel, M. L., and Jones, K. W.,** RNA-DNA hybrids at the cytological level, *Nature (London),* 223, 582, 1969.
3. **van der Ploeg, M., van Ruijn, P., Bauman, J. G., Landegent, J. E., and Raap, A. K.,** Hybridocytochemistry as a tool for the investigation of chromatic organisation, *Basic Appl. Histochem.,* 29, 181, 1985.
4. **Hudson, P., Penschow, J., Shine, J., Ryan, G., Niall,H., and Coghlan, J.,** Hybridization histochemistry: use of recombinant DNA as a 'homing probe' for tissue localization of specific mRNA populations, *Endocrinology,* 108, 353, 1981.
5. **Haase, A., Brahic, M., Stowring, L., and Blum, H.,** Detection of viral nucleic acids by *in situ* hybridization, *Methods Virol.,* 7, 189, 1984.
6. **Pardue, M. L.,** *In situ* hybridization, in *Nucleic Acid Hybridization,* Hames, B. D. and Higgins, S. J., Eds., IRL Press, Oxford, 1985, 179.
7. **Gowans, E. J., Baroudy, B. M., Negro, F., Ponzetto, A., Purcell, R. H., and Gerlin, J. L.,** Replication and stable expression of hepatitis delta virus RNA in hepatocyte nuclei after *in vivo* infection, *Virology,* in press.
8. **Gowans, E. J.,Negro, F., Baroudy, B. M., Korba, B. E., Bonino, F., Ponzetto, A., Purcell, R. H., and Gerin, J. L.,** Molecular markers of WHV and HDV in single cells, *UCLA Symp. Mol. Cell. Biol.,* 70, 591, 1987.
9. **Gowans, E. J., Burrell, C. J., Jilbert, A. R., and Marmion, B. P.,** Patterns of single- and double-stranded hepatitis B virus DNA and viral antigen accumulation in infected liver cells, *J. Gen. Virol.,* 64, 1229, 1983.
10. **Jilbert, A. R., Freiman, J. S., Burrell, C. J., Holmes, M., Gowans, E. J., Rowland, R., Hall, P., and Cossart, Y. E.,** Early virus cell interactions in duck hepatitis B Virus infection: a time course study of tissue tropism and sites of virus production, in *Viral Hepatitis and Liver Disease,* Zuckerman, A. J., Ed., Alan R. Liss, New York, 1988, 497.
11. **Rogers, A. W.,** *Techniques for Autoradiography, 3rd ed.,* Elsevier/North Holland, Amsterdam, 1979.
12. **Maddox, P. H. and Jenkins, D.,** 3-Aminopropyltriethoxysilane (APES): a new advance in section adhesion, *J. Clin. Pathol.,* 40, 1256, 1987.
13. **Lawrence, J. B. and Singer, R. H.,** Quantitative analysis of *in situ* hybridization methods for the detection of active gene expression, *Nucleic Acid Res.,* 13, 1777, 1985.
14. **Hoefler, H., Childers, H., Montminy, M. R., Lechan, R. M., Goodman, R. H., and Wolfe, H. J.,** *In situ* hybridization methods for the detection of somatostatin mRNA in tissue sections using antisense RNA probes, *Histochem. J.,* 18, 597, 1986.
15. **Kaltenbach, F. J., Hillemanns, H. G., Fettig, O., and Hilgarth, M.,** Thrombin cell block techic in gynecologic cytodiagnosis, *Acta Cytol. (Baltimore),* 17, 128, 1973.
16. **Binder, M., Tourmente, S., Roth, J., Renaud, M., and Gehring, W. J.,** *In situ* hybridization at the electron microscope level: localization of transcripts on ultra thin sections of Lowicryl K4M-embedded tissue using biotinylated probes and protein-A gold complexes, *J. Cell Biol.,* 102, 1646, 1986.
17. **Hayashi, S., Gillam, I. C., Delaney, A. D., and Tener, G. M.,** Acetylation of chromosome squashes of *Drosophila melanogaster* decreases the background in autoradiographs from hybridization with ^{125}I-labelled RNA, *J. Histochem. Cytochem.,* 26, 677, 1978.
18. **Shivers, B. D., Schachter, B. S., and Pfaff, D. W.,** *In situ* hybridization for the study of gene expression in the brain, *Methods Enzymol.,* 124, 497, 1986.
19. **Lawrence, J. B. and Singer, R. H.,** Intracellular localisation of messenger RNA for cytoskeletal proteins, *Cell,* 45, 407, 1986.
20. **Taylor, J., Mason, W., Summers, J., Goldberg, J., Aldrich, C., Coates, L., Gerin, J., and Gowans, E.,** Replication of human hepatitis delta virus in primary cultures of woodchuck hepatocytes, *J. Virol.,* 61, 2891, 1987.
21. **Arrand, J. E.,** Preparation of nucleic acid probes, in *Nucleic Acid Hybridization,* Hames, B. D. and Higgins, S. J., Eds., IRL Press, Oxford, 1985, 17.
22. **Coghlan, J. P., Alfred, P., Haralambidis, J., Niall, H. D., Penschaw, J., and Tregear, G. W.,** Hybridization histochemistry, *Anal. Biochem.,* 149, 1, 1985.
23. **Jilbert, A. R., Freiman, J. S., Gowans, E. J., Holmes, M., Cossart, Y. E., and Burrell, C. J.,** Duck hepatitis B virus DNA in liver, spleen and pancreas: analysis by *in situ* and Southern-blot hybridization, *Virology,* 158, 330, 1987.
24. **Gowans, E. J., Burrell, C. J., Jilbert, A. R., and Marmion, B. P.,** Cytoplasmic (but not nuclear) hepatitis B virus (HBV) core antigen reflects HBV DNA synthesis at the level of the infected hapatocyte, *Intervirololgy,* 24, 220, 1985.

25. Radiochemical Centre, Labelling of DNA with 32P by nick translation, Technical Bulletin 80/3, Amersham, Buckinghamshire, England, 1980.
26. **Feinberg, A. P. and Vogelstein, B.**, A technique for radiolabelling DNA restriction endonuclease fragments to high specific activity, *Anal. Biochem.*, 132, 6, 1983.
27. **Hu, N. and Messing, J.**, The making of strand-specific probes, *Gene*, 17, 271, 1982.
28. **Melton, D. A., Krieg, P. A., Rebagliati, M. R., Maniatis, T., Zinn, K., and Green, M. R.**, Efficient *in vitro* synthesis of biologically active RNA and RNA hybridization probes from plasmids containing a bacteriophage SP6 promoter, *Nucleic Acids Res.*, 12, 7035, 1984.
29. **Cox, K., De Leon, D. V., Angerer, L. M., and Angerer, R. C.**, Detection of mRNAs in sea urchin embryos by *in situ* hybridization using asymmetric RNA probes, *Dev. Biol.*, 101, 485, 1984.
30. **Cohen, G.**, personal communication, 1987.
31. **Lynn, D. A., Angerer, L. M., Bruskin, A. M., Klrein, W. H., and Angerer, R. C.**, Localization of a family of mRNAs in a single cell type and its precursors in sea urchin embryos, *Proc. Natl. Acad. Sci. U.S.A.*, 80, 2656, 1983.
32. **Crum, C. P., Nuovo, G., Friedman, D., and Silverstein, S. J.**, Restricted expression of HPV 16 open reading frames in genital precancers, Sixth Int. Papillomavirus Workshop, Washington, D.C., 82, 1987.
33. **Britten, R. J. and Davidson, E. H.**, Hybridization strategy, in *Nucleic Acid Hybridization*, Hames, B. D. and Higgins, S. J., Eds., IRL Press, Oxford, 1985. 1.
34. **Kurnit, D. M.**, DNA helical content during the C-banding procedure, *Cytogenet. Cell. Genet.*, 13, 313, 1974.
35. **Szabo, P., Elder, R., Steffensen, D. M., and Uhlenbeck, O. C.**, Quantitative *in situ* hybridization of ribosomal RNA species to polythene chromosomes of *Drosophila melanogaster*, *J. Mol. Biol.*, 115, 539, 1977.
36. **Gerhardt, D. S., Kawasaki, E. S., Bancroft, F. C., and Szabo, P.**, Localization of a unique gene by direct hybridization *in situ*, *Proc. Natl. Acad. Sci. U.S.A.*, 78, 3755, 1981.
37. **Renz, M. and Kurz, C.**, A colorimetric method for DNA hybridization, *Nucleic Acids Res.*, 12, 3435, 1984.
38. **Maniatis, T., Fritsch, E. F., and Sambrook, J.**, *Molecular Cloning, A Laboratory Manual*, CSH Laboratory, Cold Spring, Harbor, N.Y., 1982.
39. **Brahic, M. and Haase, A.**, Detection of viral sequences of low reiteration frequency by *in situ* hybridization, *Proc. Natl. Acad. Sci. U.S.A.*, 75, 6125, 1978.
40. **McAllister, L. B., Scheller, R. H., Kandel, E. R., and Axel, R.**, *In situ* hybridization to study the origin and fate of identified neurons, *Science*, 222, 800, 1983.
41. **Harper, M. E., Marselle, L. M., Gallo, R. C., and Wong-Staal, F.**, Detection of lymphocytes expressing human T-lymphotropic virus type III in lymph nodes and peripheral blood from infected individuals by *in situ* hybridization, *Proc. Natl. Acad. Sci. U.S.A.*, 83, 772, 1986.
42. **Penschaw, J. D., Haralambidis, J., Alfred, P., Tregear, G. W., and Coughlan, J. P.**, Location of gene expression in CNS using hybridization histochemistry, *Methods Enzymol.*, 124, 534, 1986.
43. **Young, W. S. and Kuhar, M. J.**, A new method for receptor autoradiography: [^{3}H] opiod receptors in rat brain, *Brain Res.*, 179, 255, 1979.
44. **Dure, B. G. M. and Salmon, S. E.**, High speed scintillation autoradiography, *Science*, 190, 1093, 1975.
45. **Rogers,A. W.**, Scintillation autoradiography at the light microscope level: a review, *Histochem. J.*, 13, 173, 1981.
46. **Williamson, D. J.**, On the specificity of riboprobes for cellular RNA in hybridization cytochemistry, *J. Histochem. Cytochem.*, 36, 811, 1988.
47. **Gowans, E. J., Burrell, C. J., Jilbert, A. R., and Marmion, B. P.**, Detection of hepatitis B virus DNA sequences in infected hepatocytes by *in situ* cytohybridization, *J. Med. Virol.*, 8,67, 1981.
48. **Jilbert, A. R., Burrell, C. J., Gowans, E. J., and Rowland, R.**, Histological aspects of *in situ* hybridization, *Histochemistry*, 85, 505, 1986.
49. **Sternberger, L. A.**, Immunofluorescence, in *Immunochemistry*, John Wiley & Sons, New York, 1979.
50. **Gendelman, H. E., Moench, T. R., Narayan, O., Griffin, D. E., and Clements, J. E.**, A double labelling technique for performing immunochemistry and *in situ* hybridization in virus infected cell cultures and tissues, *J. Virol. Methods*, 11, 93, 1985.
51. **Duncan, D. E., Stumpf, W. E., Pilgrim, C. H., and Breese, G. R.**, High resolution autoradiography at the regional topographic level with [^{14}C]2-deoxyglucose and [^{3}H]2-deoxyglucose, *J. Neurosci. Methods*, 20, 105, 1986.

Chapter 6

THE STUDY AND DIAGNOSIS OF HUMAN GENETIC DISORDERS USING NUCLEIC ACID PROBES

Grant R. Sutherland and John C. Mulley

TABLE OF CONTENTS

I. INTRODUCTION

Genetic disorders result from changes to the normal DNA ranging from single nucleotide substitutions to gross chromosomal changes visible at the level of resolution of the light microscope. There are about 4000 known diseases resulting from mutations of single genes which affect man. Further, there are many other human diseases and malformations which have a genetic component to their etiology. One in 160 live born children has a major chromosome abnormality such as an aneuploidy, a balanced rearrangement, a deletion, or a partial trisomy.

Human genetic diseases can be conveniently placed into four categories. First, there are monogenic diseases due to the mutation of a gene with a known product. For example, the α-thalassemias result from mutation of the gene for the α chain of the hemoglobin molecule. Second, there are monogenic diseases which are due to mutation of a gene of unknown function. The chromosomal location of the gene may be known but its product remains unknown. The gene for myotonic dystrophy for example, is located near the centromere on chromosome 19, but nothing is known of the pathogenesis of the disease. Third, there are polygenic and multifactorial disorders determined by a combination of more than one gene and environmental factors. Coronary heart disease and diabetes mellitus are two examples. Finally, there are disorders which are due to a visible chromosomal change. These can vary from the presence of an extra whole chromosome (trisomy) as seen in Down syndrome to small deletions of chromosome material that are barely discernible by light microscopy and which may remove only a small number of genes.

These diseases are studied and diagnosed by a variety of procedures. Direct examination of the mutation site is possible for monogenic disorders by restriction, oligonucleotide, or deletion analysis. Where this is not possible, indirect examination of the mutation site is carried out by linkage using intragenic or extragenic genetic markers. Study of polygenic and multifactorial diseases is far more difficult because few of the components responsible for a given disorder may be known. Visible chromosomal changes are more easily detected by conventional microscopy than by molecular procedures, but detailed analysis of the observed changes can often be obtained only by molecular techniques.

There are four situations in which the diagnosis of monogenic disease is important. The first, which does not involve DNA probes, is the assessment of the individual with abnormal phenotype or disease process to reach a diagnosis of the disease. Correct diagnosis is essential for the selection of probes for the remaining three diagnostic situations. The importance of this step cannot be overemphasized. A diagnosis implies which set of probes will be linked to or be within a disease producing gene. If an incorrect initial diagnosis is made, the wrong probes will be used for the other diagnoses which may be relevant to the family of the index case. The exercise will thus be futile and the results useless at best and seriously misleading at worst.

The second situation in which diagnosis of genetic disease is important is presymptomatic diagnosis. This is where an individual at risk of carrying a mutant gene which will result in the onset of a disease process in the future is identified as carrying the gene. Presymptomatic diagnosis usually applies to autosomal dominant conditions with variable or late age of onset.

Carrier testing is the third important diagnostic situation. This applies to females at risk of carrying X-linked conditions and thus at risk of having affected sons. For X-linked recessive conditions the females are not at risk of expressing the disease themselves. If the condition is X-linked dominant with incomplete penetrance, some females may also be affected. Carrier detection is also important in autosomal recessive disorders, particularly those for which the mutant gene is at fairly high frequency, e.g., Tay-Sachs disease in Jewish populations. Carrier detection in autosomal recessive diseases is usually by chemical measurement of a gene product rather than by using DNA probes.

The final situation is prenatal diagnosis. This is where a couple is at risk of having a child with a specific genetic disease. Diagnosis at an early stage of pregnancy indicates whether the fetal genotype would result in an individual who would have the disease in question.

Nucleic acid probes are having a major impact on presymptomatic diagnosis, carrier testing, and prenatal diagnosis. The indication for study is a family history based on definite clinical diagnosis. Presymptomatic diagnosis and carrier testing influence decisions on reproduction and the utilization of prenatal diagnosis. Diagnosis can be virtually definitive in many situations, but in others risk figures are expressed as probabilities.

Nucleic acid probes are used to study some aspects of genetic disease processes. These include the nature of mutation in germ cells and somatic cells, and the identification of deleted and rearranged chromosomal material beyond the resolution of the light microscope. Nucleic acid probes have been used to study malignant disorders with associated chromosomal changes within the tumor. These are another major group of human diseases which some regard as genetic, but a survey of malignancy is outside the scope of this review.

II. DEVELOPMENT OF PROBES FOR APPLICATION TO STUDY AND DIAGNOSIS

The majority of nucleic acid probes used in the diagnosis and study of human genetic disease are DNA, either cloned naturally occurring genomic sequences, cDNA copies of mRNA, or synthetic oligonucleotides. The various types of probes, their use, and the methods by which they are located in the genome will be discussed in this section.

A. Oligonucleotide Probes

Nucleic acid synthesizers are used to construct custom-made oligonucleotides. The nucleotide sequence of constructs is deduced either from that portion of the amino acid sequence of the protein of interest that is least susceptible to redundancy of the genetic code, or from a gene sequence which will distinguish normal from mutant genes. Oligonucleotides are used to probe cDNA libraries to identify and clone the corresponding gene[1] or are directly applied to diagnosis where a known base substitution is responsible for the disease process.[2,3]

B. Libraries

A library, in molecular biological terms, is a culture of a vector (plasmid, phage, cosmid) into which the DNA from a species or tissue of interest has been inserted. Ideally, each individual organism in the library will contain a different segment of cloned DNA and all the DNA of interest will be present at least once. A DNA sequence of specific interest is then selected from the library.

All probes used in human genetics (apart from synthetic oligonucleotides) are derived from libraries. Libraries are constructed by reducing genomic DNA or DNA prepared enzymatically from the processed mRNA of specific tissues (cDNA) to manageable fragment sizes and ligating these into a suitable vector.[4,5] The number of vector organisms is then increased (amplification) and the library is ready for selection of the individual vector organisms which contain required DNA inserts.

cDNA libraries are constructed from a given tissue and represent only the coding sequences which are transcribed in that tissue. These libraries are used for the isolation of cloned coding sequences of genes of known product, using oligonucleotides or complementary sequences from related species or genomic libraries.

Genomic libraries are designed to represent the coding and noncoding sequences of the original cellular DNA. They may be derived from the entire human genome, a specific chromosome, or part thereof. Chromosome-specific libraries are obtained by flow sorting metaphase chromosomes or by digesting DNA from human X rodent somatic cell hybrids. The genomic libraries contain clones with both coding and noncoding (anonymous) sequences from which flanking and intervening sequences of genes can be isolated using the cDNA clones as probes, and for detecting restriction fragment length polymorphisms (RFLPs).

Two procedures are used for the isolation of clones containing unique sequences from genomic libraries. Libraries constructed from human DNA alone are probed with nick-translated total human DNA to screen out clones containing high copy repeats. Clones that do not hybridize contain low copy or single-copy inserts which can be used to search for RFLPs. Libraries constructed from human X rodent hybrid cells are probed with pBLUR which hybridizes to Alu sequences. Clones which hybridize to pBLUR are thus identified as containing human inserts. When clones selected in this fashion are used as probes special precautions are required to overcome the technical problems caused by the repeat sequences they will contain. High stringency hybridization conditions[6] or the blocking of repeat sequences by hybridizing the probe with nonradioactively labeled sonicated genomic DNA (prereassociation technique[7]) can usually cope but are not ideal. Once a useful sequence in such a clone is identified the repeat sequences are removed and the unique sequence subcloned into a plasmid.

Finding an RFLP and demonstrating tight linkage to a disease gene by analysis of many families is laborious. To some extent, RFLPs can be targeted by isolating them from chromosome-specific libraries and mapping them against DNA panels constructed from somatic cell hybrids containing characterized segments of human chromosomes.[8,9] Subtractive hybridization is a further technique for isolating probes to very small defined regions. This is based on libraries derived from patients with minute deletions associated with the disease of interest and was first used to obtain intragenic probes to the Duchenne muscular dystrophy gene.[10] The technique of pulsed field gel electrophoresis[11] and availability of jumping and linking libraries[12] make possible the indirect cloning of disease genes from large fragments which hybridize to probes known to flank disease genes.

C. Probes to Disease Genes of Known Product

McKusick[13] lists approximately 300 genetic diseases where there is a known defective gene product. The majority (more than 200) of these are enzyme deficiencies but also included are abnormal coagulation factors, complement components, transport proteins, peptide hormones, collagens, cell surface receptors, epidermal proteins, and membrane transport defects.

Probes to genes of a known product are perhaps the easiest to obtain. A number of approaches have been successful. If a gene product has been isolated, purified, and amino acid-sequenced, candidate DNA sequences can be determined and oligonucleotide probes to these sequences synthesized and used to look for clones containing the gene in cDNA libraries. When the gene product is highly expressed in specialized tissues, the corresponding mRNA is relatively more abundant than those associated with other gene products. This increases the probability of selecting the clone of interest from a library. If the amino acid sequence is not known and the protein not highly expressed in specialized tissues, a monoclonal antibody to a gene product maybe used to probe a cDNA library constructed in an expression vector and the antibody used to identify clones expressing the product. If a gene from another mammalian species has been cloned, it will frequently have sufficient sequence homology with the human gene to allow clones containing part of the human gene to be isolated from a phage library.

Using these approaches several hundred (de la Chapelle[14] lists 249) human genes of known function have been cloned and many of these are genes which, when mutated, give rise to genetic disease. Once such genes have been cloned they can be used in various ways to study and diagnose the diseases associated with them. Gene structure is investigated by restriction enzyme mapping and DNA sequencing. Table 1 lists some of the important disease-producing genes which have been cloned.

D. Restriction Fragment Length Polymorphisms

RFLPs are polymorphic restriction sites inherited through families in a Mendelian manner. They normally have no phenotypic effect and are detectable as variations in fragment size

Table 1
SOME IMPORTANT DISEASE-PRODUCING GENES OF KNOWN PRODUCT WHICH HAVE BEEN CLONED

Gene name	Symbol	Chromosome	Disease
Antithrombin III	AT3	1	AT3 deficiency
β globin complex	HBBC	11	β-thalassemias
Phenylalanine Hydroxylase	PAH	12	Phenylketonuria
α 1-Antitrypsin	PI	14	α 1-Antitrypsin deficiency
α-Globin complex	HBA	16	α-thalassemias
Adenosine deaminase	ADA	20	Combined immune deficiency
Factor VIII	F8	X	Hemophilia A
Factor IX	F9	X	Hemophilia B
Hypoxanthine phosphoribosyl-transferase	HPRT	X	Lesch-Nyhan syndrome
Ornithine transcarbamylase	OTC	X	OTC deficiency
von Willebrand factor	F8VWF	12	von Willebrand disease
Glucose-6-phosphate dehydrogenase	G6PD	X	G6PD deficiency
Glucocerebrosidase	GBA	1	Gaucher disease
β-hexosaminidase	HEXB	5	Tay Sachs and
α-hexosaminidase	HEXA	15	Sandhoff diseases
Complement C4 genes	C4A C4B	6	C4 deficiency
Complement C2	C2	6	C2 deficiency
Collagen, type 1, 2 chain	COL1A2	7	Osteogenesis imperfecta
β-glucuronidase	GUSB	7	Mucopolysaccharidosis VII
β-glucosidase	GAA	17	Fabry disease
Low density lipoprotein	LDL	19	Familial hypercholesterolemia

of DNA arising from the presence or absence of an endonuclease recognition site or from duplication or deletion between restruction sites. Probes detecting RFLPs from genomic DNA define a locus which may be used as a genetic marker in the same way as conventional blood group, enzyme, and protein polymorphisms.

RFLPs are detectable irrespective of whether the probe is part of a cloned gene or merely an anonymous DNA fragment randomly isolated. The functional role of probe DNA, whether it is a coding sequence, an intron sequence or spacer sequence, is irrelevant other than for defining whether probes are intragenic or extragenic. The list of probes detecting RFLPs is updated biannually[15] and already far exceeds the number of conventional markers (protein polymorphisms) available to human geneticists. RFLPs originate from situations shown in Figure 1. Examples of autoradiographs arising from these situations are illustrated in Figure 2.

An RFLP allele is a marker for a gene or segment of chromosome which can allow that segment to be followed through a family. For this to be possible heterozygosity is essential. If an individual is not heterozygous the RFLP is said to be uninformative. The chance of an RFLP being informative in a carrier of a disease gene is equal to the frequency of heterozygotes in the population.This is simply $2pq$ for a two allele polymorphism, where p and q are the allele frequencies. The frequency of heterozygotes is greater when alleles are at intermediate frequencies and when more than two alleles, or haplotypes, are present. Some authors use a calculated polymorphism information content (PIC) value[16] to express the usefulness of an RFLP for linkage studies.

E. Probes to Disease of Unknown Gene Product

The cloning of genes of unknown product cannot be carried out by direct methods and

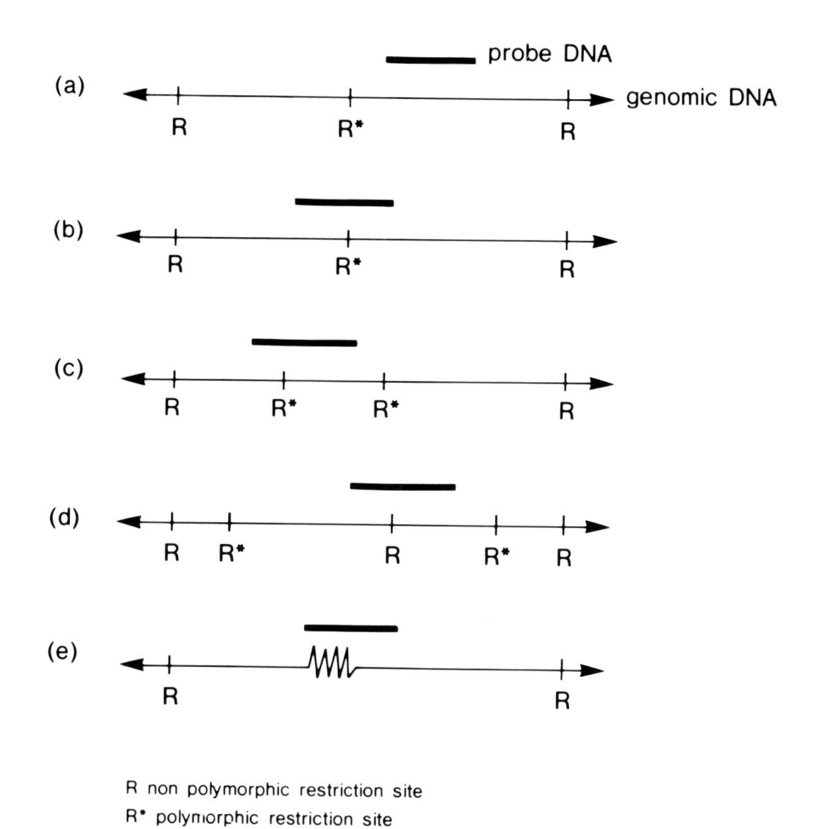

FIGURE 1. Mechanisms which determine RFLPs: (a) two-allele polymorphism with each allele represented by a single band on the autoradiograph; e.g., APOC2 (*Taq*I); (b) two-allele polymorphism with one allele represented by a double band on the autoradiograph, e.g., 52A (*DXS51*) (*Taq*I); (c) multiple allele polymorphism determined by more than one polymorphic restriction site between nonpolymorphic sites; the number of fragments depends upon whether the probe spans the polymorphic sites (a or b above) and the number of internal polymorphic sites; (d) two two-allele polymorphisms determined by the one probe on the one digest; haplotypes are scored directly from the autoradiograph, e.g., *MT2A* (*Taq*I); (e) polymorphism caused by variable length of tandem repeats; these can be two-allele polymorphisms, e.g., *AT3* (*Bam*HI), polymorphisms with a number of identifiable alleles, e.g., St14 (*Taq*I) (*DX52*) or multiple alleles, e.g., 3′HVR of alpha globin (*Pvu*II or *Rsa*I).

has become known as reverse genetics. One approach is to map the gene to a specific chromosomal region, usually by classical genetic linkage with blood groups, enzyme and protein polymorphisms, or RFLPs. These markers themselves have already been mapped to a chromosome. Another approach utilizes the rare individuals who have the disease because of a chromosomal deletion or translocation. These approaches are hampered by the limitations of existing technology to characterize vast lengths of DNA even within small physical regions of metaphase chromosomes.

The genes for chronic granulomatous disease, retinoblastoma, and Duchenne muscular dystrophy have been wholly or partly isolated using this approach (see below). Once the gene has been isolated the reverse process continues of trying to determine the product of the gene and its function. This has only been achieved for chronic granulomatous disease. The process of determining the gene product should eventually lead to some of the diseases now being diagnosed using DNA probes becoming amenable to carrier detection or presymptomatic diagnosis using regular biochemical assays of the product (if the gene is expressed in a tissue available from biopsy). Prenatal diagnosis will only be possible by non-DNA procedures if the product is expressed in chorionic villi or amniocytes.

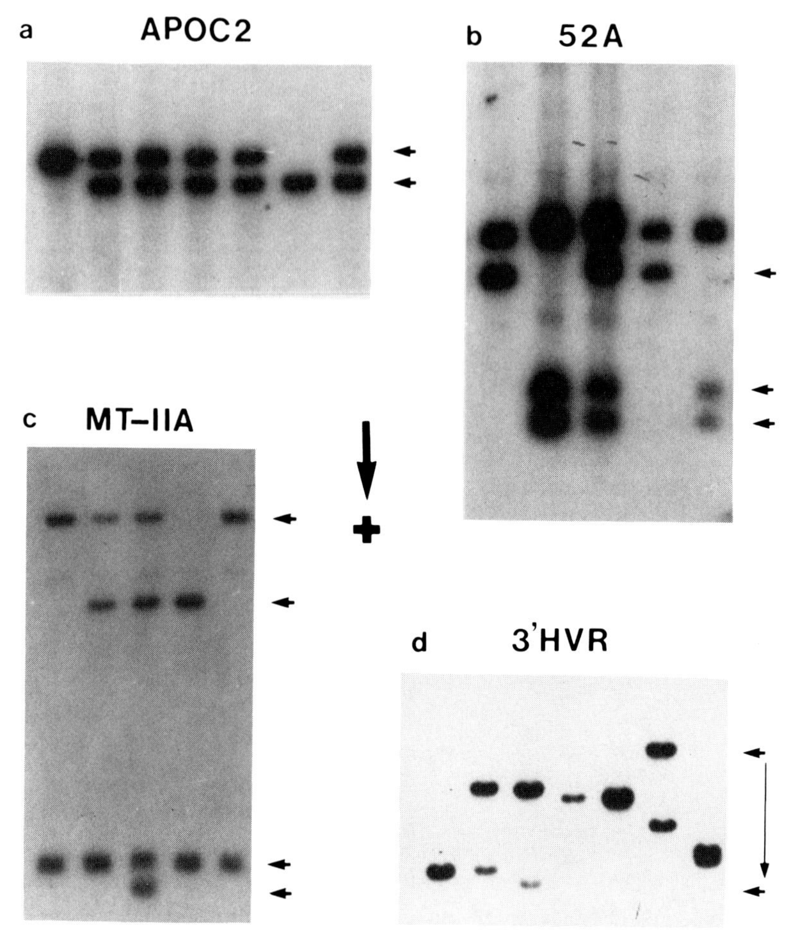

FIGURE 2. (a) *APOC2 Taq*I polymorphism; the two alleles 3.8 and 3.5 (arrowed) are examples of situation (a) of Figure 1; (b) 52A *(DXS51) Taq*I polymorphism; the two alleles 1.3 and 0.7, 0.6 (arrowed) are examples of situation (b) of Figure 1; (c) *MT2A Taq*I polymorphism; the two independent two-allele polymorphisms 7.8 and 5.3 (upper arrows), and 1.7 and 1.6 (lower arrows) are examples of situation (d) of Figure 1; haplotype coding from left to right is A/A, A/C, A/D or B/C (can only be determined by family study), C/C and A/A. (d) 3′HVR *Pvu*I polymorphism; this is an example of situation (e) of Figure 1 showing extensive genetic variation with hybridization bands ranging from 1 kb to 8 kb (arrows).

Once close linkage of RFLPs to the disease-producing genes has been established, they can be used for carrier detection, presymptomatic and prenatal diagnosis. No knowledge of the disease mechanism is necessary and although it is desirable to use fragments of the gene as probes where these are avilable, this is not essential. There are a number of important genetic diseases where this procedure is currently in use (see below). This form of indirect diagnosis will continue until the gene is eventually cloned and sufficient intragenic markers found to replace those that are extragenic. This will eventually occur for all diseases.

F. Mapping Probes

Indirect diagnosis by linkage relies heavily on gene mapping for the development of suitable probes. The X chromosome has been mapped in greater detail than the autosomes because of its unique inheritance pattern. Recessive traits are expressed in boys and every mating involving a heterozygous female is informative because of male hemizygosity.

Gene maps are constructed from physical or genetic data. Physical maps are established

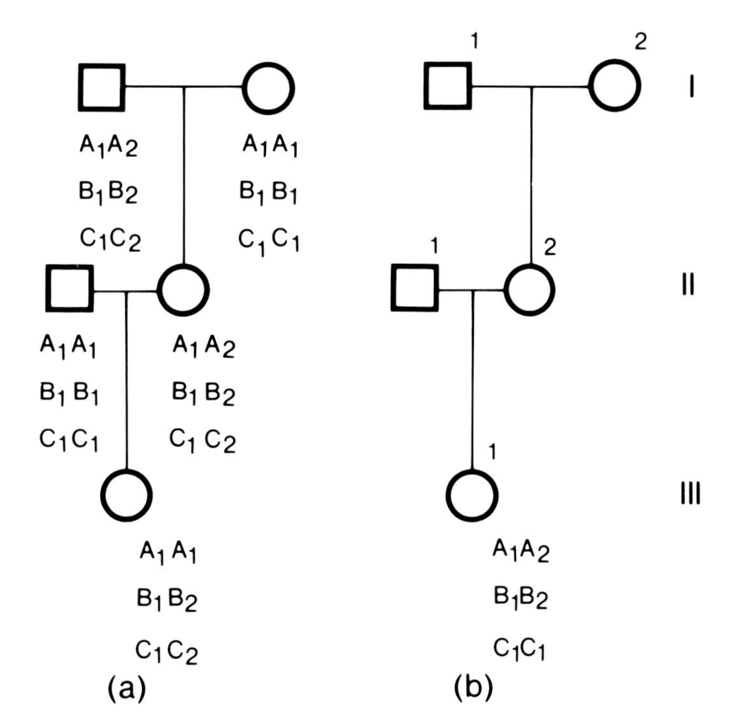

FIGURE 3. Logic for the determination of gene order from three point analysis. Three loci A, B, and C have alleles A1 and A2, B1 and B2, and C1 and C2, respectively. All three loci are informative in II-2. Genotypes are identical in pedigrees (a) and (b), except for III-1. Assume loci A, B and C have already been defined as a linkage group by two point analysis. (a) A1, B1, and C1 in III-1 came from II-1. Therefore, A1, B2, and C2 came from II-2 where linkage phase is known from the genotype of I-2. Recombination has occurred between A and both B and C. Double recombination is unlikely within a tight linkage group, there A is not between B and C. The order of B and C relative to A is unknown, and can only be determined after recombination is observed between loci B and C. (b) This pedigree demonstrates recombination between B and C. Recombination has occurred between C and both A and B. Hence C is unlikely to lie between A and B. The order must therefore be A-B-C. The chromosomal orientation of the cluster must be determined by physical means. The closer the loci are within a linkage group, the observation of recombinants necessary for determination of order become less likely. Order cannot be determined with certainty by two locus analysis because of the broad probability limits associated with estimates of recombination frequency (map distances).

from somatic cell hybrid panels characterized for fragments of human chromosomes or by *in situ* hybridization, especially to metaphase chromosomes expressing fragile sites or having characterized translocations. Genetic maps are derived from linkage analysis and give distances between loci based on recombination frequencies. Two-point analysis is carried out manually or using the computer program LIPED[17] for complex pedigrees. Analyses extended to more than two loci use the LINKAGE package.[18] Gene order is established within linkage groups by three-point informative crosses (Figure 3). Order is essential to demonstrate that markers flank a disease gene.

A genetic linkage map of man with markers evenly spread throughout the genome will soon be fully established. Any disease gene for which a product is not known may then be assigned by linkage to a marker with a known map location.[16] This permits indirect diagnosis and is the prelude to eventual indirect cloning. Huntington disease, polycistic kidney disease, and cystic fibrosis have been assigned in this way. The frequency of recombination between two markers separated by 10^6 base pairs is on average 1%. However, this relationship varies with position on the chromosome. Recombination frequency increases from a minimum around the centromere to a maximum near the telomere.

Risks based on loosely linked markers are sensitive to variations in the recombination fraction used. If the recombination fraction is underestimated it will result in overly optimistic predictions of individuals not carrying a gene or overly gloomy ones that the gene is indeed present. Sufficient numbers of large families with the disease are rarely available for accurate estimates of recombination frequency on their own. This problem can be overcome using normal families with very large sibships to firmly establish a background map of marker loci by multilocus procedures. Disease families are then used to determine the relative position of the disease gene.[19]

Once the map is firmly established, diagnosis and carrier detection by linkage is used for disorders not amenable to direct investigation either because the gene product is not known, known but not amenable to safe biopsy for testing, existing tests on the gene product are unreliable or equivocal, or there is delayed onset of clinical or other measurable symptoms.

G. Alpha Satellite Probes

Chromosome-specific satellite DNA comprises a diverse family of tandemly repeated DNA sequences. This DNA is located near the centromeres of all human chromosomes and will facilitate the development of centromere-based genetic linkage maps. The set of centromere polymorphisms on the X for example are virtually unique for each chromosome examined.[20]

Centromeres represent fixed mapping points in the same way as other cytogenetic markers, such as fragile sites. Centromere markers have the potential to scan larger segments of chromosome than other markers because, in general, frequency of recombination is less than centromeric regions than in telomeric regions of chromosomes. Hence linkage will be detected over greater physical distances in the region of the centromeres than at telomeres.

H. DNA Fingerprint (Minisatellite) Probes

Hypervariable minisatellite regions of DNA consist of tandem repeats of a few tens of bases.[21] They contain an invariant core sequence of 10 to 15 base pairs but unlike the α satellite repetitive sequences occur at many loci within the genome. Individuals differ at given loci by the number of repetitions. Digestion by restriction endonucleases *Hinf*I and/or *Sau*3A (with recognition sequences flanking the minisatellite) generates variable length fragments depending on the number of repeats present.

Multiple bands in the 2 to 20 kb range are detected after electrophoresis of digests using minisatellite sequences as probes in low stringency hybridizations.[21] Minisatellite probes 33.6 and 33.15 together detect multiple bands representing about 60 loci.[22] At least half represent resolvable unlinked loci spread over the autosomes.[22] If randomly distributed throughout the entire genome, there is approximately a 20% chance that at least one minisatellite locus is within 10 centimorgans (cM) of a given disease locus. To illustrate the extent of variability present, Wong et al.[23] cloned one selected fingerprint band. Using a fragment of this band as a probe, at least 77 different alleles were detected from only 79 individuals.

A similar but unrelated set of bands is detectable using wild-type M13 bacteriophage as the probe.[24] This surprising observation results from two 15-base pair repeats within M13 which hybridize to human DNA, in the absence of salmon sperm DNA.

The power of minisatellite probes lies in the simultaneous detection of many highly polymorphic marker loci within the genome using a single probe. This has obvious application to gene mapping and the identification of individuals. In contrast to conventional genetic markers, it is not normally possible to identify alternative alleles from the bands, and linkage data cannot be pooled from unrelated pedigrees. When linkage is suggested or demonstrated by association with a given band within a large sibship, that band needs to be excised from the gel. Single copy sequences immediately adjacent to the minisatellite portion need to be subcloned in order to confirm linkage in additional families and to obtain a chromosomal localization using *in situ* hybridization or hybrid panels. Alternatively, the entire minisatellite maybe isolated and used as the probe in high stringency hybridizations.

III. DIAGNOSTIC PROCEDURES FOR MONOGENIC DISORDERS

Once a gene has been cloned its use as a probe for the diagnosis of the disease depends upon the type of mutation which gives rise to the disease. This can vary from simple situations in which virtually every individual with the disease has the same mutation, as is seen in sickle cell anemia and in geographical isolates for β-thalassemia to more complex ones which will include the great majority of diseases, especially the rarer ones, where there will be different mutations in virtually every family in which the disease is present. An example of such a disease is hemophilia A. Some of these diseases will be presented in more detail once the general principles of the use of DNA probes for diagnosis have been outlined.

The mutations the cause disease vary and include single base changes, deletions ranging from one or more bases to entire genes, insertion of bases, and disruption by translocation with breakpoints within the gene.The gene product may be absent or abnormal. Phenotypic expression is affected by amino acid substitution or changes in transcription, processing, and translation. Single base changes have been found to be responsible for premature stop codons, frameshift mutations, prevention of splicing, alteration of splice sites, and rates of transcription. General diagnostic procedures are needed for detection of these types of mutations.

Caution is necessary for heterogeneous disorders. Clinical heterogeneity arises from different mutations at the same locus or from mutations at different loci affecting the same phenotype. Where more than one locus is involved there is more than one defective gene product and the loci maybe in different chromosomes (e.g., dominant osteogenesis imperfecta can be linked to either the *COLIA1* gene on chromosome 17 or the *COLIA2* gene on chromosome 7[25]). Careful clinical diagnosis is essential, though for some groups of disorders this may not be conclusive. A definite diagnosis is necessary before a probe is used for diagnosis in other family members. In the presence of known heterogeneity and unreliable clinical discrimination, initial studies to establish or exclude linkage to candidate markers must be carried out in each family, provided it is large enough, to establish which gene is defective in order to determine the diagnosis. If there are no candidate markers, as for heterogeneous disorders of unknown product, investigations to initially establish a linkage for subsequent diagnostic application need to be carried out on large kindreds. Pooling families in order to reach significance for a linkage comparison, which is a standard procedure for homogenous diseases, is invalid for heterogenous disorders.

DNA from family members is usually extracted from a sample of blood. Twenty ml of blood will usually yield 400 to 500 μg of DNA. For mid-trimester prenatal diagnosis, DNA can be extracted from uncultured or cultured amniotic fluid cells, but preferably DNA should be obtained much earlier by direct extraction from chorionic villus biopsy samples to allow first trimester diagnosis. Approximately 1 μg wet weight of biopsy sample yields 1 μg of DNA. Diagnosis is generally based on the principle that restriction enzymes cut DNA in a predictable and reproducible way and probes hybridize to these restriction fragments under the appropriate conditions.

Diagnostic services need to be established with the capacity to carry out DNA probing and computer-assisted genetic analysis where necessary. The use of nucleic acid probes is labor intensive. For most diseases, many probes will be needed to render families informative for at least one probe. New probes must be established as they become available. The service must be available throughout the year and able to respond to urgent situations. Ideally, families are studied prior to request for prenatal diagnosis, but unexpected pregnancy may leave little time for completion of detailed DNA investigation between confirmation of pregnancy and the cut-off time for chorionic villus biopsy (Table 2).

Table 2
THE HUMAN GESTATION PERIOD

Week	Stage of gestation
0	Day 1 of last period
2	Conception
5—6	Pregnancy confirmed
	Investigation:
	See medical geneticist
	Collect DNA from family
	Find informative marker
	Organize villus biopsy
10	Chorionic villus biopsy
13	Cut-off for first trimester termination
16	Amniocentesis
20	Cut-off for mid-trimester social termination

Note: There is limited time for determination of informative markers for linkage analysis between confirmation of pregnancy (5 to 6 weeks) and chorionic villus biopsy (10 weeks). Results of chorionic villus sampling are desirable prior to the cut-off period in which simpler termination can be performed (13 weeks). There can be difficulties in cultiring amniotic fluid cells in time for DNA analysis. Preferably, the family is studied prior to conception of an at-risk fetus.

A. Diagnosis by Direct Examination of the Mutation Site

The advantage of direct diagnosis is that family study is not required. The diagnosis is made directly from results of testing the individual.

1. Restriction Analysis

In some instances, a single base change can give rise to an altered gene product and destroy or create a new restriction enzyme recognition sequence. If all, or a large proportion of, individuals with a particular disease have it because of a common mutation of this type, restriction analysis using the appropriate cloned part of the gene involved will allow for a direct determination of whether an individual carries normal or mutant genes. In such a case the restriction enzyme provides a simple means of DNA sequencing at the site of the mutation. The only human disease in which restriction analysis is the primary method of diagnosis is sickle cell anemia (see later).

2. Oligonucleotide Analysis

When single base changes which result in abnormal gene products do not alter restriction enzyme recognition sequences they can be detected using synthetic oligonucleotides. Oligonucleotides of 16 to 19 base pairs require a perfect match otherwise they will be removed by washing after hybridization. Removal occurs with one base mismatch. The test for presence of the disease allele and for presence of the normal allele is controlled by use of accurate hybridization and washing temperatures. Positive and negative controls are necessary for both oligonucleotides. One oligonucleotide is specific for the disease allele and the other specific for the normal allele.

Oligonucleotide analysis is more technically demanding than other approaches, and where there are informative intragenic probes and a suitable family structure a linkage approach is preferred. α 1-antitrypsin deficiency and some geographical isolates of β-thalassemia have been diagnosed using oligonucleotide analysis (see later).

A new method of directly examining the mutation site has been described for detection of the sickle cell anemia allele.[26] This method involves an enzymatic amplification of the segment of the β-globin gene in which the mutation occurs, a process called a polymerase chain reaction. The amplified sequence is then hybridized with an end-labeled oligonucleotide (40-mer) and digested with *Dde*I and *Hinf*I. This liberates a labeled 8-mer from the normal allele and a 3-mer from the mutant allele which are separated by polyacrylamide gel electrophoresis. This process is called oligonucleotide restriction analysis.

3. Deletion Analysis

Small deletions may not be discernible at the level of resolution of the light microscope. These deletions are only detectable by molecular techniques so they do not fit into the category of visible chromosomal change nor, apart from intragenic deletions, do they fit into the category of monogenic disorders. For convenience, the contiguous gene syndromes in which the deletion encompasses more than one gene will be considered later.

a. Altered Restriction Fragments

This is identical to the principle of detecting a diallelic RFLP based on simple insertion or deletion. In this case the RFLP generated by the deletion is associated with the genetic disorder rather than being part of the normal variation in the population. Diagnosis is based on an alteration to the restriction fragment pattern as in some types of thalassemia and other monogenic diseases in which deletions account for a proportion of cases.

b. Absence of Restriction Fragments

A proportion of mutations for many disorders are deletions which remove the entire sequence homologous with the probe. The frequency of such deletions in boys with Duchenne muscular dystrophy (DMD) is approximately 10% using the pERT and XJ series of probes.[27] Detection of deletions is simple and definitive in males using X-linked probes and double hybridization (Figure 4). Where a family has been characterized for a deletion, unequivocal prenatal diagnosis is available for any male fetus.

There are difficulties in determining carrier status for females in X-linked disorders and for males and females in disorders which are autosomal, using band intensities to determine gene dosage. Even with careful quantitation of DNA, correlation of band intensities with gene dosage is often ambiguous.[2] A female in a DMD family which is characterized for a deletion of pERT87 can be unambiguously excluded as a carrier if she is heterozygous with any of the polymorphic probes that map within the deletion. At least nine pERT87 polymorphisms exist. If none of these are heterozygous then the female has a high probability of carrying a deletion for most or all of the pERT87 region. This may be demonstrated unequivocally if anomalous maternal inheritance is detected for any of these which would behave as null alleles.

B. Indirect Diagnosis by Linkage

This approach infers the presence or absence of a deleterious mutation using markers near to, but not at, the site of the mutation. It has general applicability to any mutation within the gene. The unit of study is the family and sufficient members of it must be studied to allow diagnostic results to be given to even a single person. Computerized risk analysis of probe results from one or more loci is often essential for interpretation especially where there is genetic recombination, the possibility that an isolated case is a new mutation, if there are missing family members, and when clinical data such as penetrance, age of onset, and quantititative variation need to be taken into account.

Except for prenatal diagnosis of males in X-linked conditions, correct paternity is essential. When in doubt, paternity or sample identification may be checked using the minisatellite fingerprint probes.[21,24] If a sample mixup between sexes is suspected the probe pDP34 (to locus DXYS1)[30] may be used for unequivocal sex determination by testing for the presence

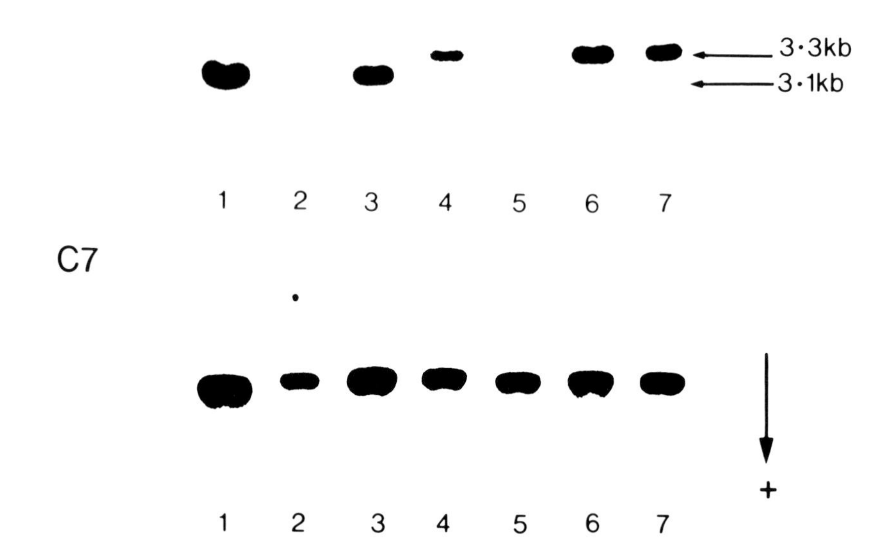

FIGURE 4. Deletion analysis using a *DMD* intragenic marker. *Taq*I digested DNA from probands as probed with pERT87-15, tghen the same filter was reprobed with C7. Individuals 2 and 5 are deleted for the pERT87-15 region. Hybridization to C7 demonstrates sufficient DNA in all tracks for hybridization unless sequences complementary to the probe are deleted (From Mulley, J. C., Gedeon, A. K., Haan, E. A., Sheffied, L. J., White S. J., Bates, L. J., Robertson, E. F., and Sutherland, G. R., *Aust. Paediatr. J.*, in press. With permission.)

or absence of a Y-linked restriction fragment (Figure 5). To minimize the chance of incomplete digestion of genomic DNA, the buffer recommended by the manufacturers with the addition of 5 m*M* spermidine can be routinely used.

Linkage, linkage phase, linkage disequilibrium, and association need to be clearly distinguished. Consider two alleles A1 and A2, with frequencies *p1* and *q1* at locus A, and two alleles B1 and B2, with frequencies *p2* and *q2* at locus B. Loci A and B are linked with reduced recombination between them, if alleles at each locus tend to be co-inherited due to their close proximity on homologs of a chromosome pair. Phase relates to the spatial distribution of alleles along the homolog (or haploid gamete) for two or more linked loci. If A1 and B1 are on the same homolog they are in coupling phase. If A1 and B1 are on different homologs they are in repulsion phase. Alleles at two linked loci are in linkage equilibrium in the population if, for example, the observed population frequency of the combination A1 and B1 on the one homolog is equal to the expected *p1p2*. Disequilibrium exists if there are allelic associations between loci such that the observed population frequency for any allelic combination significantly departs from the expected frequency. A haplotype is a group of alleles from tightly linked loci that rarely recombine because of close proximity to each other. Unfortunately, the term haplotype has often been loosely used to refer to the alleles of a group of linked markers on one homolog irrespective of the tightness of linkage. The term linkage phase is used in the context of association by linkage in family analysis where different associations occur in different families; whereas the term linkage disequilibrium will be encountered when disease associations within a population are considered later.

1. Family Analysis Using Intragenic Probes

Although diagnosis is indirect with respect to the mutation site, use of intragenic probes

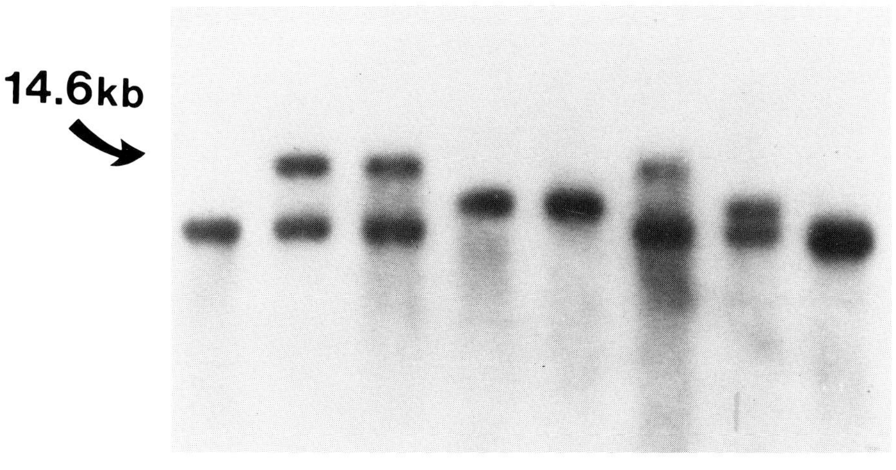

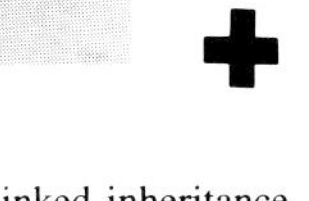

FIGURE 5. The pDP34 *Taq*I polymorphism. The two alleles 11.8 and 10.6 follow normal X-linked inheritance. The 14.6 kb band (arrow) is amel specific. Three of the eight samples are males. (Courtesy of Dr. G. Suthers).

may be classified as gene analysis because the possibility of recombination can be neglected.

Generally there is no need to be concerned about the possibility of genetic recombination between an intragenic RFLP and the disease. There is, however, a theoretical risk that this could occur. Suppose a gene is 100 kb in length and has the RFLP at one end and the disease-producing mutation at the other. One hundred kb is roughly equivalent to 0.1 of a recombination unit (centimorgan, cM); hence we could expect recombination between such an intragenic RFLP and the disease one in a thousand times. In practical genetics this risk can be ignored in most situations. The only known gene for which there is significant recombination between intragenic RFLPs and the disease is the Duchenne/Becker muscular dystrophy gene (see below).

a. X-Linked Recessive Disorders

A typical family is shown in Figure 6. Males are exclusively affected in most conditions with transmission of disease genes through unaffected female carriers. All daughters of affected males will be carriers.

All matings involving heterozygous females are informative for linkage analysis because males are hemizygous. Linkage phase can be determined from the grandfather from an affected boy, or from a normal sibling of an affected boy provided the case is familial rather than sporadic. No information is required from the father for prenatal diagnosis of a male fetus but marker genotype must be known for the father to determine the origin of X chromosomes for any female assessed for carrier status. In the above family (Figure 6) III-2 is a noncarrier female who will not pass the disorder to any male offspring whereas the fetus III-3 will be affected.

b. Autosomal Dominant Disorders

A typical family is shown in Figure 7. RFLPs are fully informative for all offspring if heterozygous in the carrier parent and homozygous in the spouse. If heterozygous in the spouse as well, then RFLPs are only informative for homozygous offspring. These are the only cases where the marker allele linked to the disease can be traced through the pedigree.

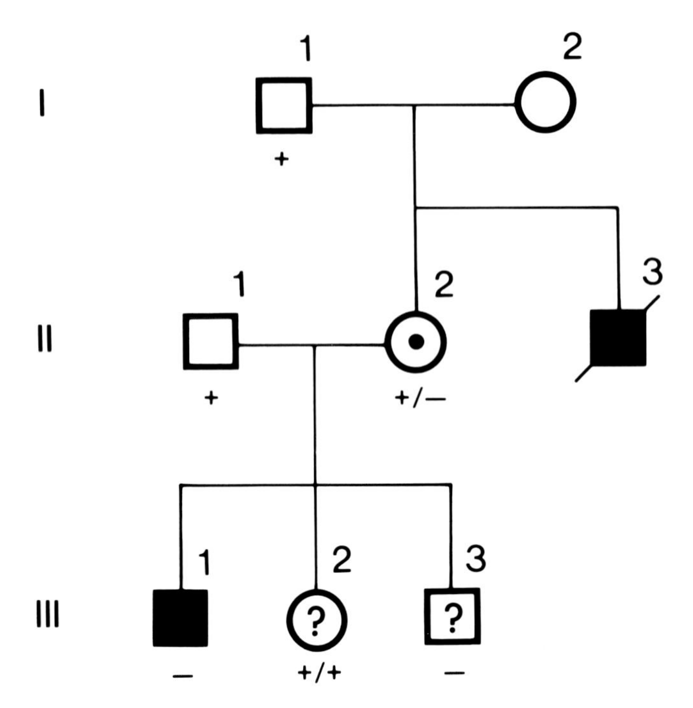

FIGURE 6. Basic principle for carrier detection and prenatal diagnosis of any X-linked disorder. Presence or absence of a restriction site is shown as + or −. II-2 is an obligate carrier. The normal allele (from I-1) is associated with the restriction site marker while the abnormal allele (from III-1) is associated with the lack of a restriction site. Hence, (in the absence of recombination) III-2 is not a carrier but III-3 would be affected.

These pedigrees are often complicated by variable age of onset, incomplete penetrance, and variable expressivity. Hence, it is not unusual for a disorder to skip a generation or have a range of severity. Linkage phase can only be determined with certainty from affected individuals because unaffected individuals may or may not carry the disease gene. For such diseases, detailed clinical evaluation for accurate diagnosis of as many individuals within the family as possible is essential prior to DNA studies. Some autosomal dominant disorders are almost always familial, some almost always new mutations, and others could be either unless more than one case has been documented within the one family.

In the family shown (Figure 7), III-1 is a carrier but III-2 will remain unaffected. Genotypes of missing individuals, such as I-1, can often be inferred.

c. Autosomal Recessive Disorders

Only the nuclear family is relevant for DNA studies. As in Figure 8, both parents are obligate carriers and marker genotypes of both parents are needed for prediction. At least one parent must be heterozygous at the marker locus, and there are four situations in which complete or partial prenatal diagnosis can be performed; these are given in Table 3. The following conditions apply to prenatal diagnosis in the situations shown in Table 3.

1. Definite diagnosis is possible for every fetus only in situation 1.
2. In situations 2 to 4 it is possible to diagnose an unaffected fetus, but it is not possible to diagnose an affected fetus.
3. If diagnosis is not possible for a fetus in situations 2 to 4, additional markers must be examined until a haplotype can be established to allow the fetus to be diagnosed as

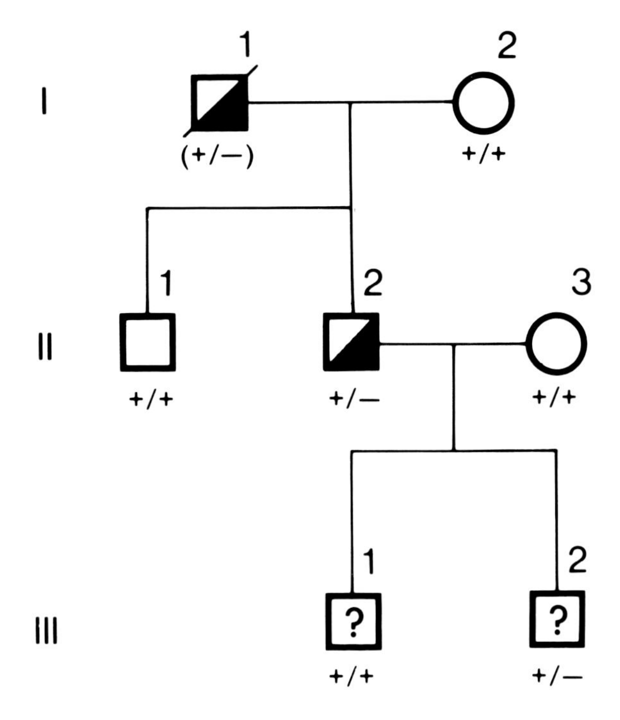

FIGURE 7. Basic principle for presymptomatic diagnosis of an autosomal dominant disease. Presence or absence of a restriction site is shown as + or −. Genotype of the deceased individual I-1 can be inferred in this family. The abnormal allele is associated with the absence of the restriction site in I-1 and II-2, hence III-1 would be unaffected but III-2 could expect to develop the disease.

unaffected, as in situations 2 to 4, or to allow parental origin of all haplotypes to be determined in order to establish that the fetus is affected, as in situation 1.

Siblings of individuals with autosomal recessive diseases have a two thirds chance of being carriers and will often be identified as such. Current nucleic acid probes will not help when they come to reproduce. For example, their spouse could also be a carrier of the same disease (one in 20 chance for cystic fibrosis), and probe could only demonstrate that a fetus is unaffected when it does not carry any allele or haplotype present in the affected uncle or aunt.

2. Family Analysis Using Extragenic Probes

The same principles apply for extragenic markers as for the analysis using intragenic probes, except for the added complication of recombination. Linkage analysis is used in both cases.

The probes are usually anonymous DNA fragments randomly picked from libraries or any cloned gene other than the disease gene. In either case, linkage and the recombination frequencies need to be firmly established prior to diagnostic application.

The risk figure determined using extragenic probes depends on the recombination fraction, Θ, and on the family structure available for phase determination. Phase may be deduced with varying degrees of certainty from grandparents or affected individuals elsewhere in the pedigree. Presence of two affected siblings can be as useful as having grandparents. Frequently only isolated cases are known. Risk analysis often needs to be carried out using computer programs such as LIPED or LINKAGE to take into account mutation rates, gene

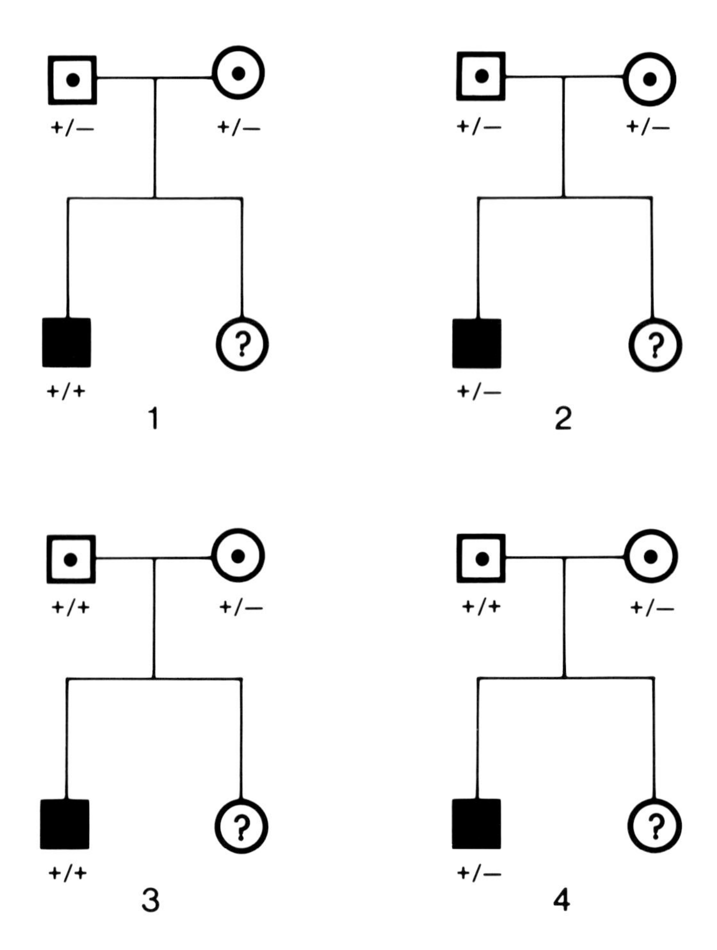

FIGURE 8. The four situations for prenatal diagnosis of an autosomal recessive disorder as described in Table 3. Presence or absence of a restriction site is shown as + or –. Parents are obligate carriers.

Table 3
THE APPLICATION OF TWO ALLELE MARKERS TO PRENATAL DIAGNOSIS OF AUTOSOMAL RECESSIVE CONDITIONS HAS FOUR INFORMATIVE OR SEMI-INFORMATIVE SITUATIONS

Situation	Marker genotype of parents	Marker genotype of affected child	Marker genotype of fetus in relation to affected child	Predicted affection of fetus
1	Both heterozygous	Homozygous	Same	Affected
			Different	Unaffected
2	Both heterozygous	Heterozygous	Same	No diagnosis[a]
			Different	Unaffected
3	One heterozygous	Homozygous	Same	No diagnosis[a]
			Different	Unaffected
4	One heterozygous	Heterozygous	Same	No diagnosis[a]
			Different	Unaffected

Note: Refer to Figure 8.

[a] In these situations there is a 50% risk that the fetus is affected.

frequencies where there are missing persons, penetrance, relevant clinical data, as well as DNA data at one or more marker loci. The program RISKDNA[31] can be used for X-linked diseases.

The interchange of genetic material, or crossing over, between the chromatids of homologous chromosomes occurs at the first meiotic division of gametogenesis. "Absolute" linkage is the situation where the chance of crossover between disease and marker loci is negligible because the marker is intragenic or in the immediate flanking region. For extragenic probes there is a finite chance of crossover depending on the recombination fraction between the probe and disease locus. Crossovers are detectable in offspring by demonstration of recombinants, which are due to an uneven number of crossovers. Extragenic probes therefore provide probabilities rather than definitive diagnoses. The precision of the probabilities depends on the accuracy of the estimate of the recombination frequency (Θ).

Extremely precise diagnoses can be made using linked markers, provided the analysis is based on informative markers flanking the disease gene. In this situation incorrect diagnosis is only possible if a rare double crossover at predictable frequency has occurred between the flanking markers. This is undetectable using flanking marker loci until clinical expression at the disease locus occurs. No useful change in probability can be offered if a single crossover occurs between the flanking markers, because it cannot be determined where the crossover occurs in relation to the disease gene; hence the efficiency of the procedure is greatly reduced with increasing map distances between flanking markers. Single crossovers within the linkage group are detectable and do not lead to misdiagnosis.

C. Demonstration of New Mutation

A major problem when dealing with X-linked disorders and some autosomal dominant disorders with incomplete penetrance is determining where in a family mutation occurred. Approximately one third of all newly ascertained males with X-linked diseases will be new mutants. In such instances there will be negligible risk of recurrence.

When DNA probes can be used to directly determine a mutation this presents little problem, for example if a boy with DMD is deleted for part of the gene detected by a probe, say pERT87-15, and his mother is heterozygous for an RFLP detected by this probe, it is clear that mutation occurred during oogenesis in his mother.

It is, however, usually not possible to directly detect the mutation and RFLPs must be used. These may be intragenic, in which case it may be possible to detect new mutation with virtual certainty, or extragenic, in which case only a probability of new mutation can be given. Only intragenic probes will be considered for simplicity but the principles are the same for extragenic probes, only computer analysis will be required to yield probability figures.

A male can be determined to be a new mutant if his mutant gene (as identified by RFLP) occurs in a disease-free brother. A boy with a disease or his mother may be the new mutant if the boy has his normal grandfather's gene or the same gene as in his mother's normal brother. A family in which mutation for hemophilia A has occurred in either the mother's ovum or grandfather's sperm is shown in Figure 9.[32] New mutations at the *DMD* locus have been demonstrated.[33]

IV. DISEASES OF KNOWN GENE PRODUCT

A. α-1-Antitrypsin Deficiency (PI)

This disorder can result in severe liver disease in childhood and the premature development of emphysema in adults. It is due to a deficiency of α-1-antitrypsin, a major α-1-protease inhibitor. This protein is coded for the *PI* locus and there are about 50 variant forms of the enzyme which can be separated in reference laboratories by a combination of isoelectric

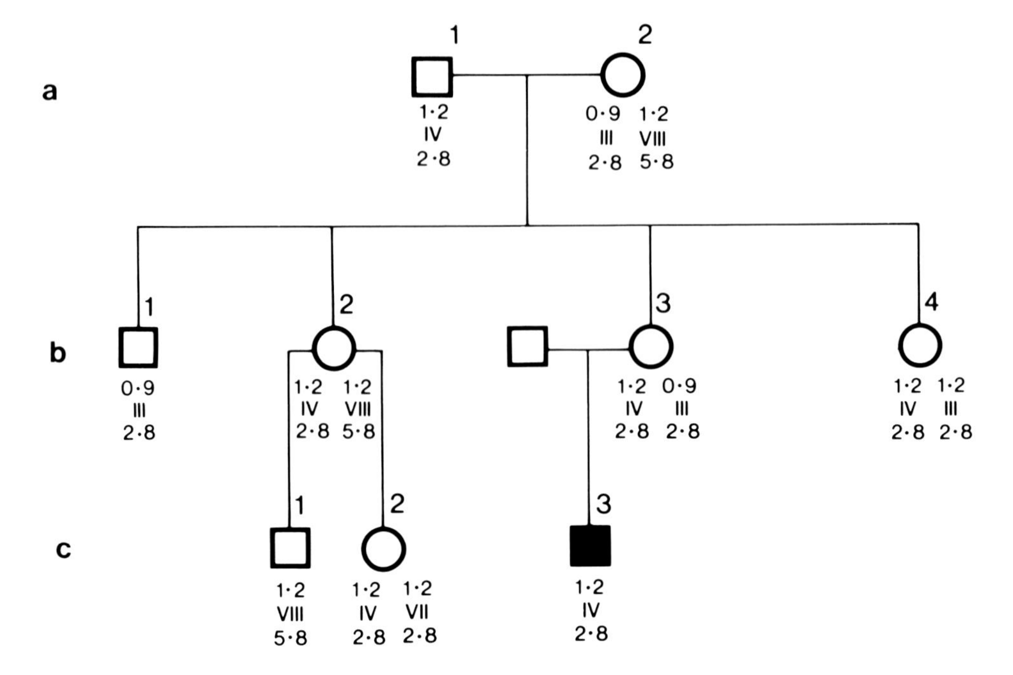

FIGURE 9. A family with hemophilia A and RFLPs (a) a *BclI* polymorphic site in exon 17-18 of the *F8* gene (alleles 0.9 and 1.2); (b) a linked polymorphism detected by the probe St14 at locus *DXS52* (alleles III, IV and VIII); and (c) a linked polymorphism detected by the probe DX13 at locus *DXS15* (alleles 2.8 and 5.8). In this family either the index case III-3 or his mother II-2 are the new mutant. This is deduced because the two chromosomes from the grandmother (I-2) are present in males who do not have hemophilia (II-1 and III-1) and the boy with hemophilia has his normal grandfather's (I-1) chromosome. It is not possible to determine whether the index case or his mother is the new mutant. The other female family members (II-2, II-4, and III-2) can be confidently informed that they are not carriers of hemophilia A. Note there has been recombination between *F8* and the other two markers in the maternal meiosis which produced the ovum from which II-4 developed. (Redrawn from Rudzki, Z., Rodgers, S. E., Casey, G., Mulley, J. C., Sutherland, G. R., and Lloyd, J. V., *Aust. N.Z. J. Med.*, 17, 609, 1987.)

focusing, acid starch gel electrophoresis, and alkaline agarose gel electrophoresis. Several variants are responsible for severe or intermediate levels of deficiency. The risk of severe liver disease in childhood is low; however, prenatal diagnosis is relevant if a sibling is severely affected because of strong intrafamilial correlation of liver disease in children homozygous for the *PI**Z* allele.[34]

The principle of prenatal diagnosis of this disorder was described under the section Oligonucleotide Analysis. Carrier detection is carried out using isoelectric focusing on serum samples. This is essential to identify the variants in carriers responsible for severe α-1-antitrypsin deficiency in their children.

Individuals with α-1-antitrypsin deficiency are usually homozygous for the *PI**Z* allele. The Z allele results from a single DNA base change which leads to the substitution of lysine in place of glutamic acid at position 342.[35] This base sequence change does not directly affect a known restriction enzyme site, hence restriction analysis as used to detect the sickle cell mutation cannot be used. It is, however, possible to directly detect the mutation using end labeled synthetic 19-mer oligonucleotides to the usual mutant alleles. This approach has been used successfully for prenatal diagnosis.[2,3]

Diagnosis using RFLPs is technically simpler.[36,37] The Z allele is part of a unique haplotype identified using *Ava*II digests and a probe extending from exon II to the 3′ flanking region of the gene.[37]

The Z allele protein also has a substitution of valine for arginine at position 213 associated with the removal of a *Mae*III restriction site. This restriction site was absent from all 31 Z alleles tested and 13/38 M1 alleles. Hence, a fetus with this restriction site will be unaffected. This is a useful polymorphism, in addition to the *Ava*II one, for prenatal diagnosis.[34] The *Ava*II and *Mae*II polymorphisms are associations resulting from strong linkage disequilibrium.

B. Familial Hypercholesterolemia (LDLR)

This disorder, which is carried by about 1 in 500 individuals predisposes to myocardial infarction after the age of 35 years and is due to a mutation in the gene coding for the low density lipoprotein reception *(LDLR)*. This gene has been cloned[38] and a 1.9 kb (base pairs 1573 to 3486) fragment of the 3′ end of the cDNA clone demonstrates a *Pvu*II polymorphism in which there is a constant 4.8 kb band and two alleles, V1 with a 19 kb band and V2 with a 16-kb and a 2.6-kb band.[39] This linkage can be used for presymptomatic diagnosis. One family has been recorded in which familial hypercholesterolemia was found to be due to a deletion of 2 kb of the *LDLR* gene.[40]

C. Hemophilia A (F8C)

Classical hemophilia is due to a deficiency of clotting factor VIII. The gene, *F8C*, is on the end of the long arm of the X chromosome and has been cloned.[41,42] There are at least three intragenic RFLPs detected using various fragments of the gene.[43] For those families which are not informative for any of the intragenic probes there are closely linked (approximately 4 cM) polymorphisms detected by the probes DX13 and St14.[44,45] These two probes are on the same side of the *F8C* locus and thus cannot be used as flanking markers.

Carrier detection is important for this disease, and although it can be carried out by direct assay of factor VIII coagulant antigen and factor VIII-related antigen, the results obtained are probabilistic. Prenatal diagnosis is requested by some couples, particularly those at risk of having sons with severe hemophilia. Considerable experience has been gained with diagnosis of this disorder[46,47] and new mutation has been demonstrated.[32,48] The most efficient strategy for diagnostic application in any particular family is to use intragenic probes. If these are uninformative, most of the remaining families will be informative for the highly polymorphic St14, or if necessary, DX13.

D. Hemophilia B (F9)

This form of hemophilia (Christmas disease) is due to deficiency of clotting factor IX. The *F9* gene is on the distal end of the long arm of the X chromosome, but closer to the centromere than that for *F8*. Carrier detection of this disorder using classical hematological techniques is less reliable than that for hemophilia A.

The *F9* gene has been cloned[49] and detects six RFLPs, although an *Mnl*I polymorphism has such small polymorphic fragments that it is best detected using synthetic oligonucleotides.[50] About 80% of families are informative when the known RFLPs are used. Unfortunately, there is considerable linkage disequilibrium at this locus so that additional polymorphisms do not greatly increase the proportion of informative families.

As with most diseases which have been studied to date, some mutants are deletions.[51,52] Dosage levels on Southern blots have been used in a deletion family for carrier detection[51] but caution has been urged in the quantitative use of Southern blots for this purpose.[53]

For the 20% or so of uninformative families it should be possible to use other linked probes as flanking markers (52A proximally and cX55.7 distally) for carrier detection.

There has been considerable experience gained in the use of the *F9* probe for carrier detection of hemophilia B.[54] Prenatal diagnosis has not been reported, possibly because the condition is not thought to be serious enough for this,[55] although an exclusion following parental request for prenatal diagnosis has been reported.[56]

E. Lesch-Nyhan Syndrome (HPRT)

This syndrome of severe mental retardation with self-mutilation results from the absence of activity of the enzyme hypoxanthine phosphoribosyl transferase *(HPRT)*. Reduced enzyme activity leads to gouty arthritis. The gene for the enzyme is on the long arm of the X chromosome. The gene has been cloned[57] and used to study the mutant gene in boys with Lesch-Nyhan syndrome.[58] Five of 28 unrelated boys had abnormal patterns on Southern analysis and their mutations were characterized as different deletions (three boys), one duplication, and one undetermined change. Northern blotting of two additional boys who had normal Southern blot patterns showed one who produced normal message and one who produced none. Hence, from 28 boys studied, the mutational change was not detected in 21 and each of the other seven had a different mutation responsible for his *HPRT* deficiency. This indicates that for rare genetic disorders each family will probably have a unique mutation (or two) and that direct analysis of the site of mutation using cloned genes as probes will be an inefficient means of diagnosis. In such diseases the use of RFLPs, either intragenic or closely linked, is more likely to be successful for this purpose.[59]

Turner et al.[60] used the *HPRT* probe to study somatic mutation (as opposed to the above study which was looking at the results of germ line mutation) by probing the DNA from clones of 6-thioguanine-resistant lymphocytes. This technique selects, from the blood of normal individuals, lymphocytes which have undergone mutation at the *HPRT* gene. A clone is then derived from each lymphocyte selected. Twelve of 21 mutant clones gave abnormal Southern blot patterns due to a variety of deletions, amplifications, and the appearance of novel bands.

F. Ornithine Transcarbamylase Deficiency (OTC)

This X-linked disorder is the commonest of the inherited urea cycle defects. It results in severe hyperammonemia and usually death in boys; females may have a protein intolerance which, in a few cases, can produce hyperammonemia and require careful management with restriction of protein intake. The variability of expression in females is usually attributed to differential X-chromosome inactivation.

Prenatal diagnosis has required fetal liver biopsy because the enzyme is not expressed in cultured amniotic fluid cells. The gene has been cloned[61,62] and 1 in 15 males tested was found to have a gene deletion.[63] At least one other gene deletion has been recorded in a boy with a visible chromosomal deletion.[64] Since deletion mutations in this condition appear to be uncommon, prenatal diagnosis relies mainly on linkage with intragenic RFLPs. About 80% of female carriers will be heterozygous for at least one such RFLP and, provided linkage phase is known, are suitable candidates for prenatal diagnosis.[64,66]

G. Phenylketonuria (PAH)

Classical phenylketonuria *(PKU)* is due to homozygosity for mutant forms of the enzyme phenylalanine hydroxylase. The disease is usually detected shortly after birth by the Guthrie test and treated. If the condition is not treated it soon results in moderate to severe mental retardation. The gene has been cloned[67] and there is a demand for prenatal diagnosis. The gene for phenylalanine hydroxylase (*PAH*) has been subcloned as a full length cDNA in the *Eco*RI site of pBR322. This probe identifies at least ten polymorphic restriction sites at the phenylalanine hydroxylase locus.

With all the RFLPs associated with this probe about 90% of families are informative[68,69] for prenatal diagnosis. In general, DNA probes in recessive diseases can be used for prenatal diagnosis only for the full siblings of affected children. Normal siblings who are shown to be carriers have about a 1 in 40 chance that they could marry a carrier. If such a spouse was found to be a carrier by phenylalanine loading[70] prenatal diagnosis could not be used to detect an affected fetus since it would not be known which of the RFLPs was associated with the mutant gene. It may be possible by RFLP study and phenylalanine loading of the spouse's parents and siblings to determine this, but there is no simple method available using DNA technology alone.

H. Sickle Cell Anemia (HBB)

Sickle cell anemia results from homozygosity for an abnormal human hemoglobin β-chain, HbS. Heterozygotes for this gene have sickle cell trait which protects against falciparum malaria and is carried by about 8% of American blacks. The sickle cell mutation is a single base change (GAG to GTG) which results in the change of glutamic acid to valine in position 6 of the β-chain. This base change destroys *Mst*II, *Sal*I, *Mnl*I, and *Dde*I restriction enzyme recognition sites. When DNA is cut with these enzymes and probed on a Southern blot with cloned fragments of the β-globin gene, altered band patterns are seen.[71,72] Hence, for this disease part of the cloned gene can be used as a direct probe for the presence or absence of a specific disease-producing mutation.

A new method, polymerase chain reaction* followed by oligonucleotide restriction analysis has been devised to detect the sickle cell mutation (see Section III.A.2).

I. α-Thalassemia (HBA)

Within the α-globin gene cluster on the short arm of chromosome 16 there are normally two functional α-globin genes, α_1 and α_2. Reduced synthesis of α-globin chains results in an excess of α and β chains leading to formation of Hb Barts (γ^4) in fetal life and Hb H (β_4) in adult life. Reduced chain synthesis is usually due to deletion of α-genes. Individuals with four α-genes are normal, three genes gives an α-thalassemia carrier (a benign condition), two genes gives α-thalassemia trait (a less benign condition), one gene gives Hb H disease and no α-genes gives Hb Barts hydrops fetalis (lethal before or shortly after birth). Different populations have different deletions which are characteristic of that population.[73-75]

Prenatal diagnosis of Hb Barts hydrops fetalis can be achieved by direct detection of the presence or absence of α-globin genes by probin DNA from uncultured amniotic fluid cells or chorion villi biopsy samples.[74,76] There are also a variety of nondeletion mutations which can give rise to α-thalassemia[77a] where prenatal diagnosis depends upon linkage using one of a number of described RFLPs.

J. β-Thalassemia (HBB)

This group of disorders is due to the production of abnormal β-globin subunits of the hemoglobin molecule. Prior to the introduction of nucleic acid probes for the prenatal diagnosis of this group of disorders, such diagnoses were achieved only by the process of fetal blood sampling for the study of β-chain synthesis. This procedure has now been superseded by the use of nucleic acid probes[77,78] DNA samples are obtained from the cells present in amniotic fluid, these cells after a period of tissue culture, or by direct DNA extraction of tissue obtained by chorion villus biopsy.

The β subunit of hemoglobin is coded for by the β-globin gene which is part of the β-globin gene complex on the short arm of chromosome 11. There are many different mutations of the β-globin gene which can result in β-thalassemia[77a] and one approach to its prenatal diagnosis has been by linkage with polymorphic restriction sites within this gene cluster.[79] However, in some populations most, if not all individuals who have a mutant form of the β-globin gene carry the identical mutation. For example, more than 95% of the β-thalassemia genes in Sardinia are due to a nonsense mutation (GAG to TAG) at the codon for amino acid 39.[80] This allows for prenatal diagnosis using synthetic oligonucleotides (usually 19-mers) to detect the DNA sequence change.[81] Also in this population the mutated

* PCR methodology 15 now developing as the basis for many rapid nonradioactive detection systems.

gene is always on a chromosome on which there is *Bam*HI polymorphic restriction site 3′ to the β-globin gene — hence, absence of this site indicates the presence of the normal allele, but the presence of the site does not indicate that the abnormal allele is present since this site is also on a proportion of chromosomes carrying the normal allele.

In the Hong Kong Chinese population the *Bam*HI site is present with the abnormal allele, and this has been used for prenatal diagnosis.[82] Similar examples of such an association resulting from linkage disequilibrium have been noted for other populations. In Mediterranean families an *Ava*III restriction site within the pseudo-β-globin gene of the β-globin cluster was found to be absent from 54/115 β-thalassemia chromosomes but from only 4/120 normal chromosomes.[83] The various restriction site polymorphisms of the β-globin gene cluster in different populations have been reviewed by Antonarakis et al.[77a] and the molecular mechanisms giving rise to the thalassemias and their detection by Kan.[84]

V. DISEASES OF UNKNOWN GENE PRODUCT

A. Adult Polycystic Kidney Disease (PKD1)

The gene for this disease is carried by about 1 in 1000 individuals in the population. It is an autosomal dominant disorder which is one of the commonest causes of renal failure requiring transplantation. The gene was mapped to the short arm of chromosome 16 by the demonstration of close linkage (5 cM), to a highly polymorphic 3′ region flanking the α-globin gene complex, the so-called 3′ hypervariable region (3′HVR). The gene is also tightly linked to that for the enzyme phosphoglycolate phosphatase *(PGP)*[85] but *3′HVR*, *PGP*, *PKD1* gene order has not been established. The 3′HVR probe has been used for prenatal diagnosis of this disorder.[86] Flanking probes are now being characterized.

B. Chronic Granulomatous Disease (CGD)

The gene for this X-linked disorder of phagocytic cells was the first human gene with an unknown product to be cloned purely on the basis of its map position. A transcript of the gene was found in phagocytic cells and was abnormal in four patients with CGD. The gene product has not yet been isolated but is predicted to be a polypeptide of at least 468 amino acids.[87]

C. Cystic Fibrosis of the Pancreas (CF)

CF is the commonest autosomal recessive disease in Caucasians of Northern European origin. Approximately 1 in 20 individuals carry the gene and one in 2500 infants are born with the disease, which is a generalized dysfunction of the exocrine glands that results in chronic lung disease as well as pancreatic insufficiency. Death usually occurs by the end of the 3rd decade although this situation is changing rapidly and longer survival can be expected in the future. There is a great demand for prenatal diagnosis and carrier detection. The pathogenetic basis of CF is unknown, carrier detection is not possible but a reasonably reliable (95%) prenatal diagnosis* is available for high risk pregnancies.[88]

The *CF* gene was first found to be linked to the gene for a classical enzyme polymorphism — paroxonase.[89] It was subsequently mapped to the long arm of chromosome 7[90-94] and RFLPs detected by the cloned oncogene *MET* and the probe PJ3.11 are closely linked to it. The gene order is uncertain since its region in the middle of the long arm of chromosome 7 is one of low genetic recombination.[95]

The use of probes for prenatal diagnosis of autosomal recessive disorders by linkage was outlined above and applies to cystic fibrosis.[96] In general, the probes can only be used for prenatal diagnosis for siblings of an affected child, but as shown by Brock et al.[97] in some families that can circumvent the need for prenatal diagnosis (Figure 9).

* RFLP analysis now allows much greater accuracy.

D. Duchenne/Becker Muscular Dystrophy (DMD/BMD)

Perhaps the best example of the application of reverse genetics has been (and continues to be) the delineation of this gene. Duchenne muscular dystrophy (DMD) affects about one in 2500 males. It is a progressive muscular degeneration which leads inevitably to death in the late teens or early twenties. Becker muscular dystrophy (BMD) is a milder form of muscular dystrophy which affects adult males. Both follow X-linked inheritance. Early studies mapped both disease genes to the middle of the short arm of the X chromosome,[98-100] and these two forms of muscular dystrophy are due to different mutations of the same gene.[101] The type of mutation which leads to one disease or the other has not been characterized.

An array of RFLPs genetically linked to the DMD locus was generated from a phage library of flow sorted human X chromosomes. Two types of patient were crucial for further progress. A few females were known to have DMD and all of these had a translocation between one of their X chromosomes and an autosome, with the break in the X chromosome in band p21.[102] The reason these girls had DMD was because the translocation disrupted the gene and their normal X chromosome was inactivated by the process of dosage compensation. Boys with a number of unrelated X-linked diseases (adrenal hypoplasia, DMD, glycerol kinase deficiency, chronic granulomatous disease, McLeod phenotype) were the second type of patient who had visible deletions of chromosome band Xp21.[103,104]

Using the boys with the deletions, Kunkel and co-workers[10] developed a technique of reassociation of denatured DNA from the deleted X and normal X chromosomes to construct a library enriched for the sequences from the deleted segment, and from this clones shown to be in the DMD gene itself were isolated — the pERT series of clones. One of the girls with a translocation had ribosomal genes from chromosome 21 translocated adjacent to the disrupted DMD gene. This allowed the cloning of the translocation point and the isolation of another series of clones within the DMD gene — the XJ series.[105]

Kunkel's group used the intragenic probes to look for evolutionarily conserved sequences and isolated a 16-kb transcript of the *DMD* gene.[106] This result has yet to make any impact of the diagnostic use of probes for *DMD*, but is potentially a major advance in the further characterization of this gene. Burmeister and Lehrach[107] have used pulse field gradient electrophoresis and the probes in the vicinity of the *DMD* locus to make a long range restriction map of this area.

There are three diagnostic problems associated with DMD. The first is the detection of females who carry this gene in families in which there is at least one boy with the disease. This can be done with a fair degree of success using serum creatinine kinase levels, but it is not precise and is elevated in only about half of carrier females. The second is the detection of new mutations. About one third of newly ascertained cases will be the result of new mutation but either the boy or his mother could be the new mutant. The third problem is prenatal diagnosis which, prior to the use of DNA probes, relied upon sexing the fetus and termination of male pregnancies. Of course half the male fetuses so terminated would have been unaffected by DMD.

A genetic map of part of the short arm of the X chromosome around the *DMD* gene is shown in Figure 10. There are more than 20 polymorphic probes (Table 4), intragenic and extragenic, which are linked to *DMD*. About 10% of boys with DMD of BMD have a deletion of one of the pERT or XJ probe sequences (Figure 4) and there may be more who have deletions which would be detectable using the many additional probe sequences now being isolated as part of the process of characterization of the DMD gene.* In these cases prenatal diagnosis is straight forward and any female family member hetrozygous for the

* Deletion frequency using intragenic cDNA probes now exceeds 50%.

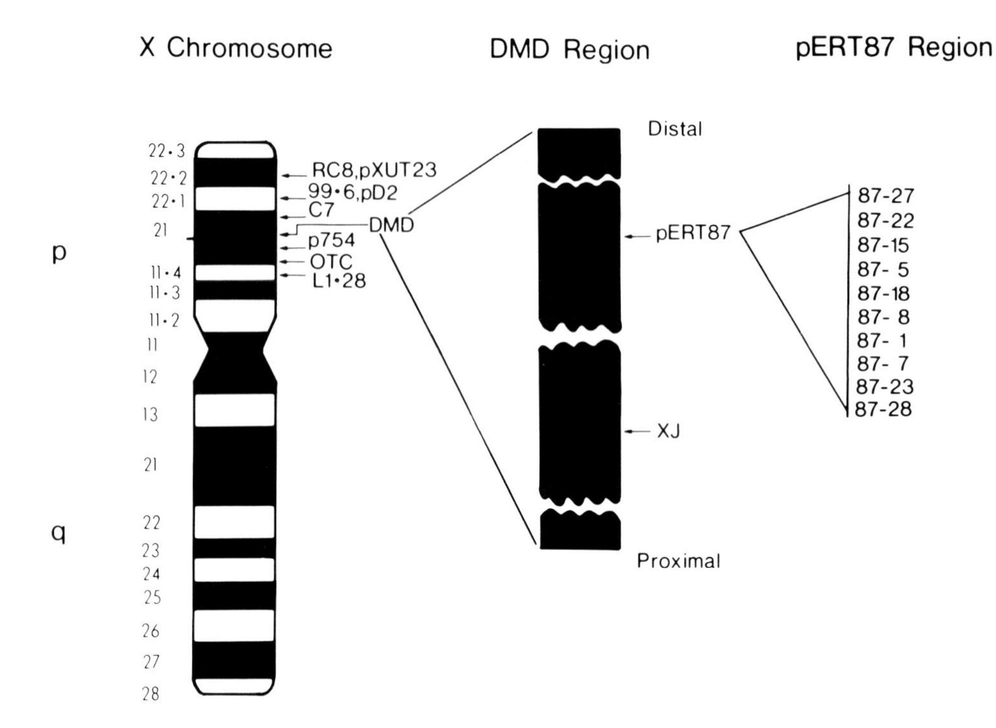

FIGURE 10. Genetic map of the short arm of the X chromosome showing order of the PERT probes in relation to the *DMD* gene.

Table 4
CLINICALLY USEFUL DNA PROBES LINKED TO *DMD* ON Xp

Clone	Restriction enzyme	Heterozygote frequency	Approximate recombination frequency
Intragenic (for rapid risk determination subject to r—5% inaccuracy)			
pERT87-1	*Bst*N1	0.47	0.05
	*Xmn*I	0.45	
	*Msp*I	0.46	
pERT87-8	*Bst*XI	0.48	0.05
	*Taq*I	0.41	
pERT87-15	*Bam*HI	0.47	0.05
	*Taq*I	0.44	
	X*mn*I	0.44	
XJ1.1	*Taq*I	0.41	0.05
XJ1.2	*Bcl*I	0.42	0.05
XJ2.3	*Taq*I	0.54	0.05
Loosely linked (for use as flanking markers for accurate risk determination)			
RC8	*Aaq*I	0.28	0.20
pXUT23	*Bgl*II	0.27	0.20
99-6	*Pst*I	0.41	0.15
pD2	*Pvu*II	0.41	0.15
C7	*Eco*RV	0.26	0.10
p754	*Pst*I	0.47	0.10
OTC	*Msp*I	0.48	0.12
L1.28	*Taq*I	0.44	0.20

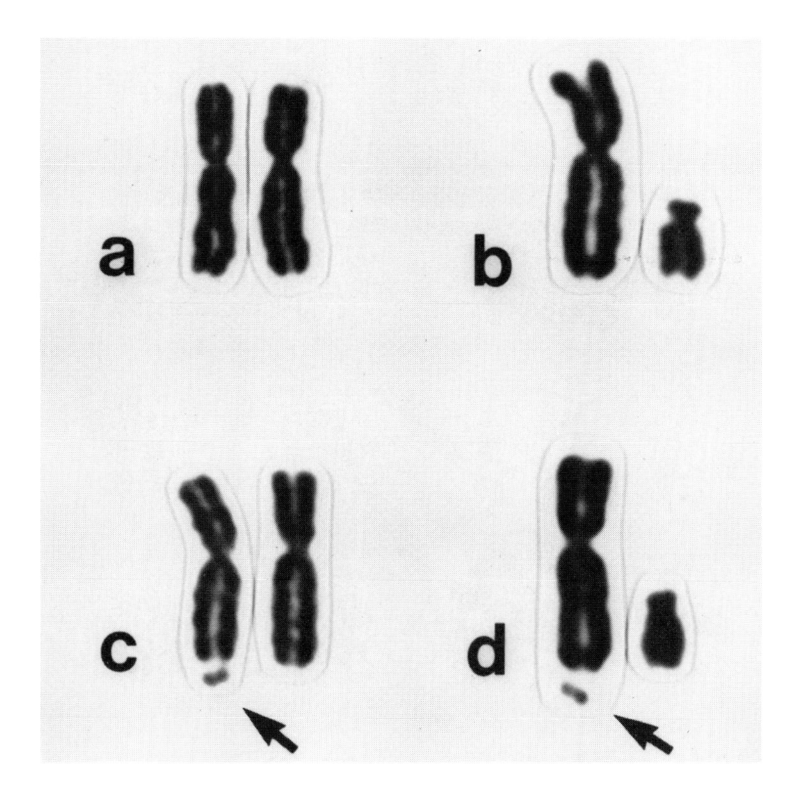

FIGURE 11. Plain stained sex chromosomes from a normal female (a) and male (b) compared with a fragile X female (c) and a fragile X male (d). The arrows indicate the appearance of the fragile site. (From Gardner, R. J. M. and Sutherland, G. R., *Genetic Advice in Chromosomal Conditions*, Oxford University Press, New York, 1988. With permission.)

deleted marker can be assured they are noncarriers.* This is one of the few areas in which DNA probes are used diagnostically to directly examine the mutation causing the disease.

When the index case has no detectable deletion, diagnosis is by linkage. A major problem here is that exons of the *DMD* gene are spread over a relatively large distance and there is 4 to 5% recombination between the intragenic probes and the disease. This is the only known disorder where significant recombination has been demonstrated between the intragenic probe and the mutation. This means that diagnosis must have a 4 to 10% inaccuracy (depending upon whether the family is linkage phase known) associated with diagnosis based on these probes. However, if linked probes flanking the disease gene are used, the accuracy of diagnosis can be greatly improved, up to 98 to 99%, depending upon exactly which of the flanking probes are informative.[108,109] The use of these flanking probes, however, greatly increases the amount of laboratory work required so that present practice is divided between these two approaches. Even though there is much recombination with intragenic probes, they are useful for the separation of potential female carriers into high or low risk classes by combining data from pedigree, CK, and DNA using LINKAGE.[28] Most families are informative for a least one polymorphism within the pERT87 and XJ series of probes.

E. Fragile (X)-Linked Mental Retardation (FRAXA)

The fragile X chromosome is associated with the commonest cause of inherited mental retardation in man.[13] The fragile X chromosome is detected cytogenetically (Figure 11) in retarded individuals but is not readily detectable in many normal male and female carriers.

* Rare exceptions are attributable to germline mosaicism.

Table 5
CLINICALLY USEFUL DNA PROBES LINKED TO FRAGILE X[a]

These Need to Be Applied as Flanking Pairs for Carrier Detection Studies

Clone	Restriction enzyme	Heterozygote frequency	Approximate recombination frequency
Proximal			
F9	*Taq*I	0.60	0.20
	*Dde*I		
cX55.7	*Taq*I	0.27	0.15[a]
4D-8	*Msp*I	0.27	0.10[a]
Distal			
St14	*Taq*I	0.90	0.15
DX13	*Bgl*II	0.47	0.15

[a] Recombination fractions are tentative and based on limited data.

The genetics of the disorder are bizarre and not fully understood, but approximately X-linked dominant with reduced penetrance.[110] Transmission can occur through apparently normal males, which is unusual for an X-linked disorder. All mothers of affected boys, including isolated cases, are regarded as carriers because new mutations are apparently restricted to sperm. There is an apparent increase in penetrance in succeeding generations, possibly associated with ascertainment bias.

The use of DNA probes for this disorder is primarily for carrier detection and to a lesser extent for prenatal diagnosis. Prenatal diagnosis using cytogenetic methods appears to be reasonably reliable.[111] The major problem in using DNA probes is that this region of the X chromosome is one of high recombination and the closest polymorphic probes are 5 to 20 cM from the fragile site. The probes of use are shown in Table 5. Even though the probes are only loosely linked to the disease, Bayesian probabilities of greater than 97 to 99% or less than 1 to 3% can be obtained for carrier status from the combined analysis of flanking RFLP results, cytogenetic results, and clinical status using LINKAGE.[112] Most families are informative for flanking markers using the highly polymorphic St14 probe distal for the fragile site and the F9, cX55.7, and 4D-8 probe proximal to the fragile site.

F. Huntington Disease (HD)

Huntington disease is an autosomal dominant neurodegenerative disorder with variable age of onset. Several RFLPs demonstrated by the probe G8 at locus *D4S10* have been shown to be about 4 cM from the Huntington disease gene.[113,144] The probe has only been recently released for presymptomatic and prenatal diagnosis. The gene is on the tip of the short arm of chromosome 4 and is deleted in patients with the Wolff-Hirschorn syndrome.[115]

There are tremendous ethical problems associated with presymptomatic diagnosis of Huntington disease using DNA probes. Many of these have not been resolved.[116] A method of avoiding the problem of presymptomatic diagnosis but of assisting families at risk involves the prenatal diagnosis of fetuses at 50% or 0% risk without determining the carrier status of the parent.[117,118] This works on the assumption that couples will elect to terminate a pregnancy at 50% risk of having Huntington disease and is outlined in Figure 12. Most, if not all, cases are familial. Careful diagnosis is essential to distinguish this disorder from other neurological disorders before application of the G8 probes.

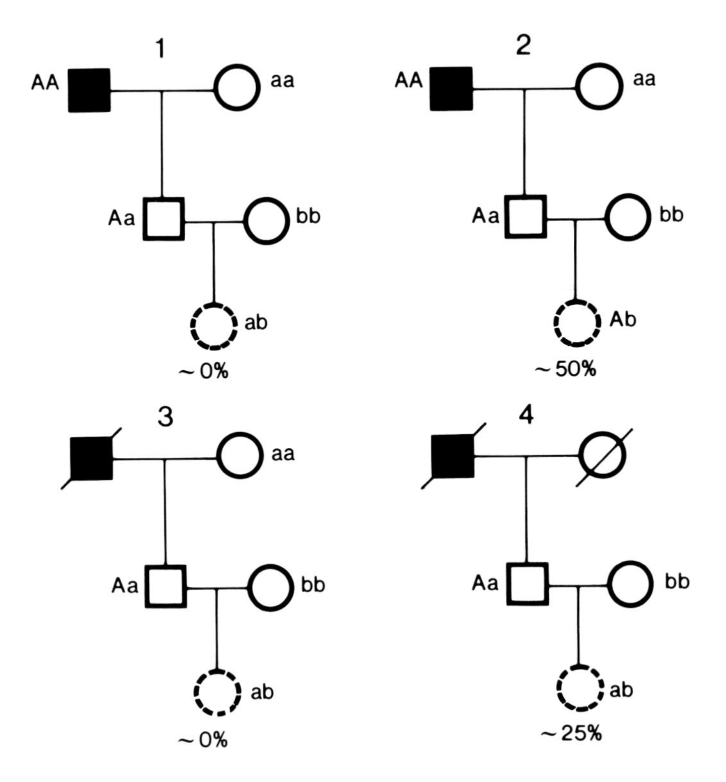

FIGURE 12. Strategy for prenatal testing for Huntington disease without informing the parent of their risk of developing the disease. Pedigree structures for prediction in pregnancies at risk: (1) no prediction can be made for the parent; the fetus has here inherited the a marker type from the healthy grandparent; Huntington disease can thus be excluded, in the absence of recombination; (2) although no prediction can be made for the parent at risk, the fetus has inherited the A marker from the affected grandparent and is thus at the same 50% risk as is the parent; (3) Huntington disease can be excluded in the fetus, since it has inherited the a marker from the healthy grandparent, as in example 1; (4) no prediction can be made for the fetus based on RFLPs since both grandparents are dead. The fetus retains its prior risk of 25%.

G. Myotonic Dystrophy (DM)

The gene for this autosomal dominant disorder has been mapped to chromosome 19, close to the centromere.[119] The clinical severity of this disease varies widely, from a congenital form (only seen when the gene is inherited from the mother) to being asymptomatic, even into old age. There is a demand for presymptomatic and prenatal diagnosis. Most, if not all, cases encountered are familial.

There are *Taq*1, *BamH*I, *Ban*I and *Bfl*I polymorphisms detected by the cloned apolipoprotein CII *(APOC2)* gene and a genomic subclone pSC11. These markers are 4 cM or less from the mytonic dystrophy lcous.[120,121] Prediction in informative families is 92 to 96% accurate using DNA criteria alone (depending upon whether linkage phase is known). The currently symptom-free members of the family should have electromyograms and slit-lamp microscope examinations of the eyes to detect presymptomatic changes. The combination of clinical results from full neurological examination (taking into account variable age of onset) with DNA information using LINKAGE improves the accuracy beyond that available using DNA alone.[122] The APOCII probe has been used for prenatal diagnosis.[120,123]

H. Other Diseases

There are a small number of dominant and many X-linked diseases for which there are

RFLPs that are more or less closely linked and which have been or could be used for diagnosis. The number is increasing rapidly and the reader is referred to the biennial reports from the International Workshops on Human Gene Mapping (the latest at the time of writing is HGM8) for an update on this. Also, Cooper and Schmidtke[124] regularly publish reviews on this topic. In some cases very few families are available for study and there may be insufficient data for the application of DNA probes to diagnosis.

There are diseases that have not been otherwise mentioned for which there are ready diagnostic tests using conventional biochemistry. For example, Hunter syndrome and adrenoleukodystrophy are closely linked to probes on the distal end of the X chromosome (St14, DX13), but these are more laborious to use than appropriate direct assays for the gene product.[125,126]

VI. STUDY OF POLYGENIC AND MULTIFACTORIAL DISORDERS BY DISEASE ASSOCIATION

Diseases which are basically familial but not simple Mendalian disorders can be studied by searching for associations with variation at candidate genes. This is distinct from linkage analysis in families where the coinheritance of the disease and the marker are examined. An association is the occurrence of a certain marker allele and the disease more often than could be attributed to chance alone. Comparisons are made between groups of affected patients and groups of carefully chosen controls from the normal population.

An association at the population level has two causes. It is direct if the marker allele is responsible for the disease. RFLPs could conceivably be directly involved if they were structural length polymorphisms affecting regulatory sequences or if they were single base changes responsible for creation or destruction of a restriction site causing the polymorphism and an altered gene product. Alternatively, the marker alleles have no effect themselves and the association is indirectly caused by linkage disequilibrium. Linkage disequilibrium occurs when the distribution of alleles among loci is nonrandom and has been explained in Section III.

Markers are of use for diagnosis only if the association is strong and all components of the etiology have been identified. Such associations would be useful for risk assessment of individuals with a family history for the disorder. A search for association is feasible only when candidate genes with a product involved in the disease process are known.

A. Coronary Heart Disease

Genetic predisposition is an important risk component of the etiology of this disorder. Genetic heterogeneity and the strong influence of environmental factors renders linkage analysis as applied to monogenic disorders inappropriate in most families. Linkage analysis is only practical using single large pedigrees in which affected and unaffected individuals can be distinguished by clinical criteria with segregation analysis compatible with the presence of a major disease gene.[127]

Candidate genes chosen because of their role in lipoprotein and triglyceride metabolism have been examined for disease associations. Certain alleles of the low density lipoprotein receptor gene *(LDLR)* on chromosome 19 and apolipoprotein genes *(APOA2* on chromosome 1, *APOB* on chromosome 2, and the *APOA1-APOC3-APOA4* cluster on chromosome 11) all have been implicated.[128,130] These markers are probably in linkage disequilibrium with mutations causing hypertriglyceridemia and hypercholesterolemia, the lipids contributing to the disease state when present in abnormal concentrations.The form of this disease due to mutation at the LDLR gene leading to familial hypercholesterolemia is a dominant monogenic disorder and has been discussed in Section IV.

B. Diabetes Mellitus

Diabetes is a heterogeneous group of diseases arising from glucose intolerance. The insulin-dependent (Type I) and non-insulin-dependent (Type 2) forms are probably polygenic and multifactorial, like coronary heart disease. Disease associations have been suggested for both forms of diabetes.[13,132]

Insulin-dependent diabetes results from lack of insulin. There is a strong association between this disorder and the serologically defined HLA alleles DR3 and DR4. New HLA RFLPs detect these same associations[132] probably by linkage disequilibrium. The same markers are said to be associated with other autoimmune diseases: myasthenia gravis, coeliac disease, and premature ovarian failure.[133]

Insulin production is reduced but not absent for the non-insulin-dependent form of diabetes. Polymorphic length variation in the 5′ flanking region of the insulin candidate gene on chromosome 11 is more prevalent in non-insulin-dependent diabetes than in non-diabetics.[131] The structural changes within the 5′ flanking region associated with this RFLP may affect the expression of the insulin gene.

C. Affective Disorders

The affective disorders (depression and/or mania) are common, approximately 1 in 200 individuals suffering from such a disorder at some time during their life. There is considerable evidence that such disorders are familial, and polygenic or multifactorial inheritance has been considered probable. It has recently been shown that in some families there appears to be a single gene of major effect linked to RFLPs on the short arm of chromosome 11. In other families such a gene is unlinked to markers on the short arm of chromosome 11.[134,135]

VII. MOLECULAR CYTOGENETICS

Chromosome anomalies can be of a structural or numerical type and can occur either at mitosis or meiosis. Those which occur at meiosis will give rise to individuals with constitutional chromosome anomalies. Those which occur at mitosis usually give rise to mosaicism which may affect the whole individual or which may affect only a single tissue or organ.

DNA probes have been used to study meiotic errors and determine the parental origin and timing of the error in trisomies, and the nature of rearrangements. Probes have been used diagnostically to detect autosomal trisomies. Some tumors appear to have loss of chromosomal heterozygosity as one step in their genesis, and this has been convincingly demonstrated for retinoblastoma and osteosarcoma.

A. Y Chromosome-Specific Probes

A number of Y chromosome- and X and Y chromosome-specific probes have been produced. Goodfellow et al.[136] list 27 single copy and 2 repeated cloned sequences with homology to the X and Y chromosomes, and 22 single and 4 repeat cloned sequences which are Y chromosome specific. Some of these have been used for a variety of diagnostic purposes and to study chromosomal rearrangements actually or potentially involving the Y chromosome.[137]

1. Fetal Sexing

Prenatal diagnosis of the fetal sex is important in the genetic management of many X-linked disorders. Y-specific probes have been applied to amniotic fluid and chorion villus biopsy cells for this purpose by several groups.[138,139] This approach to fetal sexing never became popular since direct chromosome analysis from the tissues involved, especially for chorion villus samples, is as rapid, less technically complex, and detects chromosomal abnormalities at the same time.

2. 45,X Males

Y-specific probes have been used, in conjunction with X chromosome probes, to explore the origin of 45,X males.[140] (45,X individuals are usually phenotypically female). It was concluded that the most likely origin was 45,X/46,XY mosaicism with loss of the XY cell line from accessible tissues after male differentiation. Another 45,X male studied[141] was found to have material from the short arm and proximal long arm of the Y chromosome translocated to the pericentromeric region of chromosome 15. This translocation was detected using a series of Y specific probes on Southern blots and *in situ* hybridization. Conventional cytogenetic analysis yielded no hint of this translocation.

3. 46,XX Males

The occurrence of males with apparently normal female karotypes has been difficult to explain for many years and several hypotheses have been proposed. Guellaen et al.[142] used ten unique sequence probes and the 3.4 kb *Hae*III fragment probe to study DNA from four such males. One showed no evidence of Y chromosome material and the other three had some, but not all, of the sequences probed for. None had the 3.4 kb Y-specific repeat sequence. Apparent XX males are probably a heterogenous group, some being XX/XY mosaics, some having an X/Y translocation and possibly others being the result of autosomal gene mutations.[143]

4. Isochromosomes for the Long Arm of the Y Chromosome

Schmidtke et al.[144] used the X and Y homologous probe pDP34 to locus *DXYS1* on the short arm of the Y chromosome to study a male with an apparent monocentric isochromosome for the long arm of the Y. Such a chromosomal anomaly usually results in a female phenotype, the male determining genes being on the short arm of the Y chromosome. Using pDP34 they were able to show that the patient had sequences from the short arm of the Y chromosome, although they were unable to locate them in the genome.

5. Y/Autosome Translocations

Burk et al.[145] used the cloned Y-specific 3.4 and 2.1 kb repeats liberated by *Hae*III digestion to study two female patients with extra chromosomal material on the short arm of chromosome 22. This material showed a fluorescence pattern with quinacrine compatible with origins from the Y chromosome. In one patient the DNA probe showed the presence of both repeat sequences and in the other neither was present. Hence the DNA probes were able to distinguish between unusual chromosomes of similar appearance. Lau et al.[146] studied a child with developmental delay who had an unusual chromosome 18. This chromosome had a brightly fluorescent band potentially of Y chromosome origin; however, a complication was that the child's mother also had an unusually brightly fluorescent satellite on chromosome 22. *In situ* hybridization with the Y-specific 3.4 kb repeat showed that it hybridized to the abnormal chromosome 18 and not the 22. On Southern blots the child had approximately twice the number of Y-specific repeat sequences as his father. The authors concluded that the child had a *de novo* chromosomal rearrangement between chromosome 18 and the Y which had occurred during spermatogenesis.

B. X Chromosome Probes

Hassold et al.[147] used seven RFLPs on the X chromosome to determine the parental origin of the X chromosome in ten spontaneously aborted 45,X fetuses. The X was of maternal origin in six instances, paternal in three, and could not be determined in one.

The mechanism of origin of isochromosomes for the long arm of the human X chromosome has been studied[148] using a series of unique probes to detect RFLPs in both arms of this chromosome. Of five isochromosomes studied, three were derived from a maternal and two

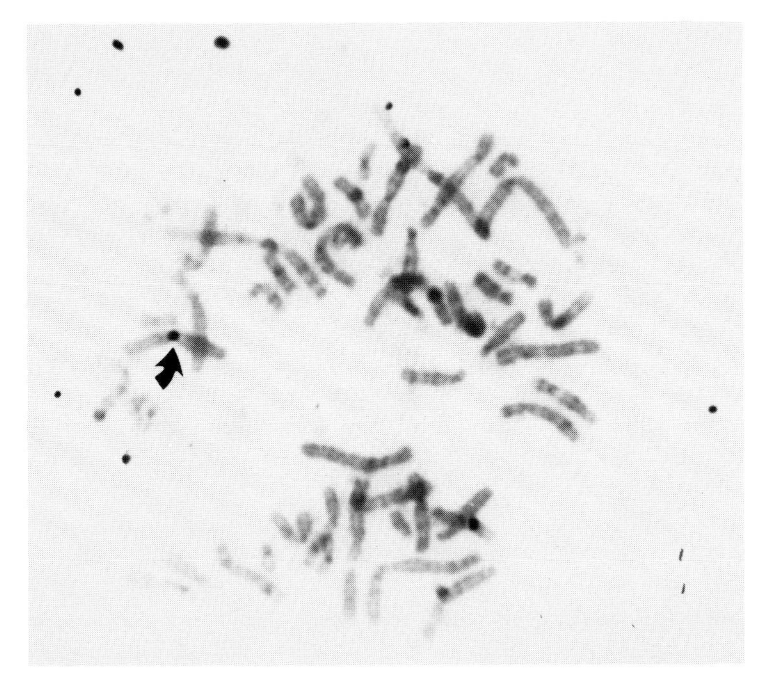

FIGURE 13. Metaphase from a patient with isochromosome for X1 heterozygous for *DXS14* showing hybridization of the probe 58.1 adjacent to the centromere of the isochromosome (arrow). (Courtesy of Dr. D. F. Callen.)

from a paternal chromosome. One patient with such a chromosome of paternal origin was found to be heterozygous at *DXS14*, (detected by probe 58.1) on the proximal short arm of the X chromosome. *In situ* hybridization of the probe showed *DXS14* to be present on the isochromosome (Figure 13). This is a situation similar to that seen for an isochromosome of the long arm of the y[144] (see above) and suggests that a mechanism for the formation of such an X isochromosome[148] — breakage and reunion between the chromatids of the two arms of a single chromosome — may have more general application to the origin of isochromosomes.

Probes to the X chromosomes have been used to determine the limits of small deletions, particularly those which include the DMD locus (see above).

C. Automal Probes

Probes to specific regions of the autosomes have been used to study a number of autosomal chromosome abnormalities.

1. Autosomal Trisomy

Davies et al.[149] used a cloned sequence from chromosome 21 (D21K9) on dosage blots to confirm trisomy 21 and used an RFLP detected by this probe to determine the parental origin of the extra chromosome in this condition. As they pointed out, cytogenetic methods are simpler for this latter purpose but not always possible. Julien et al.[150] used a chromosome 21 specific probe as a means of prenatally diagnosing Down syndrome within 24 h of amniocentesis. The biotin-labeled probe was used for *in situ* hybridization to interphase nuclei. Normal nuclei showed two foci of hybridization and trisomy 21 nuclei showed three such foci. A similar approach using aminoacetylfluorene labeling of a chromosome 18 specific α satellite probe has been used to detect trisomy 18 in interphase nuclei.[151] It is unlikely that such methods will replace cytogenetic analysis for prenatal diagnosis since such analysis will detect any chromosome abnormality and not just the one being sought by the probe.

2. Cri-du-Chat Syndrome

Preliminary studies towards the molecular characterization of the deletion of the short arm of chromosome 5 seen in the Cri-du-Chat syndrome were reported by Carlock and Wasmuth.[152]

3. Cat Eye Syndrome

A malformation syndrome known as the Cat Eye syndrome, because of the presence of colobomata, has been known for many years but the origin of the extra chromosome in such patients has remained certain. Chromosomes 22 and 13 have been suspected of being involved, but using a probe to locus *D22S9* at 22q11, McDermit et al.[153] showed that Cat Eye syndrome patients have either three or four copies of this sequence. This indicated that trisomy or tetrasomy for 22q11 is the chromosomal abnormality in this syndrome. Using *in situ* hybridization with probes on chromosome 22, Duncan et al.[154] further defined the abnormal chromosome which results from rearrangements of a chromosome 22 with a break distal to *D22S9* but proximal to the C lambda immunoglobulin genes and at least one subgroup of the V lambda genes.

4. Additional Marker Chromosomes

A major problem in clinical cytogenetics is the identification and determination of significance of extra small chromosomes present in addition to the normal complement. Such chromosomes can be apparently innocuous or can be the cause of mental handicap. The use of *in situ* hybridization with chromosome centromere-specific probes can help solve this problem.

Mattei et al.[155] used an anonymous cloned sequence which mapped to chromosome 18p11.3 (probe B78, locus *D18S3*) to identify a small marker chromosome as being derived from chromosome 18. The chromosome-specific α satellite probes (section II) should be able to be similarly used for this purpose.

5. Retinoblastoma

This is a highly malignant tumor which arises in the eyes of infants and children. The tumor appears to follow dominant inheritance in some families and in about 21% of patients there is a cytogenetically detectable deletion of band q14 of chromosome 13. There is a gene for retinoblastoma *(RBI)* at this point on the chromosome.

Using probes which detect RFLPs on chromosome 13 it has been shown that cells in the tumor itself are homozygous or hemizygous for the region of chromosome on which the retinoblastoma gene is situated[156,157] even though the normal cells of the patient are heterozygous for this region.

The tumor is thought to arise in individuals who carry the mutant form of the *RBI* gene in cells which become homozygous or hemizygous for this gene. Hemizygosity is achieved either by loss of the chromosome carrying the normal gene, by mitotic nondisjunction, or by deletion of the normal gene from the chromosome. Homozygosity can be achieved by mitotic nondisjunction followed by reduplication, somatic recombination, gene conversion, or point mutation.

Using RFLPs on chromosome 13 and gene dosage studies for the enzyme esterase D, the gene for which is frequently deleted when the retinoblastoma gene is deleted, Cavenee et al.[158] have indicated that prenatal and postnatal prediction of the likelihood of developing retinoblastoma is possible.

A DNA segment which is probably the *RBI* gene (or at least its cDNA sequence) has been cloned.[159] On Northern blot analysis no transcript of this gene can be found in retinoblastoma tumor tissue (or in one osteosarcoma) although it was present in a virally transformed retinal epithelial cell line and in a variety of tumors. Using this gene probe,

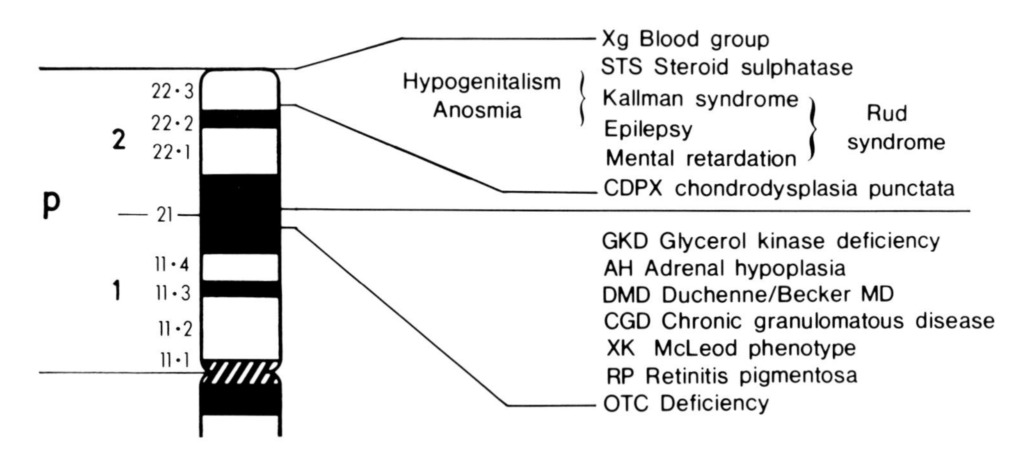

FIGURE 14. Order of genes on the tip of the X chromosome and in band Xp21 which can give rise to variable contiguous gene syndromes when involved in deletions.

p4.7R, or a subclone of it, the DNA in a series of retinoblastomas and osteosarcomas was analyzed on Southern blots and 30% of tumors had abnormal patterns indicating either heterozygous or homozygous alteration of the *RBI* gene.

Loss of heterozygosity to allow an otherwise recessive gene to become dominant appears to be the mechanism for oncogenesis in Wilms tumor,[160] the gene for which is on the short arm of chromosome 11. This mechanism may be more general and awaits the results of studies on other childhood and inherited cancers.

D. Contiguous Gene Syndromes

Schmickel[162] has used the term "contiguous gene syndromes" to describe several malformation complexes which are due to deletions of segments of chromosome, which may or may not be visible cytogenetically, which remove a small number of genes. These genes are usually unrelated in any developmental sense but simply share adjacent locations on a chromosome.

There are several malformation complexes on the X chromosome. The best defined are those which involve deletion of the *DMD* and adjacent genes (Sections III and V). Small deletions near the tip of the short arm of the X chromosome can remove genes for steroid sulfatase, and those resulting in Kallmann syndrome (anosmia and hypogonadotropic hypogonadism). A larger deletion can result in Rud syndrome which comprises features of the Kallmann syndrome plus ichthyosis (steroid sulfatase deficiency), epilepsy, mental retardation, and retinitis pigmentosa. It is uncertain why retinitis pigmentosa is part of this syndrome since the gene for this disorder maps to the proximal part of Xp. The extent of the deletion determines which clinical features are apparent (Figure 14). The deletions are not usually apparent cytogenetically but some patients have t(X;Y) translocations. The deletions can be detected and their extent mapped using nucleic acid probes.[163,164]

Some contiguous gene syndromes involve autosomal deletions and these include the trichorhinophalangeal syndromes (Langer-Giedion) associated with deletion of 8q24, Wilms; tumor/aniridia complex (11p13), retinoblastoma (13q14, see above), Prader-Willi syndrome (15q11) (Figure 15), and Miller-Dieker (17p13) and DiGeorge (22q11) syndromes.[161] As probes have been used to define the X deletions, they should be able to be used to define the autosomal ones. A start in this direction has been made for the Prader-Willi deletion[165] and that associated with the Wilms' tumor/aniridia complex.[166]

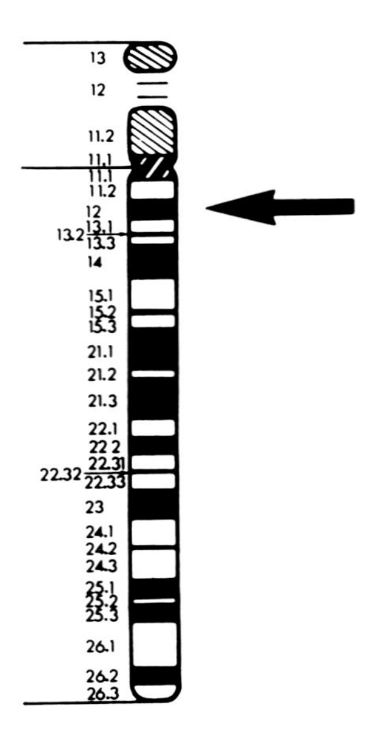

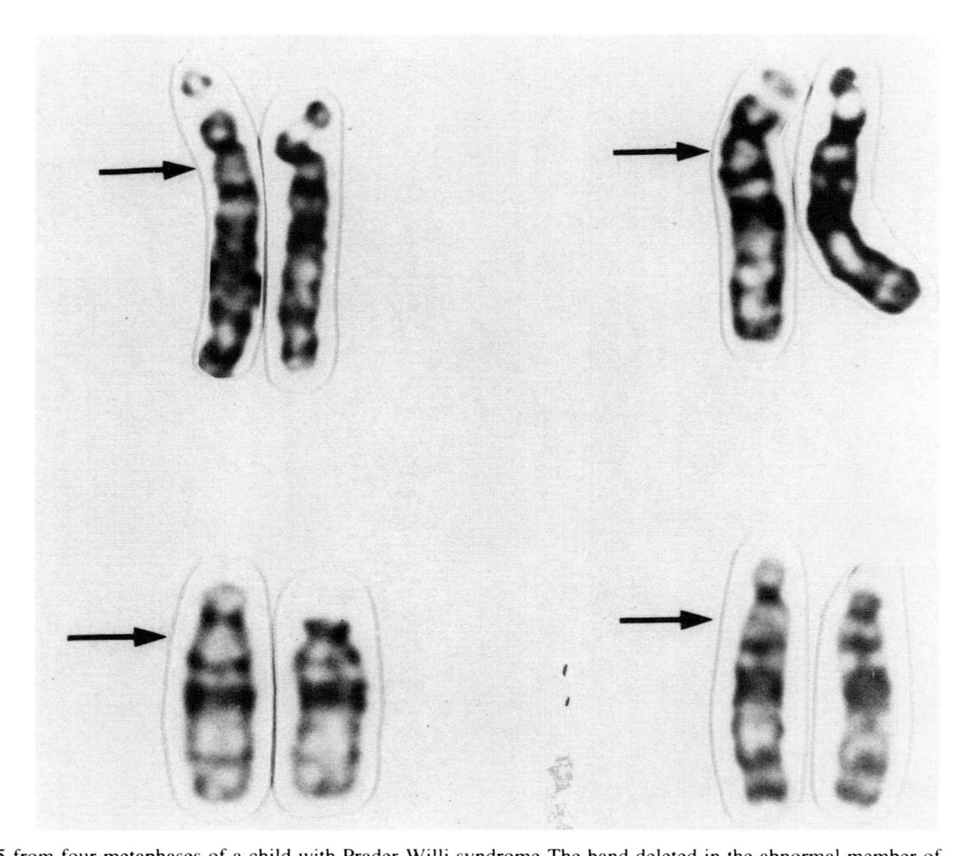

FIGURE 15. Chromosome pair 15 from four metaphases of a child with Prader-Willi syndrome. The band deleted in the abnormal member of each pair is arrowed on the normal homolog and the ideogram. (From Gardner, R. J. M. and Sutherland, G. R. *Genetic Adivce in Chromosomal Conditions*, Oxford University Press, New York, in press. With permission.)

VIII. CONCLUSIONS

Nucleic acid probes have had a dramatic effect on technology available for the diagnosis of human genetic disease. Nevertheless, the impact of this technology is only beginning to be felt. In the next few years most of the significant disease-producing genes will have been either cloned or mapped. It may take a little longer for reverse genetics to lead to the cloning of those genes which produce disease but for which the pathogenetic basis is unknown. This will, however, undoubtedly proceed at an ever increasing pace and the physiological basis of most genetic diseases will be understood. Once this has been achieved, nucleic acid probes may no longer be required for diagnosis in many instances since direct biochemical assay of the gene products maybe, if amenable to biopsy, a more efficient approach.

Given adequate laboratory resources, diagnostic procedures are now routine for a large number of monogenic disorders that have been well studied. Disorders for which only extragenic probes are presently available will ultimately be diagnosed using more accurate intragenic probes. Treatment may become available for an increasing range of genetic disorders once their pathogenetic basis has been established. This may be either by drug treatment to correct body chemistry, or by the introduction of normal genes into specific tissues.

ACKNOWLEDGMENTS

Original work reported in this chapter was supported by the National Health and Medical Research Council of Australia and the Adelaide Children's Hospital Research Foundation. We thank Dr. G. K. Suthers for critically reading the manuscript.

GLOSSARY

Alleles — different forms of homologous DNA sequences at the same locus. A gene or extragenic DNA sequence may have many alleles but in man there are only two present at any one autosomal locus (one on each homolog).

Alpha satellite — diverse family of chromosome-specific tandemly repeated DNA sequences located near the centromeres of all human chromosomes.

Autosome — any chromosome except for the sex chromosomes (X and Y in man) which can be referred to as the gonosomes.

Centromere — the point of primary constriction of the chromosome and the structure which attaches to the microtubules of the spindle during cell division.

Dominant disease gene — a disease gene which produces an abnormal phenotype when present in only one of a pair of chromosomes (heterozygote). The distinction between dominant and recessive (q.v.) is somewhat arbitrary.

Expression — the degree of phenotypic alteration caused by a gene.

Genetic marker — polymorphic locus used to tag a chromosomal region in order to trace its inheritance through a family or to distinguish between individuals within a population.

Hemizygous — describes the state of a normal diploid organism having only one copy of a gene or DNA sequence. In man, all males are hemizygous for genes on the X (and Y) chromosome. Deletion of part of an autosome results in hemizygosity on the normal homolog for the genes in the deleted region.

Heterozygous — the state of a diploid organism having different alleles at a locus on homologous chromosomes.

Homolog — the name given to each of the two members of a pair of chromosomes.

Isochromosome — a chromosome which contains two copies of one arm of the normal chromosome from which it is derived, one on either side of the centromere.

Locus — a position on a chromosome where there is only a single gene or characterized DNA sequence.

Gene mapping — assignment and regional localization of genes or DNA fragments to chromosomes.

Minisatellite region — hypervariable tandem repeat of an invariant core sequence of a few tens of bases.

Penetrance — is the manifestation of an abnormal gene as an abnormal phenotype. A gene may be nonpenetrant, i.e., its carrier is of normal phenotype when the gene usually results in an abnormal phenotype. Incomplete pentrance refers to the lack of phenotypic abnormality in all carriers of a mutant gene (c/f expression q.v.).

Polymorphism — the presence of any one of two or more genetically distinct phenotypes within a population, the least common of which occurs with a frequency greater than that which could be maintained by recurrent mutation alone.

Recessive disease gene — a gene which causes no significant phenotypic abnormality when present on only one of a pair of chromosomes (heterozygote) but results in such abnormality when present on both chromosomes (homozygote).

Sibship — a group of individuals with the same parents (i.e., a group of brothers and sisters).

Telomere — a structure at the end of all chromosome arms. The structure cannot be seen morphologically but has characteristic DNA sequences.

Translocation — is the process whereby two chromosomes break and rejoin to create morphologically different chromosomes.

X-linked disease — one for which the gene determining the disease is located on the X chromosome.

REFERENCES

1. **Noda, M., Furutani, Y., Takahashi, H., Toyosata, M., Hirose, T., Inayama, S., Nakanishi, S., and Numa, S.,** Cloning and sequence analysis of cDNA for bovine adrenal preproenkephalin, *Nature (London)*, 295, 202, 1982.
2. **Kidd, V. J., Golbus, M. S., Wallace, R. B., Itakura, K., and Woo, S. L. C.,** Prenatal diagnosis of α-anitrypsin deficiency by direct analysis of the mutation site in the gene, *N. Engl. J. Med.*, 310, 639, 1984.
3. **Hejtmancik, J. F., Ward, P. A., Mansfield, T., Sifers, R. N., Harris, S., and Cox, D. W.,** Prenatal diagnosis of α-antitrypsin deficiency by restriction fragment length polymorphisms, and comparison with oligonucleotide probe analysis, *Lancet*, 2, 767, 1986.
4. **Maniatis, T., Hardison, R. C., Lacy, E., Lauer, J., O'Connell, C., Quon, D., Sim, G. K., and Efstratiadis, A.,** The isolation of structural genes from libraries of eukaryotic DNA, *Cell*, 15, 687, 1978.
5. **Williams, J. G.,** The preparation and screening of a cDNA clone bank, in, *Genetic Engineering I*, Williamson, R., Ed., Academic Press, New York, 1981.
6. **Fisher, J. H., Guselia, J. F., and Scoggin, C. H.,** Molecular hybridization under conditions of high stringency permits cloned DNA segments containing reiterated DNA sequences to be assigned to specific chromosomal locations, *Proc. Natl. Acad. Sci. U.S.A.*, 81, 520, 1984.
7. **Sealey, P., Whittaker, P. A., and Southern, E. M.,** Removal of repeated sequences from hybridization probes, *Nucl. Acid Res.*, 13, 1905, 1985.
8. **Cavenee, W., Leach, R., Mohandas, T., Pearson, P., and White, R.,** Isolation and regional localization of DNA segments revealing polymorphic loci from human chromosome 13, *Am. J. Hum. Genet.*, 36, 10, 1984.
9. **Carlock, L. R., Skarecky, D., Dana, S. L., and Wasmuth, J. J.,** Deletion mapping of human chromosome 5 using chromosome-specific DNA probes, *Am. J. Hum. Genet.*, 37, 839, 1985.
10. **Kunkel, L. M., Monaco, A. P., Middlesworth, W., Ochs, H. D., and Latt, S. A.,** Specific cloning of DNA fragments absent from the DNA of a male patient with an X chromosome deletion, *Proc. Natl. Acad. Sci. U.S.A.*, 82, 4778, 1985.

11. **Anand, R.**, Pulsed field gel electrophoresis: a technique for fractionating large DNA molecules, *Trend. Genet.*, 2, 278, 1986.
12. **Poustka, A., Pohl, T. M., Barlow, D. P., Frischauf, A.-M., and Lerach, H.**, Construction and use of human chromosome jumping libraries from *Not*I-digested DNA, *Nature (London)*, 325, 353, 1987.
13. **McKusick, V. A.**, *Mendelian Inheritance in Man*, 7th ed., Johns Hopkins University Press, Baltimore, 1986.
14. **de la Chapelle, A.**, The 1985 human gene map and human gene mapping 1985, *Cytogenet. Cell Genet.*, 40, 1, 1985.
15. **Willard, H. F., Skolnick, M. H., Pearson, P. L., and Mandel, J.-L.**, Report of the committee on human gene mapping by recombinant DNA techniques, *Cytogenet. Cell Genet.*, 40, 360, 1985.
16. **Botstein, D., White, R. L., Skolnick, M., and Davis, R. W.**, Construction of a genetic linkage map in man using restriction fragment length polymorphisms, *Am. J. Hum. Genet.*, 32, 314, 1980.
17. **Ott, J.**, Estimation of the recombination fraction in human pedigrees: efficient computation of the likelihood for human linkage studies, *Am. J. Hum. Genet.*, 26, 588, 1974.
18. **Lathrop, G. M., Lalouel, J. M., Julier, C., and Ott, J.**, Multilocus linkage analysis in humans: detection of linkage and estimation of recombination, *Am. J. Hum. Genet.*, 37, 482, 1985.
19. **Drayna, D. and White, R.**, The genetic linkage map of the human X chromosome, *Science*, 230, 753, 1985.
20. **Willard, H. F., Waye, J. S., Skolnick, M. H., Schwartz, C. E., Powers, V. E., and England, S. B.**, Detection of restriction fragment length polymorphisms at the centromeres of human chromosomes by using chromosome-specific satellite DNA probes: implications for development of centromere-based genetic linkage maps, *Proc. Natl. Acad. Sci. U.S.A.*, 83, 5611, 1986.
21. **Jeffreys, A. J., Wilson, V., and Thein, S. L.**, Hypervariable 'minisatellite' regions in human DNA, *Nature (London)*, 314, 67, 1985.
22. **Jeffreys, A. J., Wilson, V., Thein, S. L., Weatherall, D. J., and Ponder, B. A.**, DNA "Fingerprints" and segregation analysis of multiple markers in human pedigrees, *Am. J. Hum. Genet.*, 39, 11, 1986.
23. **Wong, Z., Wilson, V., Jeffreys, A. J., and Thein, S. L.**, Cloning a selected fragment from a human DNA 'fingerprint': isolation of an extremely polymorphic minisatellite, *Nucl. Acid Res.*, 14, 4605, 1986.
24. **Vassart, G., Georges, M., Monsieur, R., Brocas, H., Lequarre, S. P., and Christophe, D.**, A sequence in M13 phage detects hypervariable minisatellites in human and animal DNA, *Science*, 235, 683, 1987.
25. **Sykes, B., Wordsworth, P., Ogilvie, D., Anderson, J., and Jones, N.**, Osteogenesis imperfecta is linked to both type I collagen structural genes, *Lancet* 2, 69, 1986.
26. **Embury, S. H., Scharf, S. J., Saiki, R. K., Gholson, M. A., Golbus, M., Arnheim, N., and Erlich, H. A.**, Rapid prenatal diagnosis of sickle cell anemia by a new method of DNA analysis, *N. Engl. J. Med.*, 316, 656, 1987.
27. **Kunkel L. M. et al.**, Analysis of deletions in DNA from patients with Becker and Duchenne muscular dystrophy, *Nature (London)*, 322, 73, 1986.
28. **Mulley, J. C., Gedeon, A. K., Haan, E. A., Sheffield, L. J., White, S. J., Bates, L. J., Robertson, E. F., and Sutherland, G.R.**, Application of DNA probes to carrier detection and prenatal diagnosis of Duchenne (and Becker) muscular dystrophy, *Aust. Paediatr. J.*, in press.
29. **Thomas, N. S. T., Ray, P. N., Worton, R. G., and Harper, P. S.**, Molecular deletion analysis in Duchenne muscular dystrophy, *J. Med. Genet.*, 23, 509, 1986.
30. **Page, D., de Martinville, B., Barker, D., Wyman, A., White, R., Francke, U., and Botstein, D.**, Single-copy sequence hybridizes to polymorphic and homologous loci on human X and Y chromosomes, *Proc. Natl. Acad. Sci. U.S.A.*, 79, 5352, 1982.
31. **Sarfarazi, M. and Williams, H.**, A computer programme for estimation of genetic risk in X linked disorders, combining pedigree and DNA probe data with other conditional information. *J. Med. Genet.*, 23, 40, 1986.
32. **Rudzki, Z., Rodgers, S. E., Casey, G., Mulley, J. C., Sutherland, G. R., and Lloyd, J. V.**, Demonstration of a recent mutation in a family with isolated haemophilia A, *Aust. N.Z. J. Med.*, 17, 609, 1987.
33. **Roncuzzi, L., Ferlini, A., Pirozzi, A., and Romeo, G.**, Origin of new mutations in Duchenne muscular dystrophy, *Hum. Genet.*, 74, 456, 1986.
34. **Cox, D. W. and Billingsley, G. D.**, Restriction enzyme MaeIII for prenatal diagnosis of α-antitrypsin deficiency, *Lancet*, 2, 741, 1986.
35. **Kidd, V. J., Wallace, R. B., Itakura, K., and Woo, S. L. C.**, α-Antitrypsin deficiency detection by direct analysis of the mutation in the gene, *Nature (London)* 304, 230, 1983.
36. **Anon.**, α-antitrypsin deficiency and prenatal diagnosis, *Lancet*, 1, 421, 1987.
37. **Cox, D. W. and Mansfield, T.**, Prenatal diagnosis of α-antitrypsin deficiency and estimates of fetal risk for disease, *J. Med. Genet.*, 24, 52, 1987.
38. **Yamamoto, T., Davis, L. G., Brown, M. S., Schneider, W. J., Casey, M. L., Goldstein, J. L., and Russell, D. W.**, The human LDL receptor: a cysteine-richprotein with multiple Alu sequences in its mRNA, *Cell*, 39, 27, 1984.

39. **Humphries, S. E., Horsthemke, B., Seed, M., Holm, M., Wynne, V., Kessling, A. M., Donald, J. A., Jowett, N., Galton, D. J., and Williamson, R.,** A common DNA polymorphism of the low-density lipoprotein (LDL) receptor gene and its use in diagnosis, *Lancet* 1, 1003, 1985.
40. **Horsthemke, B., Kessling, A. M., Seed, M., Wynn, V., Williamson, R., and Humphries, S. E.,** Identification of a deletion in the low density lipoprotein (LDL) receptor gene in a patient with familial hypercholesterolaemia, *Hum. Genet.*, 71, 75, 1985.
41. **Wood, W. I., Capon, D. J., Simonsen, C. C., Eaton, D. L., Gitschier, J., Keyt, B., Seeburg, P. H., Smith, D. H., Hollingshead, P., Wion, K. L., Delwart, E., Tuddenham, E. G. D., Vehar, G. A., and Lawn, R. M.,** Expression of active human factor VIII from recombinant DNA clones, *Nature (London)*, 312, 330, 1984.
42. **Toole, J. J., Knopf, J. L., Woznmey, J. M., Sultzman, L. A., Buecker, J. L., Pittman, D. D., Kaufman, R. J., Brown, E., Shoemaker, C., Orr, E. C., Amphlett, G. W., Foster, W. B., Coe, M. L., Knutson, G. J., Fass, D. N., and Hewick, R. M.,** Molecular cloning of a cDNA encoding human antihaemophilic factor, *Nature (London)* 312, 342, 1984.
43. **Wion, K. L., Tuddenham, G. D., and Lawn, R.M.,** A new polymorphism in the factor VIII gene for prenatal diagnosis of hemophilia A, *Nucl. Acid. Res.*, 14, 4535, 1986.
44. **Winter, R. M., Harper, K., Goldman, E., Mibashan, R. S., Warren, R. C., Rodeck, C. H., Penketh, R. J. A., Ward, R. H. T., Hardisty, R. M., and Pembrey, M. E.,** First trimester prenatal diagnosis and detection of carriers of haemophilia A using the linked DNA probe DX13, *Br. Med. J.*, 291, 765, 1985.
45. **Oberlé, I., Camerino, G., Heilig R., Grunebaum, L., Cazenave, J.-P., Crapanzano, C., Mannucci, P. M., and Mandel, J.-L.,** Genetic screening for hemophilia A (classic hemophilia) with a polymorphic DNA probe, *N. Engl. J. Med.*, 312, 682, 1985.
46. **Antonarakis, S. E., Waber, P. G., Kittur, S. D., Patel, A. S., Kazazian, H. H., Mellis, M. A., Counts, R. B., Stamatoyannapoulos, G., Bowie, E. J. W., Fass, D. N., Pittman, D. D. Wozney, J. M., and Toole, J. J.** Hemophilia A. Detection of molecular defects of and carriers by DNA analysis, *N. Engl. J. Med.*, 313, 842, 1985.
47. **Antonarakis, S. E., Carpenter, R. J., Hoyer, L. W., Toole, J. J., Copeland, K. L., Carta, C. A., Caskey, C.T., and Kazazian, H. H.,** Prenatal diagnosis of haemophilia A by factor VIII gene analysis, *Lancet*, 1, 1407, 1985.
48. **Delpech, M., Deburgrave, N., Baudis, M., Malssonneuve, P., Bardin, J. M., Sultan, Y., and Kaplan, J.-C.,** De Novo mutation in hemophilia A established by DNA haplotype analysis and precluding prenatal diagnosis, *Hum. Genet.*, 74, 316, 1986.
49. **Choo, K. H., Gould, K. G., Rees, D. J. G., and Brownlee, G. G.,** Molecular cloning of the gene for human antihaemophilic factor IX, *Nature (London)*, 299, 178, 1982.
50. **Winship, P. R. and Brownlee, G. G.,** Diagnosis of haemophilia B carriers using intragenic oligonucleotide probes, *Lancet*, 1, 218, 1986.
51. **Peake, I. R., Furlong, B. L., and Bloom, A. L.,** Carrier detection by direct gene analysis in a family with haemophilia B (factor IX deficiency), *Lancet*, 1, 242, 1984.
52. **Giannelli, F., Choo, K. H., Rees, D. J. G., Boyd, Y., Rizza, G. R., and Brownlee, G. G.,** Gene deletions in patients with haemophilia B and antifactor IX antibodies, *Nature (London)*, 303, 181, 1983.
53. **Beaudet, A.,** DNA blotting for haemophilia B carrier detection: pitfalls in dosage analysis, *Lancet*, 1, 1182, 1984.
54. **Connor, J. M., Pettigrew, A. F., Shiach, C., Hann, I. M.,Lowe, G.D. O., and Forbes, C. D.,** Application of three intragenic DNA polymorphisms for carrier detection in haemophilia B, *J. Med. Genet.*, 23, 300, 1986.
55. **Smurl, J. F., Weaver, D. D., Jarmas, A., and Padilla, L.-M.,** Ethical considerations in medical genetics — the prenatal diagnosis of hemophilia B, *Am. J. Med. Genet.*, 17, 773, 1984.
56. **Tonnesen, T., Sondergaard, F., Guttler, F., Oberle, I., Moisan, J. P., Mandel, J. L., Hauge, M., and Damsgard, E. M.,** Exclusion of haemophilia B in male fetus by chorionic villus biopsy, *Lancet*, 2, 932, 1984.
57. **Patel, P. I., Nussbaum, R. L., Oramson, P. F., Ledbetter, D. H., Caskey, C. T., and Chinault, A. C.,** Organisation of the HPRT gene and related sequences in the human genome, *Somatic Cell Mol. Genet.*, 10, 483, 1984.
58. **Yang, T. P., Patel, P. I., Chinault, A. C., Stout, J. T., Jackson, L. G., Hildebrand, B. M., and Caskey, C.T.,** Molecular evidence of new mutation at the hprt locus in Lesch-Nyhan patients, *Nature (London)*, 310, 412, 1984.
59. **Nussbaum, R. L., Crowder,W. E., Nyhan, W. L., and Caskey, C. T.,** A three-allele restriction-fragment-length polymorphism at the hypoxanthine phosphoribosyltransferase locus in man, *Proc. Natl. Acad. Sci. U.S.A.*, 80, 4035, 1983.

60. **Turner, D. R., Morley, A. A., Hallandros, M., Kutlaca, R., and Sanderson, B., J.,** *In vitro* somatic mutations in human lymphocytes frequently result from major gene alterations, *Nature (London),* 315, 343, 1985.
61. **Horwich, A. L., Fenton, W. A., Williams, K. R., Kalousek, F., Kraus, J. P., Doolittle, R. F., Konigsberg, W., and Rosenberg, L. E.,** Structure and expression of a complementary DNA for the nuclear coded precursor of human mitochondrial ornithine transcarbamylase, *Science,* 224, 1068, 1984.
62. **Davies, K. E., Briand, P., Ionasescu, V., Ionasescu, G., Williamson, R., Brown, C., Cavard, C., and Cathelineau, L.,** Gene for OTC: characterization and linkage to Duchenne muscular dystrophy, *Nucleic Acid Res.,* 13, 155, 1985.
63. **Roz, R., Fox, J., Fenton, W. A., Horwich, A. L., and Rosenberg, L. E.,** Gene deletion and restriction fragment length polymorphisms at the human ornithine transcarbamylase locus, *Nature (London),* 313, 815, 1985.
64. **Old, J. M., Purvis-Smith, S., Wilcken, B., Pearson, P., Williamson, R., Briand, P. L., Howard, N. J., Hammond, N. J., Cathelineau, L., and Davies, K. E.,** Prenatal exclusion of ornithine transcarbamylase deficiency by direct gene analysis, *Lancet,* 1, 73, 1985.
65. **Pembrey, M. E., Old, J. M., Leonard, J. V., Rodeck, C. H., Warren, R., and Davies, K. E.,** Prenatal diagnosis of ornithine carbamoyl transferase deficiency using a gene specific probe, *J. Med. Genet.,* 22, 462, 1985.
66. **Fox, J., Hack, A. M., Fenton, W. A., Golbus, M. S., Winter, S., Kalousek, F., Rozen, R., Brusilow, S. W., and Rosenberg, L. E.,** Prenatal diagnosis of ornithine transcarbamylase deficiency with use of DNA polymorphisms, *N. Engl. J. Med.,* 315, 1205, 1986.
67. **Woo, S. L., Lidsky, A. S., Guttler, F., Chandra, T., and Robson, K. J. J.,** Cloned human phenylalanine hydroxylase gene allows prenatal diagnosis and carrier detection of classical phenylketonuria, *Nature (London),* 306, 151, 1983.
68. **Lidsky, A.S., Guttler, F., and Woo, S. L. C.,** Prenatal diagnosis of classic phenylketonuria by DNA analysis, *Lancet* , 1, 549, 1985.
69. **Dalger, S. P., Chakraborty, R., Guttler, F., Lidsky, A. S., Koch, R., and Woo, S. L. C.,** Polymorphic DNA haplotypes at the phenylalanine hydroxylase locus in prenatal diagnosis of phenylketonuria, *Lancet,* 1, 229, 1986.
70. **Blitzer, M. G., Bailey-Wilson, J. E., and Shapira, E.,** Discrimination of heterozygotes for phenylketonuria, persistent hyperphenylalaninemia and controls by phenylalanine loading, *Clin. Chim. Acta,* 161, 347, 1986.
71. **Chang, J. C. and Kan, Y. W.,** A sensitive new prenatal test for sickle cell anemia, *N. Engl. J. Med.,* 307, 30, 1982.
72. **Orkin, S. H., Little, P. F. R., Kazazlan, H. H., and Boehm, C. D.,** Improved detection of the sickle mutation by DNA analysis, application to prenatal diagnosis, *N. Engl. J. Med.,* 307, 32, 1984.
73. **Higgs, D. R., Ayyub, H., Clegg, J. B., Hill, A. V. S., Nicholls, R. D., Teal, H., Wainscoat, J. S., and Weatherall, D. J.,** Thalassaemia in British people, *Br. Med. J.,* 290, 1303, 1985.
74. **Zeng, Y.-T. and Huang, S.-Z.,** β-Globin gene organisation and prenatal diagnosis of β-thalassaemia in Chinese, *Lancet,* 1, 304, 1985.
75. **Hill, A. V. S. and Wainscoat, J. S.,** The evolution of the α- and β-globin gene clusters in human populations, *Hum. Genet.,* 74, 16, 1986.
76. **Chan, V., Ghosh, A., Chan, T. K., Wong, V., and Todd, D.,** Prenatal diagnosis of homozygous thalassaemia by direct DNA analysis of uncultured amniotic fluid cells, *Br. Med. J.,* 288, 1327, 1984.
77. **Boehm, C. D., Antonarakis, S. E., Phillips, J. A., Stetten, G., and Kazazian, H., H.,** Prenatal diagnosis using DNA polymorphisms. Report on 95 pregnancies at risk for sickle-cell disease of β-thalassemia, *N. Eng. J. Med.,* 308, 1054, 1983.
77a. **Antonarakis. S. E., Kazazlan, H. H., and Orkin, S. H.,** DNA polymorphism and molecular pathology of the human globin gene clusters, *Hum. Genet.,* 69, 1, 1985.
78. **Old, J. M., Heath, C., Fitches, A., Thein, S. L., Weatherall, D. J., Warren, R., McKenzie, C., Rodeck, C. H., Modell, B., Petrou, M., and Ward, R. H.,** First-trimester fetal diagnosis for haemoglobinopathies: report on 200 cases, *Lancet,* 2, 763, 1986.
79. **Kazazian, H. H., Phillips, J. A., Boehm, C. D., Vik, T. A., Mahoney, M. J., and Ritchye, A. K.,** Prenatal diagnosis of β-thalassaemia by amniocentesis: linkage analysis using multiple polymorphic restriction endonuclease sites, *Blood,* 567, 926, 1980.
80. **Trecartin, R. F., Liebhaber, S. A., Chang, J. C., Lee, K. Y., and Kan, Y. W.,** β°-thalassaemia in Sardinia is caused by a nonsense mutation, *J. Clin. Invest.,* 68, 1012, 1981.
81. **Rosatelli, C., Tuveri, T., DiTucci, A., Falchi, A. M., Scalas, M. T., Monni, G., and Cao, A.,** Prenatal diagnosis of beta-thalassaemia with the synthetic-oligomer technique, *Lancet,* 1, 241, 1985.
82. **Chan, V., Leung, N. K., Chan, T. K., Ghosh, A., Kan, Y. W., and Todd, D.,** BamH I polymorphism in the Chinese: its potential usefulness in prenatal diagnosis of a thalassaemia, *Brit. Med. J.,* 289, 947, 1984.

83. **Wainscoat, J. S., Thein, S. L., Old, J. M., and Weatherall, D. J.,** A new DNA polymorphism for prenatal diagnosis of β-thalassaemia in Mediterranean populations, *Lancet,* 2, 1299, 1984.
84. **Kan, Y. W.,** Thalassemia: molecular mechanism and detection, *Am. J. Hum. Genet.,* 38, 4, 1986.
85. **Reeders, S. T., Breuning, M. H., Corney, G., Jeremiah, S. J., Khan, P. M., Davies, K. E., Hopkinson, D. A., Pearson, P. L., and Weatherall, D. J.,** Two genetic markers closely linked to adult polycystic kidney disease on chromosome 16, *Br. Med. J.,* 292, 851, 1986.
86. **Reeders, S. T., Gal, A., Propping, P., Waldherr, R., Davies, K. E., Zerres, K., Hogenkamp, T., Schmidt, W., Dolata, M. M., and Weatherall, D. J.,** Prenatal diagnosis of autosomal dominant polycystic kidney disease with a DNA probe, *Lancet,* 2, 6, 1986.
87. **Royer-Pokora, B., Kunkel, L. M., Monaco, A. P., Goff, S. C., Newburger, P. E., Baehner, R. L., Cole, F. S., Curnutte, J. T., and Orkin, S. H.,** Cloning the gene for an inherited human disorder — chronic granulomatous disease — on the basis of its chromosomal location, *Nature (London),* 322, 32, 1986.
88. **Brock, D. J. H. and van Heyningen, V.,** The facts on cystic fibrosis testing, *Nature (London),* 319, 184, 1986.
89. **Eiberg, H., Schmeigelow, K., Tsui, L.-C., Buchwald, M., Niebuhr, E.,Phelan, P. D., Williamson, R., Warwick, W., Koch, C., and Mohr, J.,** Cystic fibrosis, linkage with PON, *Cytogenet. Cell Genet.,* 40, 623, 1985.
90. **Scambler, P. J., Farrall, M., Stanier, P., Bell, G., Ramirez, F., Wainwright, B. J., Bell, J., Lench, N. J., Kruyer, H., and Williamson, R.,** Linkage of COL1A2 collagen gene to cystic fibrosis, and its clinical implications, *Lancet,* 2, 1241, 1985.
91. **Knowlton, R. G., Cohen-Haguenauer, O., Cong, N. V., Frezal, J., Brown, V. A., Barker, D., Braman, J. C., Schumm, J. W., Tsui, L.-C., Buchwald, M., and Donis-Keller, H.,** A polymorphic DNA marker linked to cystic fibrosis is located on chromosome 7, *Nature (London),* 318, 380, 1985.
92. **White, R., Woodward, S., Leppert, M., O'Connell, P., Hoff, M., Herbst, J., Lalouel, J.-M., Dean, M., and Woude, G. V.,** A closely linked genetic marker for cystic fibrosis, *Nature (London),* 318, 382, 1985.
93. **Wainwright, B. J., Scambler, P. J., Schmidtke, J., Watson, E. A., Law, H.-Y., Farrall, M., Cooke, H. J., Eiberg, H., and Williamson, R.,** Localization of cystic fibrosis locus to human chromosome 7cen-q22, *Nature (London),* 318, 384, 1985.
94. **Tsui, L.-C., Buchwald, M., Barker, D., Braman, J. C., Knowlton, R., Schjumm, J. W., Eiberg, H., Mohr, J., Kennedy, D., Plasvic, N., Zsiga, M., Markiewicz, D., Akots, G., Brown, V., Helms, C., Gravius, T., Parker, C., Rediker, K., and Donis-Keller, H.,** Cystic fibrosis locus defined by a genetically linked polymorphic DNA marker, *Science,* 230, 1054, 1985.
95. **Porteous, D. J. and van Heyningen, V.,** Cystic fibrosis: from linked markers to the gene, *Trend. Genet.,* 2, 149, 1986.
96. **Farrall, M., Rodeck, C. H., Stanier, P., Lissens, W., Watson, E., Law, H.-Y., Warren, R., Super, M., Scambler, P., Wainwright, B., and Williamson, R.,** First-trimester prenatal diagnosis of cystic fibrosis with linked DNA probes, *Lancet,* 1, 1402,1986.
97. **Brock, D., Curtis, A., Holloway, S., Burn, J., and Nelson, R.,** DNA typing to avoid need for prenatal diagnosis of cystic fibrosis, *Lancet* , 2, 393, 1986.
98. **O'Brien, T., Harper, P. S., Davies, K. E., Murray, J. M., Sarfarazi, M., and Williamson, R.,** Absence of genetic heterogeneity in Duchenne muscular dystrophy shown by a linkage study using two cloned DNA sequences, *J. Med. Genet.,* 20, 249, 1983.
99. **Sarfarazi, M., Harper, P. S., Kingston, H. M., Murray, J. M., O'Brien, T., Davies, K. E., Williamson, R., Tippett, P., and Sanger, R.,** Genetic linkage relationships between the Xg blood group system and two X chromosome DNA polymorphisms in families with Duchenne and Becker muscular dystrophy, *Hum. Genet,* 65, 169, 1983.
100. **de Martinville, B., Kunkel, L. M., Bruns, G., Morie, F., Koenig, M., Mandel, J. L., Horwich, A., Latt, S. A., Gusella, J. F., Housman, D., and Francke, U.,** Localization of DNA sequences in region Xp21 of the human X chromosome: search for molecular markers close to the Duchenne muscular dystrophy locus, *Am. J. Hum. Genet.,* 37, 235, 1985.
101. **Kingston, H. M., Thomas, N. S. T., Pearson, P. L., Sarfarazi, M., and Harper, P. S.,** Genetic linkage between Becker muscular dystrophy and a polymorphic DNA sequence on the short arm of the X chromosome, *J. Med. Genet.,* 20, 255, 1983.
102. **Emanuel, B. S., Zackai, E. H., and Tucker, S. H.,** Further evidence for Xp21 location of Duchenne muscular dystrophy (DMD) locus: X;9 translocation in a female with DMD, *J. Med. Genet.,* 20, 461, 1984.
103. **Francke, U., Ochs, H. D., de Martinville, B., Giacalone, J., Lindgren, V., Disteche, C., Pagon, R. A., Hofker, M. H., van Ommen, G.-J. B.,Pearson, P. L., and Wedgewood, R. J.,** Minor Xp21 chromosome deletion in a male associated with expression of Duchenne muscular dystrophy, chronic granulamatous disease, retinitis pigmentosa, and McLeod syndrome, *Am. J. Hum. Genet.,* 37, 250,1985.

104. **Dunger, D. B., Pembrey, M., Pearson, P., Whitfield, A., Davies, K. E., Lake, B., Williams, D., and Dillon, M. J. D.,** Deletion on the X chromosome detected by direct DNA analysis in one of two unrelated boys with glycerol kinase deficiency, adrenal hypoplasia, and Duchenne muscular dystrophy, *Lancet* , 1, 585, 1986.
105. **Ray, P. N., Belfall, B., Duff, C., Logan, C., Kean, V., Thompson, M. W., Sylvester, J. E., Gorski, J. L., Schmickel, R. D., and Worton, R. G.,** Cloning of the breakpoint of an X;21 translocation associated with Duchenne muscular dystrophy, *Nature (London)* , 318, 672, 1985.
106. **Monaco, A. P., Neve, R. L., Colletti-Feener, C., Bertelson, C. J., Kurnit, D. M., and Kunkel, L. M.,** Isolation of candidate cDNAs for portions of the Duchenne muscular dystrophy gene, *Nature (London),* 323, 646, 1986.
107. **Burmeister, M. and Lehrach, H.,** Long-range restriction map around the Duchenne muscular dystrophy gene, *Nature (London),* 324, 582, 1986.
108. **Bakker, E., Goor, N., Wrogemann, K., Kunkel, L. M., Fenton, W. A., Majoor-Krakauer, D., Jahoda, M. G. J., van Ommen, G. J. B., Hofker, M. H., Mandel, J. L., Davies, K. E., Willard, H. F., Sandkuyl, L., Essen, A. J. V., Sachs, E. S., and Pearson, P. L.,** Prenatal diagnosis and carrier detection of duchenne muscular dystrophy with closely linked RFLPs, *Lancet,* 1, 655, 1985.
109. **Williams, H., Sarfarazi, M., Brown, C., Thomas, N., and Harper, P. S.,** The use of flanking markers in prediction for Duchenne muscular dystrophy, *Arch., Dis. Childh.,* 61, 218, 1986.
110. **Mulley, J. C. and Sutherland, G. R.,** Fragile X transmission and the determination of carrier probabilities for genetic counselling, *Am. J. Med. Genet.,* 26, 987, 1987.
111. **Sutherland, G. R., Baker, E., Purvis-Smith, S., Hockey, A., Krumins,E., and Eichenbaum, S. Z.,** Prenatal diagnosis of the fragile X using thymidine induction, *Prenat. Diagn.* 7, 197, 1987.
112. **Mulley, J. C., Gedeon, A. K., Thorn, K. A., Bates, L. J., and Sutherland, G. R.,** Linkage and genetic counselling for the fragile X using DNa probes 52A, F9,DX13 and St14, *Am. J. Med. Genet.,* 27, 435, 1987.
113. **Gusella, J. F., Wexler, N. S., Conneally, P. M., Naylor, S. L., Anderson, M. A., Tanzi, R. E., Watkins, P. C., Ottina, K., Wallace, M. R., Sakaguchi, A. Y., Young, A. B., Shoulson, I., Bonilla, E., and Martin, J. B.,** A polymorphic DNA marker genetically linked to Huntington's disease, *Nature (London),* 306, 234, 1983.
114. **Youngman, S., Sarfarazi, M., Quarrell, O. W. J., Conneally, P. M., Gibbons, K., Harper, P. S., Shaw, D. J., Tanzi, R. E., Wallace, M. R., and Gusella, J. F.,** Studies of a DNA marker (G8) genetically linked to Huntington disease in British families, *Hum. Genet.,* 73, 333, 1986.
115. **Gusella, J. F., Tanzi, R. E., Bader, P. I., Phelan, M. C., Stevenson, R., Hayden, M. R., Hofman, K. J., Faryniarz, A. G., and Gibbons, K.,** Deletion of Huntingtons disease-linked G8 (D4S10) locus in Wolf-Hirschhorn syndrome, *Nature (London),* 318, 75, 1985.
116. **Craufurd, D. I. O. and Harris,R.,** Ethics of predictive testing for Huntington's chorea: the need for more information, *Br. Med. J.,* 293, 249, 1986.
117. **Harper, P. S.,** The prevention of Huntington's chorea, *J. R. Coll. Physicians London,* 20, 7, 1986.
118. **Harper, P. S. and Sarfarazi, M.,** Genetic prediction and family structure in Huntington's chorea, *Br. Med. J.,* 290, 1929, 1985.
119. **Shaw, D. J., Meredith, A. L., Sarfarazi, M., Harley, H. G., Huson, S. M., Brook, J. D., Bufton, L., Litt, M., Mohandas, T., and Harper, P. S.,** Regional localisations and linkage relationships of seven 123RFLPs and myotonic dystrophy on chromosome 19, *Hum. Genet.,* 74, 262, 1986.
120. **Meredith, A. L., Huson S. M., Lunt, P. W., Sarfarazi, M., Harley, H. G., Brook, J. D., Shaw, D. J., and Harper, P. S.,** Application of a closely linked polymorphism of restriction fragment length to counselling and prenatal testing in families with myotonic dystrophy, *Br. Med. J.,* 293, 1353, 1986.
121. **Pericak-Vance, M. A., Yamaoka, L. H., Assinder, R. I. F., Hung, W.-Y., Bartlett, R. J., Stajich, J. M., Gaskell, P. C., Ross, D. A., Sherman, S., Fey, G. H., Humphries, S., Williamson, R., and Roses, A. D.,** Tight linkage of apolipoprotein C2 to myotonic dystrophy on chromosome 19, *Neurology,* 36, 1418, 1986.
122. **Haan, E. A., Mulley, J. C., Gedeon, A. K., Sheffield, L. J., Sutherland, G. R.,** Presymptomatic testing for myotonic dystrophy using the linked DNA marker APOC2, *Med. J. Aust.,* in press.
123. **Lunt, P. W., Meredith, A. L., and Harper, P. S.,** First-trimester prediction in fetus at risk for myotonic dystrophy, *Lancet,* 2, 350-351, 1986.
124. **Cooper, D. N. and Schmidtke, J.,** Diagnosis of genetic disease using recombinant DNA, *Hum. Genet.,* 73, 1, 1986.
125. **Boué, J., Oberlé, I., Heilig, R., Mandel, J. L., Moser, A., Moser, H., Larsen, J. W., Dumez, Y., and Boué, A.,** First trimester prenatal diagnosis of adrenoleukodystrophy by determination of very long chain fatty acid levels and by linkage analysis to a DNA probe, *Hum. Genet.,* 69, 272, 1985.
126. **Upadhaya, M., Sarfarazi, M., Bamforth, J. S., Thomas, N. S. T., Oberle, I., Young, I., and Harper, P. S.,** Localisation of the gene for Hunter syndrome on the long arm of X chromosome, *Hum. Genet.,* 74, 391, 1986

127. **Leppert, M. F., Hasstedt, S. J., Holm, T., O'Connell, P., Wu, L., Ash, O., Williams, R. R., and White, R.,** A DNA probe for the LDL receptor gene is tightly linked to hypercholesterolemia in a pedigree with early coronary disease. *Am. J. Hum. Genet.*, 39, 300, 1986.
128. **Ferns, G. A. A., Shelley, C. S., Stocks, J., Rees, A., Paul, H., Baralle, F., and Galton, D. J.,** A DNA polymorphism of the aproprotein AII gene in hypertriglyceridaemia, *Hum. Genet.*, 74, 302, 1986.
129. **Hegele, R. A., Huang, L.-S., Herbert, P. N., Blum, C. B., Buring, J. E., Hennekens, C. H., and Breslow, J. L.,** Apolipoprotein B-gene DNA polymorphisms associated with myocardial infarction, *N. Engl. J. Med.*, 515, 1509, 1986.
130. **Ordovas, J. M., Schaefer, E. J., Salem, D., Ward, R. H., Gluec, C. J., Vergani, C., Wilson, P. W. F., and Karathanasis, S. K.,** Apoloprotein A-1 gene polymorphism associated with premature coronary artery disease and familial hypoalphalipoproteinemia, *N. Engl. J. Med.*, 314, 671, 1986.
131. **Rotwein, P. A., Chirgwin, J., Province, M., Knowler, W. C., Pettitt, D. J., Cordell, B., Goodman, H. M., and Permutt, M. A.,** Polymorphism in the 5′ flanking region of the human insulin gene: a genetic marker for non-insulin-dependent diabetes, *N. Engl. J. Med.*, 308, 65, 1983.
132. **Festenstein, H., Awad, J., Hitman, G. A., Cutbush, S.,Groves, A. V., Cassell, P., Ollier, W., and Sachs, J. A.,** New HLA DNA polymorphisms associated with autoimmune diseases, *Nature (London)*, 322, 64, 1986.
133. **Bell, J., Smoot, S., Newby, C., Toyka, K., Rassenti, L., Smith, K., Hohlfeld, R., McDevitt, H., and Steinman, L.,** HLA-DQ Beta-chain polymorphism linked to myasthenia gravis, *Lancet*, 1, 1058, 1986.
134. **Hodgkinson, S., Sherrington, R., Gurling H., Marchbanks, R., Reeders, S., Mallet, J., McInnis, M., Petursson, H., and Brynjolfsson, J.,** Molecular genetic evidence for heterogeneity in manic depression, *Nature (London)*, 325, 805, 1987.
135. **Detera-Wadleigh, S. D., Berrettini,W. H., Goldin, L. R., Boorman, D., Anderson, S., and Gershon, E. S.,** Close linkage of c-Harvey-ras-1 and the insulin gene to affective disorder is ruled out in three North American pedigrees, *Nature (London)*, 325, 806, 1987.
136. **Goodfellow, P. N., Davies, K. E., and Ropers, H.-H.,** Report of the committee on the genetic constitution of the X and Y chromosomes, *Cytogenet. Cell Genet.*, 40, 296, 1985.
137. **McDonough, P. G., Tho, S. P., Trill, J. J., Byrd, J. R., Reindollar, R. H., and Tischfield, J. A.,** Use of two different deoxyribonucleic acid probes to detect Y chromosome deoxyribonucleic acid in subjects with normal and altered Y chromosomes, *Am. J.Obstet. Gynecol.*, 154, 737, 1986.
138. **Lau, Y.-F., Dozy, A. M., Huang, J. C., and Kan, Y. W.,** A rapid screening test for antenatal sex determination, *Lancet*, 1, 14, 1984.
139. **Gosden, J. R., Gosden, C. M., Christie, S., Cooke, H. J., Morsman, J. M., and Rodeck, C. H.,** The use of cloned Y chromosome-specific DNA probes for fetal sex determination in first trimester prenatal diagnosis, *Hum. Genet.*, 66, 347, 1984.
140. **de la Chapelle, A., Page, D. C., Brown, L., Kaski, U., Parvinen, T., and Tippett, P. A.,** The origin of 45,X males, *Am. J. Hum.Genet.*, 38, 330, 1986.
141. **Disteche, D. M., Brown, L., Saal, H., Friedman, C., Thuline, H. C., Hoar, D. L., Pagon, R. A., and Page, D. C.,** Molecular detection of a translocation (Y; 15) in a 45.X male, *Hum. Genet.*, 74, 372, 1986.
142. **Guellaen, G., Casanova, M., Bishop, C., Geldwerth, D., Andre, G., Fellous, M., and Weissenbach, J.,** Human XX males with Y single-copy DNA fragments, *Nature (London)*, 307, 172, 1984.
143. **de la Chapelle, A.,** The etiology of maleness in XX men, *Hum. Genet.*, 58, 105, 1981.
144. **Schmidtke, J., Arnemann, J., Schmid, M., Baum, F., Mayerova, A., Langenbeck, U., and Hansmann, I.,** A male with a monocentric Yq isochromosome and presence of Yp-specific DNA sequence, *Hum. Genet.*, 69, 135, 1985.
145. **Burk, R. D., Stamberg, J., Young, K. E., and Smith, K. D.,** Use of repetitive DNA for diagnosis of chromosomal rearrangements, *Hum. Genet.*, 64, 339, 1983.
146. **Lau, Y.-F., Ying, K. L., and Donnell, G. N.,** Identification of a case of Y:18 translocation using a Y-specific repetitive DNA probe, *Hum. Genet.*, 69, 102, 1985.
147. **Hassold, T., Kumlin, E., Takaesu, N., and Leppert, M.,** Determination of the parental origin of sex-chromosome monosomy using restriction fragment length polymorphisms, *Am. J. Hum. Genet.*, 37, 965, 1985.
148. **Callen, D. F., Mulley, J. C., Baker, E. G., and Sutherland, G. R.,** Determining the origin of human X isochromosomes by the use of DNA sequence polymorphisms and detection of an apparent I(xq) with xp sequences, *Hum. Genet.*, 77, 236, 1987.
149. **Davies, K. E., Harper, K., Bonthron D., Krumlauf, R., Polkey, A., Pembrey, M. E., and Williamson, R.,** Use of a chromosome 21 cloned DNA probe for the analysis of non-disjunction in Down syndrome, *Hum. Genet.*, 66, 54, 1984.
150. **Julien, C., Bazin, A., Guyot, B., Forestier, F., and Daffos, F.,** Rapid prenatal diagnosis of Down's syndrome with *in situ* hybridisation of fluorescent DNA probes, *Lancet*, 2, 864, 1986.

151. **Cremer, T., Landegent, J., Bruckner, A., Scholl, H. P., Schardin, M., Hager, H. D., Dewvilee, P., Pearson, P., and van der Ploeg, M.**, Detection of chromosome aberrations in the human interphase nucleus by visualization of specific target DNAs with radioactive and non-radioactive *in situ* hybridization techniques: diagnosis of trisomy 18 with probe L1.84, *Hum. Genet.*, 74, 346, 1986.
152. **Carlock, L. R. and Wasmuth, J. J.**, Molecular approach to analyzing the human 5p deletion syndrome, Cri du Chat, *Somat. Cell Mol. Genet.*, 11, 267, 1985.
153. **McDermid, H. E., Duncan, A. M. V., Brasch, K. R., Holden, J. J. A., Magenis, E., Sheehy, R., Burn,. J., Kardon, N., Noel, B., Schinzel, A., Teshima, I., and White, B. N.**, Characterization of the supernumerary chromosome in Cat Eye syndrome, *Science*, 232, 646, 1986.
154. **Duncan, A. M. V., Hough, C. A., White, B. N., and McDermid, H. E.**, Breakpoint localization of the marker chromosome associated with the cat eye syndrome, *Am. J. Hum. Genet.*, 38, 978, 1986.
155. **Mattei, M. G., Philip, N., Passarge, E., Molsan, J. P., Mandel, J. L., and Mattei, J. F.**, DNA probe localisation at 18pl13 band by *in situ* hybridization and identification of a small supernumerary chromosome, *Hum. Genet.*, 69, 268, 1985.
156. **Dryja, T. P., Cavenee, W., White, R., Rapaport, J. M., Petersen, R., Albert, D. M., and Bruns, G. A. P.**, Homozygosity of chromosome 13 in retinoblastoma, *N. Engl. J. Med.*, 310, 550, 1984.
157. **Cavenee, W. K., Dryja, T. P., Phillips,R. A., Benedict, W.F., Godbout, R., Gallie,B. L., Murphree, A. L., Strong, L. C., and White, R. L.**, Expression of recessive alleles by chromosomal mechanisms in retinoblastoma, *Nature (London)*, 305, 779, 1983.
158. **Cavenee, W. K., Murphree, A. L., Shull, M. M., Benedict, W. F., Sparkes, R. S., Kock, E., and Nordenskjold, M.**, Prediction of familial predisposition to retinoblastoma, *N. Engl. J. Med.*, 314, 1201, 1986.
159. **Friend, S. H., Bernards, R., Rogelj, S., Weinberg, R. A., Rapaport, J. M., Albert, D. M., and Dryja, T. P.**, A human DNA segment with properties of the gene that predisposes to retinoblastoma and osteosarcoma, *Nature (London)*, 323, 643, 1986.
160. **Solomon, E.**, Recessive mutation in aetiology of Wilms' tumour, *Nature (London)*, 309, 11, 1984.
161. **Gardner, R. J. M. and Sutherland, G. R.**, Genetic Advice in Chromosomal Conditions, Oxford University Press, New York, in press, 1987.
162. **Schmickel, R. D.**, Contiguous gene syndromes: a component of recognizable syndromes, *J. Pediatr.*, 109, 231, 1986.
163. **Ballabio, A., Parenti, G., Tippett, P., Mondello, C., Di Maio, S., Tenore, A., and Andria, G.**, X-linked ichthyosis, due to steroid sulphatase deficiency, associated with Kallmann syndrome (hypogonadotropic hypogonadism and anosmia): linkage relationships with Xg and cloned DNA sequences from the distal short arm of the X chromosome, *Hum. Genet.*, 72, 237, 1986.
164. **Francke, U.**, Microdeletions and Mendelian phenotypes, 7th Int. Congr. Hum. Genet., Abstracts, Berlin, 1986. 39.
165. **Donlon, T. A., Lalande, M., Wyman, A., Bruns, G., and Latt, S. A.**, Isolation of molecular probes associated with the chromosome 15 instability in the Prader-Willi syndrome, *Proc. Natl. Acad. Sci. U.S.A.*, 83, 4408, 1986.
166. **Van Heyningen, V., Boyd, P. A., Seawright, A., Fletcher, J. M., Fantes, J. A., Buckton, K. E., Spowart, G., Porteous, D. J., Hill, R. E., Newton, M. S., and Hastie, N. D.**, Molecular analysis of chromosome 11 deletions in aniridia-Wilms tumor syndrome, *Proc. Natl. Acad. Sci. U.S.A.*, 82, 8592, 1985.

INDEX

A

B

C

G

H

I

S

T

U

V

W

X

Y